Problem Book in Quantum Field Theory

Voja Radovanović

Problem Book in Quantum Field Theory

Second Edition

Springer

Voja Radovanović
Faculty of Physics
University of Belgrade
Studentski trg 12-16
11000 Belgrade
Serbia and Montenegro
E-mail: rvoja@phy.bg.ac.yu

Library of Congress Control Number: 2007940156

ISBN 978-3-540-77013-8 Springer Berlin Heidelberg New York
ISBN 978-3-540-29062-9 1st ed. Springer Berlin Heidelberg New York

Springer is a part of Springer Science+Business Media
springer.com

Typesetting: by the author using a Springer LaTeX macro package
Cover design: eStudio Calamar, Spain

Printed on acid-free paper SPIN: 12197873 5 4 3 2 1 0

To my daughter Natalija

Preface

This Problem Book is based on the exercises and lectures which I have given to undergraduate and graduate students of the Faculty of Physics, University of Belgrade over many years. Nowadays, there are a lot of excellent Quantum Field Theory textbooks. Unfortunately, there is a shortage of Problem Books in this field, one of the exceptions being the Problem Book of Cheng and Li [7]. The overlap between this Problem Book and [7] is very small, since the latter mostly deals with gauge field theory and particle physics. Textbooks usually contain problems without solutions. As in other areas of physics doing more problems in full details improves both understanding and efficiency. So, I feel that the absence of such a book in Quantum Field Theory is a gap in the literature. This was my main motivation for writing this Problem Book.

To students: You cannot start to do problems without previous studying your lecture notes and textbooks. Try to solve problems without using solutions; they should help you to check your results. The level of this Problem Book corresponds to the textbooks of Mandl and Show [15]; Greiner and Reinhardt [11] and Peskin and Schroeder [16]. Each Chapter begins with a short introduction aimed to define notation. The first Chapter is devoted to the Lorentz and Poincaré symmetries. Chapters 2, 3 and 4 deal with the relativistic quantum mechanics with a special emphasis on the Dirac equation. In Chapter 5 we present problems related to the Euler-Lagrange equations and the Noether theorem. The following Chapters concern the canonical quantization of scalar, Dirac and electromagnetic fields. In Chapter 10 we consider tree level processes, while the last Chapter deals with renormalization and regularization.

There are many colleagues whom I would like to thank for their support and help. Professors Milutin Blagojević and Maja Burić gave many useful ideas concerning problems and solutions. I am grateful to the Assistants at the Faculty of Physics, University of Belgrade: Marija Dimitrijević, Duško Latas and Antun Balaž who checked many of the solutions. Duško Latas also drew all the figures in the Problem Book. I would like to mention the contribution of the students: Branislav Cvetković, Bojan Nikolić, Mihailo Vanević, Marko

Vojinović, Aleksandra Stojaković, Boris Grbić, Igor Salom, Irena Knežević, Zoran Ristivojević and Vladimir Juričić. Branislav Cvetković, Maja Burić, Milutin Blagojević and Dejan Stojković have corrected my English translation of the Problem Book. I thank them all, but it goes without saying that all the errors that have crept in are my own. I would be grateful for any readers' comments.

Belgrade
August 2005

Voja Radovanović

Contents

Part I Problems

1 Lorentz and Poincaré symmetries 3
2 The Klein–Gordon equation 9
3 The γ–matrices 13
4 The Dirac equation 17
5 Classical field theory and symmetries 25
6 Green functions 31
7 Canonical quantization of the scalar field 35
8 Canonical quantization of the Dirac field 43
9 Canonical quantization of the electromagnetic field 49
10 Processes in the lowest order of perturbation theory 55
11 Renormalization and regularization 61

Part II Solutions

1 Lorentz and Poincaré symmetries 67
2 The Klein–Gordon equation 77
3 The γ–matrices 85

4 The Dirac equation 93

5 Classical fields and symmetries 121

6 Green functions 131

7 Canonical quantization of the scalar field 141

8 Canonical quantization of the Dirac field 161

9 Canonical quantization of the electromagnetic field 179

10 Processes in the lowest order of the perturbation theory 191

11 Renormalization and regularization 211

References 239

Index 241

Part I

Problems

1

Lorentz and Poincaré symmetries

- Minkowski space, M_4 is a real 4-dimensional vector space with metric tensor defined by
$$g_{\mu\nu} = \begin{pmatrix} 1 & 0 & 0 & 0 \\ 0 & -1 & 0 & 0 \\ 0 & 0 & -1 & 0 \\ 0 & 0 & 0 & -1 \end{pmatrix} . \tag{1.A}$$
Vectors can be written in the form $\boldsymbol{x} = x^\mu \boldsymbol{e}_\mu$, where x^μ are *the contravariant components* of the vector $\boldsymbol{x}$ in the basis
$$\boldsymbol{e}_0 = \begin{pmatrix} 1 \\ 0 \\ 0 \\ 0 \end{pmatrix}, \quad \boldsymbol{e}_1 = \begin{pmatrix} 0 \\ 1 \\ 0 \\ 0 \end{pmatrix}, \quad \boldsymbol{e}_2 = \begin{pmatrix} 0 \\ 0 \\ 1 \\ 0 \end{pmatrix}, \quad \boldsymbol{e}_3 = \begin{pmatrix} 0 \\ 0 \\ 0 \\ 1 \end{pmatrix} .$$
The square of the length of a vector in M_4 is $x^2 = g_{\mu\nu}x^\mu x^\nu$. The square of the line element between two neighboring points x^μ and $x^\mu + \mathrm{d}x^\mu$ takes the form
$$\mathrm{d}s^2 = g_{\mu\nu}\mathrm{d}x^\mu \mathrm{d}x^\nu = c^2\mathrm{d}t^2 - \mathrm{d}\boldsymbol{x}^2 . \tag{1.B}$$
The space M_4 is also a manifold; x^μ are global (inertial) coordinates. *The covariant components* of a vector are defined by $x_\mu = g_{\mu\nu}x^\nu$.
- *Lorentz transformations,*
$$x'^\mu = \Lambda^\mu{}_\nu x^\nu , \tag{1.C}$$
leave the square of the length of a vector invariant, i.e. $x'^2 = x^2$. The matrix Λ is a constant matrix[1]; x^μ and x'^μ are the coordinates of the same event in two different inertial frames. In Problem 1.1 we shall show that from the previous definition it follows that the matrix Λ must satisfy the condition $\Lambda^T g \Lambda = g$. The transformation law of the covariant components is given by
$$x'_\mu = (\Lambda^{-1})^\nu{}_\mu x_\nu = \Lambda_\mu{}^\nu x_\nu . \tag{1.D}$$

[1] The first index in $\Lambda^\mu{}_\nu$ is the row index, the second index the column index.

- Let $\boldsymbol{u} = u^\mu \boldsymbol{e}_\mu$ be an arbitrary vector in tangent space[2], where u^μ are its contravariant components. A dual space can be associated to the vector space in the following way. The dual basis, $\boldsymbol{\theta}^\mu$ is determined by $\boldsymbol{\theta}^\mu(\boldsymbol{e}_\nu) = \delta^\mu_\nu$. The vectors in the dual space, $\boldsymbol{\omega} = \omega_\mu \boldsymbol{\theta}^\mu$ are called dual vectors or one–forms. The components of the dual vector transform like (1.D). The scalar (inner) product of vectors $\boldsymbol{u}$ and $\boldsymbol{v}$ is given by

$$\boldsymbol{u} \cdot \boldsymbol{v} = g_{\mu\nu} u^\mu v^\nu = u^\mu v_\mu \ .$$

A tensor of rank (m, n) in Minkowski spacetime is

$$\mathsf{T} = T^{\mu_1 \ldots \mu_m}{}_{\nu_1 \ldots \nu_n}(x) \boldsymbol{e}_{\mu_1} \otimes \ldots \otimes \boldsymbol{e}_{\mu_m} \otimes \boldsymbol{\theta}^{\nu_1} \otimes \ldots \otimes \boldsymbol{\theta}^{\nu_n} \ .$$

The components of this tensor transform in the following way

$$T'^{\mu_1 \ldots \mu_m}{}_{n_1 \ldots \nu_\nu}(x') = \Lambda^{\mu_1}{}_{\rho_1} \ldots \Lambda^{\mu_m}{}_{\rho_m} (\Lambda^{-1})^{\sigma_1}{}_{\nu_1} \ldots (\Lambda^{-1})^{\sigma_n}_{\nu_n} T^{\rho_1 \ldots \rho_m}{}_{\sigma_1 \ldots \sigma_n}(x) \ ,$$

under Lorentz transformations. A contravariant vector is tensor of rank $(1, 0)$, while the rank of a covariant vector (one-form) is $(0, 1)$. The metric tensor is a symmetric tensor of rank $(0, 2)$.
- *Poincaré transformations,*[3] (Λ, a) consist of Lorentz transformations and translations, i.e.

$$(\Lambda, a)x = \Lambda x + a \ . \tag{1.E}$$

These are the most general transformations of Minkowski space which do not change the interval between any two vectors, i.e.

$$(y' - x')^2 = (y - x)^2 \ .$$

- In a certain representation the elements of the Poincaré group near the identity are

$$U(\omega, \epsilon) = e^{-\frac{i}{2} M_{\mu\nu} \omega^{\mu\nu} + i P_\mu \epsilon^\mu} \ , \tag{1.F}$$

where $\omega^{\mu\nu}$ and $M_{\mu\nu}$ are parameters and generators of the Lorentz subgroup respectively, while ϵ^μ and P_μ are the parameters and generators of the translation subgroup. The Poincaré algebra is given in Problem 1.11.
- The Levi-Civita tensor, $\epsilon^{\mu\nu\rho\sigma}$ is a totaly antisymmetric tensor. We will use the convention that $\epsilon^{0123} = +1$.

---

[2] The tangent space is a vector space of tangent vectors associated to each point of spacetime.

[3] Poincaré transformations are very often called inhomogeneous Lorentz transformations.

1.1. Show that Lorentz transformations satisfy the condition $\Lambda^T g \Lambda = g$. Also, prove that they form a group.

1.2. Given an infinitesimal Lorentz transformation

$$\Lambda^\mu{}_\nu = \delta^\mu{}_\nu + \omega^\mu{}_\nu \,,$$

show that the infinitesimal parameters $\omega_{\mu\nu}$ are antisymmetric.

1.3. Prove the following relation

$$\epsilon_{\alpha\beta\gamma\delta} A^\alpha{}_\mu A^\beta{}_\nu A^\gamma{}_\lambda A^\delta{}_\sigma = \epsilon_{\mu\nu\lambda\sigma} \det A \,,$$

where $A^\alpha{}_\mu$ are matrix elements of the matrix A.

1.4. Show that the Kronecker δ symbol and Levi-Civita ϵ symbol are form invariant under Lorentz transformations.

1.5. Prove that

$$\epsilon^{\mu\nu\rho\sigma}\epsilon_{\alpha\beta\gamma\delta} = - \begin{vmatrix} \delta^\mu{}_\alpha & \delta^\mu{}_\beta & \delta^\mu{}_\gamma & \delta^\mu{}_\delta \\ \delta^\nu{}_\alpha & \delta^\nu{}_\beta & \delta^\nu{}_\gamma & \delta^\nu{}_\delta \\ \delta^\rho{}_\alpha & \delta^\rho{}_\beta & \delta^\rho{}_\gamma & \delta^\rho{}_\delta \\ \delta^\sigma{}_\alpha & \delta^\sigma{}_\beta & \delta^\sigma{}_\gamma & \delta^\sigma{}_\delta \end{vmatrix} ,$$

and calculate the following contractions $\epsilon^{\mu\nu\rho\sigma}\epsilon_{\mu\beta\gamma\delta}$, $\epsilon^{\mu\nu\rho\sigma}\epsilon_{\mu\nu\gamma\delta}$, $\epsilon^{\mu\nu\rho\sigma}\epsilon_{\mu\nu\rho\delta}$, $\epsilon^{\mu\nu\rho\sigma}\epsilon_{\mu\nu\rho\sigma}$.

1.6. Let us introduce the notations $\sigma^\mu = (I, \boldsymbol{\sigma})$; $\bar{\sigma}^\mu = (I, -\boldsymbol{\sigma})$, where I is a unit matrix, while $\boldsymbol{\sigma}$ are Pauli matrices[4] and define the matrix $X = x_\mu \sigma^\mu$.

(a) Show that the transformation

$$X \to X' = SXS^\dagger,$$

where $S \in \mathrm{SL}(2,\mathrm{C})$[5], describes the Lorentz transformation $x^\mu \to \Lambda^\mu{}_\nu x^\nu$. This is a homomorphism between proper orthochronous Lorentz transformations[6] and the $\mathrm{SL}(2,\mathrm{C})$ group.

(b) Show that $x^\mu = \frac{1}{2}\mathrm{tr}(\bar{\sigma}^\mu X)$.

1.7. Prove that $\Lambda^\mu{}_\nu = \frac{1}{2}\mathrm{tr}(\bar{\sigma}^\mu S \sigma_\nu S^\dagger)$, and $\Lambda(S) = \Lambda(-S)$. The last relation shows that the map is not unique.

---

[4] The Pauli matrices are

$$\sigma_1 = \begin{pmatrix} 0 & 1 \\ 1 & 0 \end{pmatrix} , \quad \sigma_2 = \begin{pmatrix} 0 & -\mathrm{i} \\ \mathrm{i} & 0 \end{pmatrix} \text{ and } \quad \sigma_3 = \begin{pmatrix} 1 & 0 \\ 0 & -1 \end{pmatrix} .$$

[5] $\mathrm{SL}(2,\mathrm{C})$ matrices are 2×2 complex matrices of unit determinant.

[6] The proper orthochronous Lorentz transformations satisfy the conditions: $\Lambda^0{}_0 \geq 1$, $\det\Lambda = 1$.

1.8. Find the matrix elements of generators of the Lorentz group $M_{\mu\nu}$ in its natural (defining) representation (1.C).

1.9. Prove that the commutation relations of the Lorentz algebra

$$[M_{\mu\nu}, M_{\rho\sigma}] = \mathrm{i}(g_{\mu\sigma}M_{\nu\rho} + g_{\nu\rho}M_{\mu\sigma} - g_{\mu\rho}M_{\nu\sigma} - g_{\nu\sigma}M_{\mu\rho})$$

lead to

$$[M_i, M_j] = \mathrm{i}\epsilon_{ijl}M_l, \quad [N_i, N_j] = -\mathrm{i}\epsilon_{ijl}N_l, \quad [M_i, N_j] = \mathrm{i}\epsilon_{ijl}N_l \ ,$$

where $M_i = \frac{1}{2}\epsilon_{ijk}M_{jk}$ and $N_k = M_{k0}$. Further, one can introduce the following linear combinations $A_i = \frac{1}{2}(M_i + \mathrm{i}N_i)$ and $B_i = \frac{1}{2}(M_i - \mathrm{i}N_i)$. Prove that

$$[A_i, A_j] = \mathrm{i}\epsilon_{ijl}A_l, \quad [B_i, B_j] = \mathrm{i}\epsilon_{ijl}B_l, \quad [A_i, B_j] = 0 \ .$$

This is a well known result which gives a connection between the Lorentz algebra and "two" SU(2) algebras. Irreducible representations of the Lorentz group are classified by two quantum numbers (j_1, j_2) which come from above two SU(2) groups.

1.10. The Poincaré transformation (Λ, a) is defined by:

$$x'^{\mu} = \Lambda^{\mu}{}_{\nu}x^{\nu} + a^{\mu} \ .$$

Determine the multiplication rule i.e. the product $(\Lambda_1, a_1)(\Lambda_2, a_2)$, as well as the unit and inverse element in the group.

1.11. (a) Verify the multiplication rule

$$U^{-1}(\Lambda, 0)U(1, \epsilon)U(\Lambda, 0) = U(1, \Lambda^{-1}\epsilon) \ ,$$

in the Poincaré group. In addition, show that from the previous relation follows:

$$U^{-1}(\Lambda, 0)P_{\mu}U(\Lambda, 0) = (\Lambda^{-1})^{\nu}{}_{\mu}P_{\nu} \ .$$

Calculate the commutator $[M_{\mu\nu}, P_{\rho}]$.

(b) Show that

$$U^{-1}(\Lambda, 0)U(\Lambda', 0)U(\Lambda, 0) = U(\Lambda^{-1}\Lambda'\Lambda, 0) \ ,$$

and find the commutator $[M_{\mu\nu}, M_{\rho\sigma}]$.

(c) Finally show that the generators of translations commute between themselves, i.e. $[P_{\mu}, P_{\nu}] = 0$.

1.12. Consider the representation in which the vectors x of Minkowski space are $(x, \ 1)^T$, while the element of the Poincaré group, (Λ, a) are 5×5 matrices given by

$$\begin{pmatrix} \Lambda & a \\ 0 & 1 \end{pmatrix} .$$

Check that the generators in this representation satisfy the commutation relations from the previous problem.

1.13. Find the generators of the Poincaré group in the representation of a classical scalar field[7]. Prove that they satisfy the commutation relations obtained in Problem 1.11.

1.14. *The Pauli–Lubanski vector* is defined by $W_\mu = \frac{1}{2}\epsilon_{\mu\nu\lambda\sigma}M^{\nu\lambda}P^\sigma$.

(a) Show that $W_\mu P^\mu = 0$ and $[W_\mu,\ P_\nu] = 0$.
(b) Show that $W^2 = -\frac{1}{2}M_{\mu\nu}M^{\mu\nu}P^2 + M_{\mu\sigma}M^{\nu\sigma}P^\mu P_\nu$.
(c) Prove that the operators W^2 and P^2 commute with the generators of the Poincaré group. These operators are *Casimir operators.* They are used to classify the irreducible representations of the Poincaré group.

1.15. Show that

$$W^2|\boldsymbol{p} = 0, m, s, \sigma\rangle = -m^2 s(s+1)|\boldsymbol{p} = 0, m, s, \sigma\rangle\ ,$$

where $|\boldsymbol{p} = 0, m, s, \sigma\rangle$ is a state vector for a particle of mass m, momentum $\boldsymbol{p}$, spin s while σ is the z–component of the spin. The mass and spin classify the irreducible representations of the Poincaré group.

1.16. Verify the following relations

(a) $[M_{\mu\nu}, W_\sigma] = \mathrm{i}(g_{\nu\sigma}W_\mu - g_{\mu\sigma}W_\nu)$
(b) $[W_\mu, W_\nu] = -\mathrm{i}\epsilon_{\mu\nu\sigma\rho}W^\sigma P^\rho$.

1.17. Calculate the commutators

(a) $[W_\mu, M^2]$,
(b) $[M_{\mu\nu}, W^\mu W^\nu]$,
(c) $[M^2,\ P_\mu]$,
(d) $[\epsilon^{\mu\nu\rho\sigma}M_{\mu\nu}M_{\rho\sigma}, M_{\alpha\beta}]$.

1.18. The standard momentum for a massive particle is $(m, 0, 0, 0)$, while for a massless particle it is $(k, 0, 0, k)$. Show that the little group in the first case is SU(2), while in the second case it is E(2) group[8].

1.19. Show that conformal transformations consisting of dilations:

$$x^\mu \to x'^\mu = \mathrm{e}^{-\rho}x^\mu\ ,$$

special conformal transformations (SCT):

$$x^\mu \to x'^\mu = \frac{x^\mu + c^\mu x^2}{1 + 2c\cdot x + c^2x^2}\ ,$$

and usual Poincaré transformations form a group. Find the commutation relations in this group.

---

[7] Scalar field transforms as $\phi'(\Lambda x + a) = \phi(x)$
[8] E(2) is the group of rotations and translations in a plane.

2

The Klein–Gordon equation

- The Klein–Gordon equation,

$$(\Box + m^2)\phi(x) = 0, \tag{2.A}$$

is an equation for a free relativistic particle with zero spin. The transformation law of a scalar field $\phi(x)$ under Lorentz transformations is given by $\phi'(\Lambda x) = \phi(x)$.
- The equation for the spinless particle in an electromagnetic field, A^μ is obtained by changing $\partial_\mu \to \partial_\mu + iqA_\mu$ in equation (2.A), where q is the charge of the particle.

2.1. Solve the Klein–Gordon equation.

2.2. If ϕ is a solution of the Klein–Gordon equation calculate the quantity

$$Q = iq \int d^3\boldsymbol{x} \left(\phi^* \frac{\partial \phi}{\partial t} - \phi \frac{\partial \phi^*}{\partial t} \right) .$$

2.3. The Hamiltonian for a free real scalar field is

$$H = \frac{1}{2} \int d^3\boldsymbol{x} [(\partial_0 \phi)^2 + (\nabla \phi)^2 + m^2 \phi^2] .$$

Calculate the Hamiltonian H for a general solution of the Klein–Gordon equation.

2.4. The momentum for a real scalar field is given by

$$\boldsymbol{P} = - \int d^3\boldsymbol{x} \partial_0 \phi \nabla \phi .$$

Calculate the momentum $\boldsymbol{P}$ for a general solution of the Klein–Gordon equation.

2.5. Show that the current[1]

$$j_\mu = -\frac{\mathrm{i}}{2}(\phi\partial_\mu\phi^* - \phi^*\partial_\mu\phi)$$

satisfies the continuity equation, $\partial^\mu j_\mu = 0$.

2.6. Show that the continuity equation $\partial_\mu j^\mu = 0$ is satisfied for the current

$$j_\mu = -\frac{\mathrm{i}}{2}(\phi\partial_\mu\phi^* - \phi^*\partial_\mu\phi) - qA_\mu\phi^*\phi\,,$$

where ϕ is a solution of Klein–Gordon equation in external electromagnetic potential A_μ.

2.7. A scalar particle in the s–state is moving in the potential

$$qA^0 = \begin{cases} -V, & r < a \\ 0, & r > a \end{cases},$$

where V is a positive constant. Find the dispersion relation, i.e. the relation between energy and momentum, for discrete particle states. Which condition has to be satisfied so that there is only one bound state in the case $V < 2m$?

2.8. Find the energy spectrum and the eigenfunctions for a scalar particle in a constant magnetic field, $\boldsymbol{B} = B\boldsymbol{e}_z$.

2.9. Calculate the reflection and the transmission coefficients of a Klein–Gordon particle with energy E, at the potential

$$A^0 = \begin{cases} 0, & z < 0 \\ U_0, & z > 0 \end{cases},$$

where U_0 is a positive constant.

2.10. A particle of charge q and mass m is incident on a potential barrier

$$A^0 = \begin{cases} 0, & z < 0, z > a \\ U_0, & 0 < z < a \end{cases},$$

where U_0 is a positive constant. Find the transmission coefficient. Also, find the energy of particle for which the transmission coefficient is equal to one.

2.11. A scalar particle of mass m and charge $-e$ moves in the Coulomb field of a nucleus. Find the energy spectrum of the bounded states for this system if the charge of the nucleus is Ze.

2.12. Using the two-component wave function $\begin{pmatrix} \theta \\ \chi \end{pmatrix}$, where $\theta = \frac{1}{2}(\phi + \frac{\mathrm{i}}{m}\frac{\partial\phi}{\partial t})$ and $\chi = \frac{1}{2}(\phi - \frac{\mathrm{i}}{m}\frac{\partial\phi}{\partial t})$, instead of ϕ rewrite the Klein–Gordon equation in the Schrödinger form.

[1] Actually this is current density.

2.13. Find the eigenvalues of the Hamiltonian from the previous problem. Find the nonrelativistic limit of this Hamiltonian.

2.14. Determine the velocity operator $\boldsymbol{v} = \mathrm{i}[H, \boldsymbol{x}]$, where H is the Hamiltonian obtained in Problem 2.12. Solve the eigenvalue problem for $\boldsymbol{v}$.

2.15. In the space of two–component wave functions the scalar product is defined by

$$\langle \psi_1 | \psi_2 \rangle = \frac{1}{2} \int \mathrm{d}^3 \boldsymbol{x} \psi_1^\dagger \sigma_3 \psi_2 \ .$$

(a) Show that the Hamiltonian H obtained in Problem 2.12 is Hermitian.
(b) Find expectation values of the Hamiltonian $\langle H \rangle$, and the velocity $\langle \boldsymbol{v} \rangle$ in the state $\begin{pmatrix} 1 \\ 0 \end{pmatrix} \mathrm{e}^{-\mathrm{i}p\cdot x}$.

3

The γ–matrices

- In Minkowski space M_4, the γ–matrices satisfy *the anticommutation relations*[1]

$$\{\gamma^\mu, \gamma^\nu\} = 2g^{\mu\nu} \ . \tag{3.A}$$

- In *the Dirac representation* γ–matrices take the form

$$\gamma_0 = \begin{pmatrix} I & 0 \\ 0 & -I \end{pmatrix} , \quad \boldsymbol{\gamma} = \begin{pmatrix} 0 & \boldsymbol{\sigma} \\ -\boldsymbol{\sigma} & 0 \end{pmatrix} \ . \tag{3.B}$$

Other representations of the γ–matrices can be obtained by similarity transformation $\gamma'_\mu = S\gamma_\mu S^{-1}$. The transformation matrix S need to be unitary if the transformed matrices are to satisfy the Hermicity condition: $(\gamma'^\mu)^\dagger = \gamma'^0\gamma'^\mu\gamma'^0$. *The Weyl representation* of the γ–matrices is given by

$$\gamma_0 = \begin{pmatrix} 0 & I \\ I & 0 \end{pmatrix} , \quad \boldsymbol{\gamma} = \begin{pmatrix} 0 & \boldsymbol{\sigma} \\ -\boldsymbol{\sigma} & 0 \end{pmatrix} , \tag{3.C}$$

while in *the Majorana representation* we have

$$\begin{aligned} \gamma_0 &= \begin{pmatrix} 0 & \sigma_2 \\ \sigma_2 & 0 \end{pmatrix} , \quad \gamma^1 = \begin{pmatrix} i\sigma_3 & 0 \\ 0 & i\sigma_3 \end{pmatrix} , \\ \gamma^2 &= \begin{pmatrix} 0 & -\sigma_2 \\ \sigma_2 & 0 \end{pmatrix} , \quad \gamma^3 = \begin{pmatrix} -i\sigma_1 & 0 \\ 0 & -i\sigma_1 \end{pmatrix} . \end{aligned} \tag{3.D}$$

- The matrix γ^5 is defined by $\gamma^5 = i\gamma^0\gamma^1\gamma^2\gamma^3$, while $\gamma_5 = -i\gamma_0\gamma_1\gamma_2\gamma_3$. In the Dirac representation, γ_5 has the form

$$\gamma_5 = \begin{pmatrix} 0 & I \\ I & 0 \end{pmatrix} \ .$$

[1] The same type of relations hold in M_d, where d is the dimension of spacetime.

- $\sigma_{\mu\nu}$ matrices are defined by

$$\sigma_{\mu\nu} = \frac{\mathrm{i}}{2}[\gamma_\mu, \gamma_\nu] \,. \tag{3.E}$$

- Slash is defined as

$$\not{a} = a^\mu \gamma_\mu \,. \tag{3.F}$$

- Sometimes we use the notation: $\beta = \gamma^0$, $\boldsymbol{\alpha} = \gamma^0 \boldsymbol{\gamma}$. The anticommutation relations (3.A) become

$$\{\alpha^i, \alpha^j\} = 2\delta^{ij},\ \{\alpha^i, \beta\} = 0 \,.$$

3.1. Prove:

(a) $\gamma_\mu^\dagger = \gamma^0 \gamma_\mu \gamma^0$,
(b) $\sigma_{\mu\nu}^\dagger = \gamma^0 \sigma_{\mu\nu} \gamma^0$.

3.2. Show that:

(a) $\gamma_5^\dagger = \gamma_5 = \gamma^5 = \gamma_5^{-1}$,
(b) $\gamma_5 = -\frac{\mathrm{i}}{4!}\epsilon_{\mu\nu\rho\sigma}\gamma^\mu\gamma^\nu\gamma^\rho\gamma^\sigma$,
(c) $(\gamma_5)^2 = 1$,
(d) $(\gamma_5\gamma_\mu)^\dagger = \gamma^0\gamma_5\gamma_\mu\gamma^0$.

3.3. Show that:

(a) $\{\gamma_5, \gamma^\mu\} = 0$,
(b) $[\gamma_5, \sigma^{\mu\nu}] = 0$.

3.4. Prove $\not{a}^2 = a^2$.

3.5. Derive the following identities with contractions of the γ–matrices:

(a) $\gamma_\mu\gamma^\mu = 4$,
(b) $\gamma_\mu\gamma^\nu\gamma^\mu = -2\gamma^\nu$,
(c) $\gamma_\mu\gamma^\alpha\gamma^\beta\gamma^\mu = 4g^{\alpha\beta}$,
(d) $\gamma_\mu\gamma^\alpha\gamma^\beta\gamma^\gamma\gamma^\mu = -2\gamma^\gamma\gamma^\beta\gamma^\alpha$,
(e) $\sigma^{\mu\nu}\sigma_{\mu\nu} = 12$,
(f) $\gamma_\mu\gamma_5\gamma^\mu\gamma^5 = -4$,
(g) $\sigma_{\alpha\beta}\gamma_\mu\sigma^{\alpha\beta} = 0$,
(h) $\sigma_{\alpha\beta}\sigma^{\mu\nu}\sigma^{\alpha\beta} = -4\sigma^{\mu\nu}$,
(i) $\sigma^{\alpha\beta}\gamma^5\gamma^\mu\sigma_{\alpha\beta} = 0$,
(j) $\sigma^{\alpha\beta}\gamma^5\sigma_{\alpha\beta} = 12\gamma^5$.

3.6. Prove the following identities with traces of γ–matrices:

(a) $\text{tr}\gamma_\mu = 0$,
(b) $\text{tr}(\gamma_\mu\gamma_\nu) = 4g_{\mu\nu}$,
(c) $\text{tr}(\gamma_\mu\gamma_\nu\gamma_\rho\gamma_\sigma) = 4(g_{\mu\nu}g_{\rho\sigma} - g_{\mu\rho}g_{\nu\sigma} + g_{\mu\sigma}g_{\nu\rho})$,
(d) $\text{tr}\gamma_5 = 0$,
(e) $\text{tr}(\gamma_5\gamma_\mu\gamma_\nu) = 0$,
(f) $\text{tr}(\gamma_5\gamma_\mu\gamma_\nu\gamma_\rho\gamma_\sigma) = -4\text{i}\epsilon_{\mu\nu\rho\sigma}$,
(g) $\text{tr}(\not{a}_1 \cdots \not{a}_{2n+1}) = 0$,
(h) $\text{tr}(\not{a}_1 \cdots \not{a}_{2n}) = \text{tr}(\not{a}_{2n} \cdots \not{a}_1)$,
(i) $\text{tr}(\gamma_5\gamma_\mu) = 0$.

3.7. Calculate $\text{tr}(\not{a}_1\not{a}_2 \cdots \not{a}_6)$.

3.8. Calculate $\text{tr}[(\not{p} - m)\gamma_\mu(1 - \gamma_5)(\not{q} + m)\gamma_\nu]$.

3.9. Calculate $\gamma_\mu(1 - \gamma_5)(\not{p} - m)\gamma^\mu$.

3.10. Verify the identity

$$\exp(\gamma_5\not{a}) = \cos\sqrt{a_\mu a^\mu} + \frac{1}{\sqrt{a_\mu a^\mu}}\gamma_5\not{a}\sin\sqrt{a_\mu a^\mu} \ ,$$

where $a^2 > 0$.

3.11. Show that the set

$$\Gamma^a = \{I,\ \gamma^\mu,\ \gamma^5,\ \gamma^\mu\gamma^5, \sigma^{\mu\nu}\} \ ,$$

is made of linearly independent 4×4 matrices. Also, show that the product of any two of them is again one of the matrices Γ^a, up to ± 1, $\pm \text{i}$.

3.12. Show that any matrix $A \in C^{44}$ can be written in terms of $\Gamma^a = \{I,\ \gamma^\mu,\ \gamma^5,\ \gamma^\mu\gamma^5, \sigma^{\mu\nu}\}$, i.e. $A = \sum_a c_a\Gamma^a$ where $c_a = \frac{1}{4}\text{tr}(A\Gamma_a)$.

3.13. Expand the following products of γ–matrices in terms of Γ^a:

(a) $\gamma_\mu\gamma_\nu\gamma_\rho$,
(b) $\gamma_5\gamma_\mu\gamma_\nu$,
(c) $\sigma_{\mu\nu}\gamma_\rho\gamma_5$.

3.14. Expand the anticommutator $\{\gamma^\mu, \sigma^{\nu\rho}\}$ in terms of Γ–matrices.

3.15. Calculate $\text{tr}(\gamma_\mu\gamma_\nu\gamma_\rho\gamma_\sigma\gamma_a\gamma_\beta\gamma_5)$.

3.16. Verify the relation $\gamma_5\sigma^{\mu\nu} = \frac{\text{i}}{2}\epsilon^{\mu\nu\rho\sigma}\sigma_{\rho\sigma}$.

3.17. Show that the commutator $[\sigma_{\mu\nu}, \sigma_{\rho\sigma}]$ can be rewritten in terms of $\sigma_{\mu\nu}$. Find the coefficients in this expansion.

3.18. Show that if a matrix commutes with all gamma matrices γ^μ, then it is proportional to the unit matrix.

3.19. Let $U = \exp(\beta \boldsymbol{\alpha} \cdot \boldsymbol{n})$, where β and $\boldsymbol{\alpha}$ are Dirac matrices; $\boldsymbol{n}$ is a unit vector. Verify the following relation:

$$\boldsymbol{\alpha}' \equiv U \boldsymbol{\alpha} U^\dagger = \boldsymbol{\alpha} - (\mathrm{I} - U^2)(\boldsymbol{\alpha} \cdot \boldsymbol{n})\boldsymbol{n} \ .$$

3.20. Show that the set of matrices (3.C) is a representation of γ–matrices. Find the unitary matrix which transforms this representation into the Dirac one. Calculate $\sigma_{\mu\nu}$, and γ_5 in this representation.

3.21. Find Dirac matrices in two dimensional spacetime. Define γ_5 and calculate

$$\mathrm{tr}(\gamma^5 \gamma^\mu \gamma^\nu) \ .$$

Simplify the product $\gamma^5 \gamma^\mu$.

4

The Dirac equation

- *The Dirac equation,*

$$(\mathrm{i}\gamma^\mu \partial_\mu - m)\psi(x) = 0 \ , \tag{4.A}$$

is an equation of the free relativistic particle with spin 1/2. The general solution of this equation is given by

$$\psi(x) = \frac{1}{(2\pi)^{\frac{3}{2}}} \sum_{r=1}^{2} \int \mathrm{d}^3\boldsymbol{p} \sqrt{\frac{m}{E_{\boldsymbol{p}}}} \left(u_r(\boldsymbol{p}) c_r(\boldsymbol{p}) \mathrm{e}^{-\mathrm{i}p\cdot x} + v_r(\boldsymbol{p}) d_r^\dagger(\boldsymbol{p}) \mathrm{e}^{\mathrm{i}p\cdot x} \right) \ , \tag{4.B}$$

where $u_r(\boldsymbol{p})$ and $v_r(\boldsymbol{p})$ are the basic bispinors which satisfy equations

$$\begin{aligned} (\not p - m) u_r(\boldsymbol{p}) &= 0 \ , \\ (\not p + m) v_r(\boldsymbol{p}) &= 0 \ . \end{aligned} \tag{4.C}$$

We use the normalization

$$\begin{aligned} \bar{u}_r(\boldsymbol{p}) u_s(\boldsymbol{p}) &= -\bar{v}_r(\boldsymbol{p}) v_s(\boldsymbol{p}) = \delta_{rs} \ , \\ \bar{u}_r(\boldsymbol{p}) v_s(\boldsymbol{p}) &= \bar{v}_r(\boldsymbol{p}) u_s(\boldsymbol{p}) = 0. \end{aligned} \tag{4.D}$$

The coefficients $c_r(\boldsymbol{p})$ and $d_r(\boldsymbol{p})$ in (4.B) being given determined by boundary conditions. Equation (4.A) can be rewritten in the form

$$\mathrm{i}\frac{\partial \psi}{\partial t} = H_D \psi \ ,$$

where $H_D = \boldsymbol{\alpha} \cdot \boldsymbol{p} + \beta m$ is the so–called Dirac Hamiltonian.

- Under the Lorentz transformation, $x'^\mu = \Lambda^\mu{}_\nu x^\nu$, Dirac spinor, $\psi(x)$ transforms as

$$\psi'(x') = S(\Lambda)\psi(x) = \mathrm{e}^{-\frac{\mathrm{i}}{4}\sigma^{\mu\nu}\omega_{\mu\nu}} \psi(x) \ . \tag{4.E}$$

$S(\Lambda)$ is the Lorentz transformation matrix in spinor representation, and it satisfies the equations:

$$S^{-1}(\Lambda) = \gamma_0 S^\dagger(\Lambda) \gamma_0 \ ,$$

$$S^{-1}(\Lambda)\gamma^\mu S(\Lambda) = \Lambda^\mu{}_\nu \gamma^\nu .$$

- The equation for an electron with charge $-e$ in an electromagnetic field A^μ is given by

$$[\mathrm{i}\gamma^\mu(\partial_\mu - \mathrm{i}eA_\mu) - m]\,\psi(x) = 0 . \tag{4.F}$$

- Under parity, Dirac spinors transform as

$$\psi(t, \boldsymbol{x}) \to \psi'(t, -\boldsymbol{x}) = \gamma_0\psi(t, \boldsymbol{x}) . \tag{4.G}$$

- Time reversal is an antiunitary operation:

$$\psi(t, \boldsymbol{x}) \to \psi'(-t, \boldsymbol{x}) = T\psi^*(t, \boldsymbol{x}) . \tag{4.H}$$

The matrix T, satisfies

$$T\gamma_\mu T^{-1} = \gamma^{\mu *} = \gamma_\mu^T . \tag{4.I}$$

The solution of the above condition is $T = \mathrm{i}\gamma^1\gamma^3$, in the Dirac representation of γ–matrices. It is easy to see that $T^\dagger = T^{-1} = T = -T^*$.

- Under charge conjugation, spinors $\psi(x)$ transform as follows

$$\psi(x) \to \psi_c(x) = C\bar{\psi}^T . \tag{4.J}$$

The matrix C satisfies the relations:

$$C\gamma_\mu C^{-1} = -\gamma_\mu^T, \quad C^{-1} = C^T = C^\dagger = -C . \tag{4.K}$$

In the Dirac representation, the matrix C is given by $C = \mathrm{i}\gamma^2\gamma^0$.

4.1. Find which of the operators given below commute with the Dirac Hamiltonian:

(a) $\boldsymbol{p} = -\mathrm{i}\nabla$,
(b) $\boldsymbol{L} = \boldsymbol{r} \times \boldsymbol{p}$,
(c) $\boldsymbol{L}^2$,
(d) $\boldsymbol{S} = \frac{1}{2}\boldsymbol{\Sigma}$, where $\boldsymbol{\Sigma} = \frac{\mathrm{i}}{2}\boldsymbol{\gamma} \times \boldsymbol{\gamma}$,
(e) $\boldsymbol{J} = \boldsymbol{L} + \boldsymbol{S}$,
(f) $\boldsymbol{J}^2$,
(g) $\boldsymbol{\Sigma} \cdot \frac{\boldsymbol{p}}{|\boldsymbol{p}|}$,
(h) $\boldsymbol{\Sigma} \cdot \boldsymbol{n}$, where $\boldsymbol{n}$ is a unit vector.

4.2. Solve the Dirac equation for a free particle, i.e. derived (4.B).

4.3. Find the energy of the states $u_s(\boldsymbol{p})\mathrm{e}^{-\mathrm{i}p\cdot x}$ and $v_s(\boldsymbol{p})\mathrm{e}^{\mathrm{i}p\cdot x}$ for the Dirac particle.

4.4. Using the solution of Problem 4.2 show that

$$\sum_{r=1}^{2} u_r(p)\bar{u}_r(p) = \frac{\not{p}+m}{2m} \equiv \Lambda_+(p) \, ,$$

$$-\sum_{r=1}^{2} v_r(p)\bar{v}_r(p) = -\frac{\not{p}-m}{2m} \equiv \Lambda_-(p) \, .$$

The quantities $\Lambda_+(p)$ and $\Lambda_-(p)$ are energy projection operators.

4.5. Show that $\Lambda_\pm^2 = \Lambda_\pm$, and $\Lambda_+\Lambda_- = 0$. How do these projectors act on the basic spinors $u_r(\boldsymbol{p})$ and $v_r(\boldsymbol{p})$? Derive these results with and without using explicit expressions for spinors.

4.6. The spin operator in the rest frame for a Dirac particle is defined by $\boldsymbol{S} = \frac{1}{2}\boldsymbol{\Sigma}$. Prove that:

(a) $\boldsymbol{\Sigma} = \gamma_5\gamma_0\boldsymbol{\gamma}$,
(b) $[S^i,\ S^j] = \mathrm{i}\epsilon^{ijk}S^k$,
(c) $S^2 = -\frac{3}{4}$.

4.7. Prove that:

$$\frac{\boldsymbol{\Sigma}\cdot\boldsymbol{p}}{|\boldsymbol{p}|}u_r(\boldsymbol{p}) = (-1)^{r+1}u_r(\boldsymbol{p}) \, ,$$

$$\frac{\boldsymbol{\Sigma}\cdot\boldsymbol{p}}{|\boldsymbol{p}|}v_r(\boldsymbol{p}) = (-1)^{r}v_r(\boldsymbol{p}) \, .$$

Are spinors $u_r(\boldsymbol{p})$ and $v_r(\boldsymbol{p})$ eigenstates of the operator $\boldsymbol{\Sigma}\cdot\boldsymbol{n}$, where $\boldsymbol{n}$ is a unit vector? Check the same property for the spinors in the rest frame.

4.8. Find the boost operator for the transition from the rest frame to the frame moving with velocity v along the z–axis, in the spinor representation. Is this operator unitary?

4.9. Solve the previous problem upon transformation to the system rotated around the z–axis for an angle θ. Is this operator a unitary one?

4.10. The Pauli–Lubanski vector is defined by $W_\mu = \frac{1}{2}\epsilon_{\mu\nu\rho\sigma}M^{\nu\rho}P^\sigma$, where $M^{\nu\rho} = \frac{1}{2}\sigma^{\nu\rho} + \mathrm{i}(x^\nu\partial^\rho - x^\rho\partial^\nu)$ is angular momentum, while P^μ is linear momentum. Show that

$$W^2\psi(x) = -\frac{1}{2}(1+\frac{1}{2})m^2\psi(x) \, ,$$

where $\psi(x)$ is a solution of the Dirac equation.

4.11. The covariant operator which projects the spin operator onto an arbitrary normalized four–vector s^μ ($s^2 = -1$) is given by $W_\mu s^\mu$, where $s \cdot p = 0$, i.e. the vector polarization s^μ is orthogonal to the momentum vector. Show that

$$\frac{W_\mu s^\mu}{m} = \frac{1}{2m}\gamma_5 \not{s}\not{p} \ .$$

Find this operator in the rest frame.

4.12. In addition to the spinor basis, one often uses the helicity basis. The helicity basis is obtained by taking $\boldsymbol{n} = \boldsymbol{p}/|\boldsymbol{p}|$ in the rest frame. Find the equations for the spin in this case.

4.13. Find the form of the equations for the spin, defined in Problem 4.12 in the ultrarelativistic limit.

4.14. Show that the operator $\gamma_5 \not{s}$ commutes with the operator $\not{p}$, and that the eigenvalues of this operator are ± 1. Find the eigen–projectors of the operator $\gamma_5 \not{s}$. Prove that these projectors commute with projectors onto positive and negative energy states, $\Lambda_\pm(\boldsymbol{p})$.

4.15. Consider a Dirac's particle moving along the z–axis with momentum $\boldsymbol{p}$. The nonrelativistic spin wave function is given by

$$\varphi = \frac{1}{\sqrt{|a|^2 + |b|^2}} \begin{pmatrix} a \\ b \end{pmatrix} \ .$$

Calculate the expectation value of the spin projection onto a unit vector $\boldsymbol{n}$, i.e. $\langle \boldsymbol{\Sigma} \cdot \boldsymbol{n} \rangle$. Find the nonrelativistic limit.

4.16. Find the Dirac spinor for an electron moving along the $z-$axis with momentum $\boldsymbol{p}$. The electron is polarized along the direction $\boldsymbol{n} = (\theta, \phi = \frac{\pi}{2})$. Calculate the expectation value of the projection spin on the polarization vector in that state.

4.17. Is the operator γ_5 a constant of motion for the free Dirac particle? Find the eigenvalues and projectors for this operator.

4.18. Let us introduce

$$\psi_L = \frac{1}{2}(1 - \gamma_5)\psi \ ,$$

$$\psi_R = \frac{1}{2}(1 + \gamma_5)\psi \ ,$$

where ψ is a Dirac spinor. Derive the equations of motion for these fields. Show that they are decoupled in the case of a massless spinor. The fields ψ_L ψ_R are known as Weyl fields.

4.19. Let us consider the system of the following two–component equations:

$$i\sigma^\mu \frac{\partial \psi_R(x)}{\partial x^\mu} = m\psi_L(x) ,$$

$$i\bar{\sigma}^\mu \frac{\partial \psi_L(x)}{\partial x^\mu} = m\psi_R(x) ,$$

where $\sigma^\mu = (I, \boldsymbol{\sigma})$; $\bar{\sigma}^\mu = (I, \ -\boldsymbol{\sigma})$.

(a) Is it possible to rewrite this system of equations as a Dirac equation? If this is possible, find a unitary matrix which relates the new set of γ–matrices with the Dirac ones.

(b) Prove that the system of equations given above is relativistically covariant. Find 2×2 matrices S_R and S_L, which satisfy $\psi'_{R,L}(x') = S_{R,L}\psi_{R,L}(x)$, where $\psi'_{R,L}$ is a wave function obtained from $\psi_{R,L}(x)$ by a boost along the x–axis.

4.20. Prove that the operator $K = \beta(\boldsymbol{\Sigma} \cdot \boldsymbol{L} + 1)$, where $\boldsymbol{\Sigma} = -\frac{i}{2}\boldsymbol{\alpha} \times \boldsymbol{\alpha}$ is the spin operator and $\boldsymbol{L}$ is orbital momentum, commutes with the Dirac Hamiltonian.

4.21. Prove the Gordon identities:

$$2m\bar{u}(\boldsymbol{p}_1)\gamma_\mu u(\boldsymbol{p}_2) = \bar{u}(\boldsymbol{p}_1)[(p_1 + p_2)_\mu + i\sigma_{\mu\nu}(p_1 - p_2)^\nu]u(\boldsymbol{p}_2) ,$$

$$2m\bar{v}(\boldsymbol{p}_1)\gamma_\mu v(\boldsymbol{p}_2) = -\bar{v}(\boldsymbol{p}_1)[(p_1 + p_2)_\mu + i\sigma_{\mu\nu}(p_1 - p_2)^\nu]v(\boldsymbol{p}_2) .$$

Do not use any particular representation of Dirac spinors.

4.22. Prove the following identity:

$$\bar{u}(\boldsymbol{p}')\sigma_{\mu\nu}(p + p')^\nu u(\boldsymbol{p}) = i\bar{u}(\boldsymbol{p}')(p' - p)_\mu u(\boldsymbol{p}) .$$

4.23. The current J_μ is given by $J_\mu = \bar{u}(\boldsymbol{p}_2)\not{p}_1\gamma_\mu\not{p}_2 u(\boldsymbol{p}_1)$, where $u(\boldsymbol{p})$ and $\bar{u}(\boldsymbol{p})$ are Dirac spinors. Show that J_μ can be written in the following form:

$$J_\mu = \bar{u}(\boldsymbol{p}_2)[F_1(m, q^2)\gamma_\mu + F_2(m, q^2)\sigma_{\mu\nu}q^\nu]u(\boldsymbol{p}_1) ,$$

where $q = p_2 - p_1$. Determine the functions F_1 and F_2.

4.24. Rewrite the expression

$$\bar{u}(\boldsymbol{p})\frac{1}{2}(1 - \gamma_5)u(\boldsymbol{p})$$

as a function of the normalization factor $N = u^\dagger(\boldsymbol{p})u(\boldsymbol{p})$.

4.25. Consider the current

$$J_\mu = \bar{u}(\boldsymbol{p}_2)p^\rho q^\lambda \sigma_{\mu\rho}\gamma_\lambda u(\boldsymbol{p}_1) ,$$

where $u(\boldsymbol{p}_1)$ and $u(\boldsymbol{p}_2)$ are Dirac spinors; $p = p_1 + p_2$ and $q = p_2 - p_1$. Show that J_μ has the following form:

$$J_\mu = \bar{u}(\boldsymbol{p}_2)(F_1\gamma_\mu + F_2 q_\mu + F_3\sigma_{\mu\rho}q^\rho)u(\boldsymbol{p}_1) ,$$

and determine the functions $F_i = F_i(q^2, m)$, $(i = 1, 2, 3)$.

4.26. Prove that if $\psi(x)$ is a solution of the Dirac equation, that it is also a solution of the Klein-Gordon equation.

4.27. Determine the probability density $\rho = \bar{\psi}\gamma^0\psi$ and the current density $\boldsymbol{j} = \bar{\psi}\boldsymbol{\gamma}\psi$, for an electron with momentum $\boldsymbol{p}$ and in an arbitrary spin state.

4.28. Find the time dependence of the position operator $\boldsymbol{r}_{\mathrm{H}}(t) = \mathrm{e}^{\mathrm{i}Ht}\boldsymbol{r}\mathrm{e}^{-\mathrm{i}Ht}$ for a free Dirac particle.

4.29. The state of the free electron at time $t = 0$ is given by

$$\psi(t = 0, \boldsymbol{x}) = \delta^{(3)}(\boldsymbol{x}) \begin{pmatrix} 1 \\ 0 \\ 0 \\ 0 \end{pmatrix} .$$

Find $\psi(t > 0, \boldsymbol{x})$.

4.30. Determine the time evolution of the wave packet

$$\psi(t = 0,\ \boldsymbol{x}) = \frac{1}{(\pi d^2)^{\frac{3}{4}}} \exp\left(-\frac{\boldsymbol{x}^2}{2d^2}\right) \begin{pmatrix} 1 \\ 0 \\ 0 \\ 0 \end{pmatrix} ,$$

for the Dirac equation.

4.31. An electron with momentum $\boldsymbol{p} = p\boldsymbol{e}_z$ and positive helicity meets a potential barrier

$$-eA^0 = \begin{cases} 0, & z < 0 \\ V, & z > 0 \end{cases} .$$

Calculate the coefficients of reflection and transmission.

4.32. Find the coefficients of reflection and transmission for an electron moving in a potential barrier:

$$-eA^0 = \begin{cases} 0, & z < 0, z > a \\ V, & 0 < z < a \end{cases} .$$

The energy of the electron is E, while its helicity is $1/2$.

4.33. Let an electron move in a potential hole $2a$ wide and V deep. Consider only bound states of the electron.

(a) Find the dispersion relations.
(b) Determine the relation between V and a if there are N bound states. Take $V < 2m$. If there is only one bound state present in the spectrum, is it odd or even?
(c) Give a rough description of the dispersion relations for $V > 2m$.

4.34. Determine the energy spectrum of an electron in a constant magnetic field $\boldsymbol{B} = B\boldsymbol{e}_z$.

4.35. Show that if $\psi(x)$ is a solution of the Dirac equation in an electromagnetic field, then it satisfies the "generalize" Klein-Gordon equation:

$$[(\partial_\mu - \mathrm{i}eA_\mu)(\partial^\mu - \mathrm{i}eA^\mu) - \frac{e}{2}\sigma_{\mu\nu}F^{\mu\nu} + m^2]\psi(x) = 0 ,$$

where $F^{\mu\nu} = \partial^\mu A^\nu - \partial^\nu A^\mu$ is the field strength tensor.

4.36. Find the nonrelativistic approximation of the Dirac Hamiltonian $H = \boldsymbol{\alpha} \cdot (\boldsymbol{p} + e\boldsymbol{A}) - eA^0 + m\beta$, including terms of order $\frac{v^2}{c^2}$.

4.37. If $V_\mu(x) = \bar{\psi}(x)\gamma_\mu\psi(x)$ is a vector field, show that V_μ is a real quantity. Find the transformation properties of this quantity under proper orthochronous Lorentz transformations, charge conjugation C, parity P and time reversal T.

4.38. Investigate the transformation properties of the quantity $A^\mu(x) = \bar{\psi}(x)\gamma^\mu\gamma_5\psi(x)$, under proper orthochronous Lorentz transformations and the discrete transformations C, P and T.

4.39. Prove that the quantity $\bar{\psi}(x)\gamma_\mu\partial^\mu\psi(x)$ is a Lorentz scalar. Find its transformation rules under the discrete transformations.

4.40. Using the Dirac equation, show that $C\bar{u}^T(p,s) = v(p,s)$, where C is charge conjugation. Also, prove the above relation in a concrete representation.

4.41. The matrix C is defined by

$$C\gamma_\mu C^{-1} = -\gamma_\mu^T .$$

Prove that if matrices C' and C'' satisfy the above relation, then $C' = kC''$, where k is a constant.

4.42. If

$$\psi(x) = N_p \begin{pmatrix} \begin{pmatrix} 1 \\ 0 \end{pmatrix} \\ \frac{\sigma_3 p}{E_p + m} \begin{pmatrix} 1 \\ 0 \end{pmatrix} \end{pmatrix} \mathrm{e}^{-\mathrm{i}Et + \mathrm{i}pz} ,$$

is the wave function in frame S of the relativistic particle whose spin is 1/2, find:

(a) the wave function $\psi_c(x) = C\bar{\psi}^T(x)$ of the antiparticle,
(b) the wave function of this particle for an observer moving with momentum $\boldsymbol{p} = p\boldsymbol{e}_z$,
(c) the wave functions which are obtained after space and time inversion,

(d) the wave function in a frame which is obtained from S by a rotation about the x–axis through θ.

4.43. Find the matrices C and P in the Weyl representation of the γ–matrices.

4.44. Prove that the helicity of the Dirac particle changes sign under space inversion, but not under time reversal.

4.45. The Dirac Hamiltonian is $H = \boldsymbol{\alpha} \cdot \boldsymbol{p} + \beta m$. Determine the parameter θ from the condition that the new Hamiltonian $H' = UHU^{\dagger}$, where $U = \mathrm{e}^{\beta \boldsymbol{\alpha} \cdot \boldsymbol{p} \theta(p)}$ has even form, i.e. $H' \sim \beta$. (Foldy–Wouthuysen transformation).

4.46. Show that the spin operator $\boldsymbol{\Sigma} = \frac{\mathrm{i}}{2}\boldsymbol{\gamma} \times \boldsymbol{\gamma}$ and the angular momentum $\boldsymbol{L} = \boldsymbol{r} \times \boldsymbol{p}$, in Foldy–Wouthuysen representation, have the following form:

$$\boldsymbol{\Sigma}_{\mathrm{FW}} = \frac{m}{E_p}\boldsymbol{\Sigma} + \frac{\boldsymbol{p}(\boldsymbol{p} \cdot \boldsymbol{\Sigma})}{2E_p(m + E_p)} + \frac{\mathrm{i}\beta(\boldsymbol{\alpha} \times \boldsymbol{p})}{2E_p} ,$$

$$\boldsymbol{L}_{\mathrm{FW}} = \boldsymbol{L} - \frac{\boldsymbol{p}(\boldsymbol{p} \cdot \boldsymbol{\Sigma})}{2E_p(m + E_p)} + \frac{\boldsymbol{p}^2 \boldsymbol{\Sigma}}{2E_p(m + E_p)} - \frac{\mathrm{i}\beta(\boldsymbol{\alpha} \times \boldsymbol{p})}{2E_p} .$$

4.47. Find the Foldy–Wouthuysen transform of the position operator $\boldsymbol{x}$ and the momentum operator $\boldsymbol{p}$. Calculate the commutator $[\boldsymbol{x}_{\mathrm{FW}}, \boldsymbol{p}_{\mathrm{FW}}]$.

5

Classical field theory and symmetries

- If $f(x)$ is a function and $F[f(x)]$ a functional, *the functional derivative*, $\frac{\delta F[f(x)]}{\delta f(y)}$ is defined by the relation

$$\delta F = \int \mathrm{d}y \frac{\delta F[f(x)]}{\delta f(y)} \delta f(y) \, , \tag{5.A}$$

where δF is a variation of the functional.

- *The action* is given by

$$S = \int \mathrm{d}^4 x \mathcal{L}(\phi_r, \partial_\mu \phi_r), \tag{5.B}$$

where $\mathcal{L}$ is the Lagrangian density, which is a function of the fields $\phi_r(x)$, $r = 1, \ldots, n$ and their first derivatives. *The Euler–Lagrange equations* of motion are

$$\partial_\mu \left(\frac{\partial \mathcal{L}}{\partial(\partial_\mu \phi_r)} \right) - \frac{\partial \mathcal{L}}{\partial \phi_r} = 0 \, . \tag{5.C}$$

- *The canonical momentum* conjugate to the field variable ϕ_r is

$$\pi_r(x) = \frac{\partial \mathcal{L}}{\partial \dot{\phi}_r} \, . \tag{5.D}$$

The canonical Hamiltonian is

$$H = \int \mathrm{d}^3 x \mathcal{H} = \int \mathrm{d}^3 x (\dot{\phi}_r \pi_r - \mathcal{L}) \, . \tag{5.E}$$

- *Noether theorem*: If the action is invariant with respect to the continous infinitesimal transformations:

$$x_\mu \to x'_\mu = x_\mu + \delta x_\mu \, ,$$

$$\phi_r(x) \to \phi'_r(x') = \phi_r(x) + \delta\phi_r(x) \, ,$$

then the divergence of the current

$$j^\mu = \frac{\partial \mathcal{L}}{\partial(\partial_\mu \phi_r)} \delta\phi_r(x) - T^{\mu\nu}\delta x_\nu \ , \tag{5.F}$$

is equal to zero, i.e. $\partial_\mu j^\mu = 0$. The quantity

$$T_{\mu\nu} = \frac{\partial \mathcal{L}}{\partial(\partial^\mu \phi_r)} \partial_\nu \phi_r - \mathcal{L} g_{\mu\nu} \ , \tag{5.G}$$

is the energy–momentum tensor . The Noether charges $Q^a = \int \mathrm{d}^3 x j_0^a(x)$ are constants of motion under suitable asymptotic conditions. The index a is related to a symmetry group.

5.1. Let

(a) $F_\mu = \partial_\mu \phi$,
(b) $S = \int \mathrm{d}^4 x \left[\frac{1}{2}(\partial_\mu \phi)^2 - V(\phi)\right]$,

be functionals. Calculate the functional derivatives $\frac{\delta F_\mu}{\delta \phi}$ in the first case, and $\frac{\delta^2 S}{\delta\phi(x)\delta\phi(y)}$ in the second case.

5.2. Find the Euler–Lagrange equations for the following Lagrangian densities:

(a) $\mathcal{L} = -(\partial_\mu A^\nu)(\partial_\nu A^\mu) + \frac{1}{2}m^2 A_\mu A^\mu + \frac{\lambda}{2}(\partial_\mu A^\mu)^2$,
(b) $\mathcal{L} = -\frac{1}{4}F_{\mu\nu}F^{\mu\nu} + \frac{1}{2}m^2 A_\mu A^\mu$, where $F_{\mu\nu} = \partial_\mu A_\nu - \partial_\nu A_\mu$,
(c) $\mathcal{L} = \frac{1}{2}(\partial_\mu \phi)(\partial^\mu \phi) - \frac{1}{2}m^2\phi^2 - \frac{1}{4}\lambda\phi^4$,
(d) $\mathcal{L} = (\partial_\mu \phi - \mathrm{i}eA_\mu \phi)(\partial^\mu \phi^* + \mathrm{i}eA^\mu \phi^*) - m^2\phi^*\phi - \frac{1}{4}F_{\mu\nu}F^{\mu\nu}$,
(e) $\mathcal{L} = \bar{\psi}(\mathrm{i}\gamma_\mu \partial^\mu - m)\psi + \frac{1}{2}(\partial_\mu \phi)^2 - \frac{1}{2}m^2\phi^2 + \frac{1}{4}\lambda\phi^4 - \mathrm{i}g\bar{\psi}\gamma_5\psi\phi$.

5.3. The action of a free scalar field in two–dimensional spacetime is

$$S = \int_{-\infty}^{\infty} \mathrm{d}t \int_0^L \mathrm{d}x \left(\frac{1}{2}\partial_\mu \phi \partial^\mu \phi - \frac{m^2}{2}\phi^2\right) .$$

The spatial coordinate x varies in the region $0 < x < L$. Find the equation of motion and discuss the importance of the boundary term.

5.4. Prove that the equations of motion remain unchanged if the divergence of an arbitrary field function is added to the Lagrangian density.

5.5. Show that the Lagrangian density of a real scalar field can be taken as $\mathcal{L} = -\frac{1}{2}\phi(\Box + m^2)\phi$.

5.6. Show that the Lagrangian density of a free spinor field can be taken in the form $\mathcal{L} = \frac{\mathrm{i}}{2}(\bar{\psi}\partial\!\!\!/\psi - (\partial_\mu\bar{\psi})\gamma^\mu\psi) - m\bar{\psi}\psi$.

5.7. The Lagrangian density for a massive vector field A^μ is given by

$$\mathcal{L} = -\frac{1}{4}F_{\mu\nu}F^{\mu\nu} + \frac{1}{2}m^2 A_\mu A^\mu .$$

Prove that the equation $\partial_\mu A^\mu = 0$ is a consequence of the equations of motion.

5.8. Prove that the Lagrangian density of a massless vector field is invariant under the gauge transformation: $A_\mu \to A_\mu + \partial_\mu \Lambda(x)$, where $\Lambda = \Lambda(x)$ is an arbitrary function. Is the relation $\partial_\mu A^\mu = 0$ a consequence of the equations of motion?

5.9. The Einstein–Hilbert gravitation action is

$$S = \kappa \int \mathrm{d}^4x\sqrt{-g}R ,$$

where $g_{\mu\nu}$ is the metric of four–dimensional curved spacetime; R is scalar curvature and κ is a constant. In the weak–field approximation the metric is small perturbation around the flat metric $g^{(0)}_{\mu\nu}$, i.e.

$$g_{\mu\nu}(x) = g^{(0)}_{\mu\nu} + h_{\mu\nu}(x) .$$

The perturbation $h_{\mu\nu}(x)$ is a symmetric second rank tensor field. The Einstein–Hilbert action in this approximation becomes an action in flat spacetime (anyone familiar with general relativity can easily prove this):

$$S = \int \mathrm{d}^4x \left(\frac{1}{2}\partial_\sigma h_{\mu\nu}\partial^\sigma h^{\mu\nu} - \partial_\sigma h_{\mu\nu}\partial^\nu h^{\mu\sigma} + \partial_\sigma h^{\mu\sigma}\partial_\mu h - \frac{1}{2}\partial_\mu h \partial^\mu h\right) ,$$

where $h = h^\mu{}_\mu$. Derive the equations of motion for $h_{\mu\nu}$. These are the linearized Einstein equations. Show that the linearized theory is invariant under the gauge symmetry:

$$h_{\mu\nu} \to h_{\mu\nu} + \partial_\mu\Lambda_\nu + \partial_\nu\Lambda_\mu ,$$

where $\Lambda_\mu(x)$ is any four–vector field.

5.10. Find the canonical Hamiltonian for free scalar and spinor fields.

5.11. Show that the Lagrangian density

$$\mathcal{L} = \frac{1}{2}[(\partial\phi_1)^2 + (\partial\phi_2)^2] - \frac{m^2}{2}(\phi_1^2 + \phi_2^2) - \frac{\lambda}{4}(\phi_1^2 + \phi_2^2)^2 ,$$

is invariant under the transformation

$$\phi_1 \to \phi_1' = \phi_1\cos\theta - \phi_2\sin\theta ,$$

$$\phi_2 \to \phi_2' = \phi_1\sin\theta + \phi_2\cos\theta .$$

Find the corresponding Noether current and charge.

5.12. Consider the Lagrangian density

$$\mathcal{L} = (\partial_\mu \phi^\dagger)(\partial^\mu \phi) - m^2 \phi^\dagger \phi \,,$$

where $\begin{pmatrix} \phi_1 \\ \phi_2 \end{pmatrix}$ is an SU(2) doublet. Show that the Lagrangian density has SU(2) symmetry. Find the related Noether currents and charges.

5.13. The Lagrangian density is given by

$$\mathcal{L} = \bar{\psi}(\mathrm{i}\gamma^\mu \partial_\mu - m)\psi \,,$$

where $\psi = \begin{pmatrix} \psi_1 \\ \psi_2 \end{pmatrix}$ is a doublet of SU(2) group. Show that $\mathcal{L}$ has SU(2) symmetry. Find Noether currents and charges. Derive the equations of motion for spinor fields ψ_i, where $i = 1, 2$.

5.14. Prove that the following Lagrangian densities are invariant under phase transformations

(a) $\mathcal{L} = \bar{\psi}(\mathrm{i}\gamma_\mu \partial^\mu - m)\psi$,
(b) $\mathcal{L} = (\partial_\mu \phi^\dagger)(\partial^\mu \phi) - m^2 \phi^\dagger \phi$.

Find the Noether currents.

5.15. The Lagrangian density of a real three–component scalar field is given by

$$\mathcal{L} = \frac{1}{2}\partial_\mu \phi^T \partial^\mu \phi - \frac{m^2}{2}\phi^T \phi \,,$$

where $\phi = \begin{pmatrix} \phi_1 \\ \phi_2 \\ \phi_3 \end{pmatrix}$. Find the equations of motions for the scalar fields ϕ_i. Prove that the Lagrangian density is SO(3) invariant and find the Noether currents.

5.16. Investigate the invariance property of the Dirac Lagrangian density under chiral transformations

$$\psi(x) \to \psi'(x) = \mathrm{e}^{\mathrm{i}\alpha\gamma_5}\psi(x) \,,$$

where α is a constant. Find the Noether current and its four-divergence.

5.17. The Lagrangian density of a σ-model is given by

$$\begin{aligned}\mathcal{L} = &\frac{1}{2}[(\partial_\mu \sigma)(\partial^\mu \sigma) + (\partial_\mu \boldsymbol{\pi}) \cdot (\partial^\mu \boldsymbol{\pi})] + \mathrm{i}\bar{\Psi}\partial\!\!\!/ \Psi \\ &+ g\bar{\Psi}(\sigma + \mathrm{i}\boldsymbol{\tau} \cdot \boldsymbol{\pi}\gamma_5)\Psi - \frac{m^2}{2}(\sigma^2 + \boldsymbol{\pi}^2) + \frac{\lambda}{4}(\sigma^2 + \boldsymbol{\pi}^2)^2 \,,\end{aligned}$$

where σ is a scalar field, $\boldsymbol{\pi}$ is a three–component scalar field, Ψ a doublet of spinor fields, while $\boldsymbol{\tau}$ are Pauli matrices. Prove that the Lagrangian density $\mathcal{L}$ has the symmetry:

$$\begin{aligned}
\sigma(x) &\to \sigma(x), \\
\boldsymbol{\pi}(x) &\to \boldsymbol{\pi}(x) - \boldsymbol{\alpha} \times \boldsymbol{\pi}(x), \\
\Psi(x) &\to \Psi(x) + \mathrm{i}\frac{\boldsymbol{\alpha}\cdot\boldsymbol{\tau}}{2}\Psi(x) ,
\end{aligned}$$

where $\boldsymbol{\alpha}$ is an infinitesimal constant vector. Find the corresponding conserved current.

5.18. In general, the canonical energy–momentum tensor is not symmetric under the permutation of indices. The energy–momentum tensor is not unique: a new equivalent energy–momentum tensor can be defined by adding a four-divergence

$$\tilde{T}_{\mu\nu} = T_{\mu\nu} + \partial^\rho \chi_{\rho\mu\nu} ,$$

where $\chi_{\rho\mu\nu} = -\chi_{\mu\rho\nu}$. The two energy–momentum tensors are equivalent since they lead to the same conserved charges, i.e. both satisfy the continuity equation. If we take that the tensor $\chi_{\mu\nu\rho}$ is given by[1]

$$\chi_{\mu\nu\rho} = \frac{1}{2}\left(-\frac{\partial\mathcal{L}}{\partial(\partial^\mu\phi_r)}(I_{\rho\nu})_{rs} + \frac{\partial\mathcal{L}}{\partial(\partial^\rho\phi_r)}(I_{\mu\nu})_{rs} + \frac{\partial\mathcal{L}}{\partial(\partial^\nu\phi_r)}(I_{\mu\rho})_{rs} \right)$$

then $\tilde{T}_{\mu\nu}$ is symmetric[2]. The quantities $(I_{\rho\nu})_{rs}$ in the previous formula are defined by the transformation law of fields under Lorentz transformations:

$$\delta\phi_r \equiv \phi'_r(x') - \phi_r(x) = \frac{1}{2}\omega^{\mu\nu}(I_{\mu\nu})_{rs}\phi_s(x) .$$

(a) Find the energy–momentum and angular momentum tensors for scalar, Dirac and electromagnetic fields employing the Noether theorem.
(b) Applying the previously described procedure, find the symmetrized (or Belinfante) energy–momentum tensors for the Dirac and the electromagnetic field.

5.19. Under dilatations the coordinates are transformed as

$$x \to x' = \mathrm{e}^{-\rho} x .$$

The corresponding transformation rule for a scalar field is given by

$$\phi(x) \to \phi'(x') = \mathrm{e}^{\rho}\phi(x) ,$$

---

[1] Belinfante, Physica 6, 887 (1939)

[2] Symmetric energy–momentum tensors are not only simpler to work with but give the correct coupling to gravity.

where ρ is a constant parameter. Determine the infinitesimal form variation[3] of the scalar field ϕ. Does the action for the scalar field possess dilatation invariance? Find the Noether current.

5.20. Prove that the action for the massless Dirac field is invariant under the dilatations:

$$x \to x' = e^{-\rho}x, \quad \psi(x) \to \psi'(x') = e^{3\rho/2}\psi(x) .$$

Calculate the Noether current and charge.

[3] A form variation is defined by $\delta_0\phi_r(x) = \phi'_r(x) - \phi_r(x)$; total variation is $\delta\phi_r(x) = \phi'_r(x') - \phi_r(x)$.

6

Green functions

- The Green function (or propagator) of the Klein-Gordon equation, $\Delta(x-y)$ satisfies the equation

$$(\Box_x + m^2)\Delta(x-y) = -\delta^{(4)}(x-y) . \tag{6.A}$$

To define the Green function entirely, one also needs to fix the boundary condition.
- The Green function (or propagator) $S(x-y)$ of the Dirac equation is defined by

$$(\mathrm{i}\gamma^\mu \partial_\mu^x - m)S(x-y) = \delta^{(4)}(x-y) , \tag{6.B}$$

naturally, again with the appropriate boundary conditions fixed.
- The retarded (advanced) Green function is defined to be nonvanishing for positive (negative) values of time $x_0 - y_0$. The boundary conditions for the Feynman propagator are causal, i.e. positive (negative) energy solutions propagate forward (backward) in time. The Dyson propagator is anticausal.

6.1. Using Fourier transform determine the Green functions for the Klein–Gordon equation. Discus how one goes around singularities.

6.2. If Δ_F is the Feynman propagator, and Δ_R is the retarded propagator of the Klein–Gordon equation, prove that the difference between them, $\Delta_F - \Delta_R$ is a solution of the homogeneous Klein–Gordon equation.

6.3. Show that

$$\int \mathrm{d}^4 k \delta(k^2 - m^2)\theta(k_0) f(k) = \int \frac{\mathrm{d}^3 \boldsymbol{k}}{2\omega_k} f(k) ,$$

where $\omega_k = \sqrt{\boldsymbol{k}^2 + m^2}$.

6.4. Prove the following properties:

$$\Delta_\mathrm{R}(-x) = \Delta_\mathrm{A}(x) \ ,$$

$$\Delta_\mathrm{F}(-x) = \Delta_\mathrm{F}(x) \ .$$

Δ_A and Δ_R are the advanced and retarded Green functions; Δ_F is the Feynman propagator.

6.5. If the Green function $\bar{\Delta}(x)$ of the Klein–Gordon equation is defined as[1]

$$\bar{\Delta}(x) = \mathrm{P} \int \frac{\mathrm{d}^4 k}{(2\pi)^4} \frac{\mathrm{e}^{-ik\cdot x}}{k^2 - m^2} \ ,$$

prove the relations:

$$\bar{\Delta}(x) = \frac{1}{2}(\Delta_\mathrm{R}(x) + \Delta_\mathrm{A}(x)) \ ,$$

$$\bar{\Delta}(-x) = \bar{\Delta}(x) \ .$$

P denotes the principal value.

6.6. Write

$$\Delta(x) = -\frac{1}{(2\pi)^4} \oint_C \mathrm{d}^4 k \frac{\mathrm{e}^{-ik\cdot x}}{k^2 - m^2} \ ,$$

and

$$\Delta_\pm(x) = -\frac{1}{(2\pi)^4} \oint_{C_\pm} \mathrm{d}^4 k \frac{\mathrm{e}^{-ik\cdot x}}{k^2 - m^2}$$

in terms of integrals over three momentum, $\boldsymbol{k}$. The integration contours are given in Fig. 6.1.

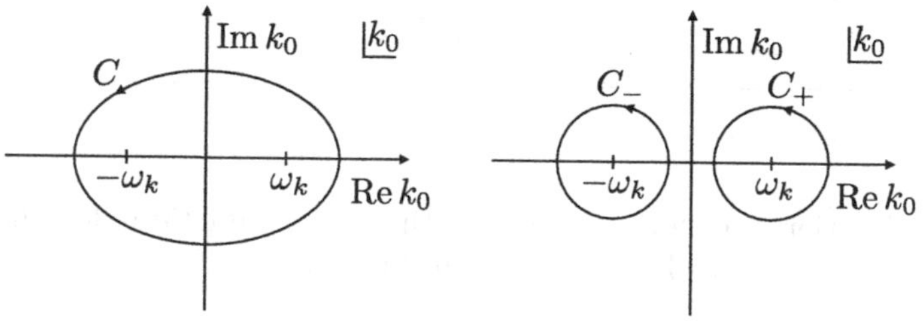

Fig. 6.1. The integration contours C and $C_\pm$.

In addition, prove that $\Delta(x) = \Delta_+(x) + \Delta_-(x)$.

[1] $\bar{\Delta}(x)$ is also called the principal-part propagator.

6.7. Show that

$$\left.\frac{\partial \Delta(x)}{\partial x^i}\right|_{x^0=0} = 0 ,$$

$$\left.\frac{\partial \Delta(x)}{\partial x^0}\right|_{x^0=0} = -\delta^{(3)}(\boldsymbol{x}) .$$

6.8. Prove that $\Delta(x)$ is a solution of the homogeneous Klein–Gordon equation.

6.9. Prove the following relation:

$$\Delta_{\mathrm{F}}(x)|_{m=0} = -\frac{1}{4\pi}\delta(x^2) + \frac{\mathrm{i}}{4\pi^2}\mathrm{P}\frac{1}{x^2} ,$$

where Δ_{F} is the Feynman propagator of the Klein–Gordon equation.

6.10. Prove that

$$\Delta_{\mathrm{R,A}}|_{m^2=0} = -\frac{1}{2\pi}\theta(\pm t)\delta(x^2) .$$

6.11. If the source ρ is given by $\rho(y) = g\delta^{(3)}(\boldsymbol{y})$, show that

$$\phi_{\mathrm{R}} = \frac{g}{4\pi}\frac{\exp(-m|\boldsymbol{x}|)}{|\boldsymbol{x}|} ,$$

where $\phi_{\mathrm{R}}(x) = -\int \mathrm{d}^4 y \Delta_{\mathrm{R}}(x-y)\rho(y)$.

6.12. Show that the Green function of the Dirac equation, $S(x)$ has the following form

$$S(x) = (\mathrm{i}\partial\!\!\!/ + m)\Delta(x) ,$$

where $\Delta(x)$ is the Green function of the Klein–Gordon equation with corresponding boundary conditions.

6.13. Starting from definition (6.B), determine the retarded, advanced, Feynman and Dyson propagators of the Dirac equation. Also, prove that the difference between any two of them is a solution of the homogenous Dirac equation.

6.14. If the source is given by $j(y) = g\delta(y_0)\mathrm{e}^{\mathrm{i}\boldsymbol{q}\cdot\boldsymbol{y}}(1,0,0,0)^{\mathrm{T}}$, where g is a constant while $\boldsymbol{q}$ is a constant vector, calculate

$$\psi(x) = \int \mathrm{d}^4 y S_{\mathrm{F}}(x-y)j(y) .$$

S_{F} is the Feynman propagator of the Dirac field.

6.15. Calculate the Green function in momentum space for a massive vector field, described by the Lagrangian density

$$\mathcal{L} = -\frac{1}{4}F_{\mu\nu}F^{\mu\nu} + \frac{1}{2}m^2 A_\mu A^\mu .$$

$F_{\mu\nu} = \partial_\mu A_\nu - \partial_\nu A_\mu$ is the field strength.

6.16. Calculate the Green function of a massless vector field for which the Lagrangian density is given by

$$\mathcal{L} = -\frac{1}{4}F_{\mu\nu}F^{\mu\nu} + \frac{1}{2}\lambda(\partial A)^2 \; .$$

The second term is known as the gauge fixing term; λ is a constant.

7

Canonical quantization of the scalar field

- The operators of a complex free scalar field are given by

$$\phi(x) = \frac{1}{(2\pi)^{\frac{3}{2}}} \int \frac{\mathrm{d}^3\boldsymbol{k}}{\sqrt{2\omega_k}} (a(\boldsymbol{k})\mathrm{e}^{-\mathrm{i}k\cdot x} + b^\dagger(\boldsymbol{k})\mathrm{e}^{\mathrm{i}k\cdot x}) , \tag{7.A}$$

$$\phi^\dagger(x) = \frac{1}{(2\pi)^{\frac{3}{2}}} \int \frac{\mathrm{d}^3\boldsymbol{k}}{\sqrt{2\omega_k}} (b(\boldsymbol{k})\mathrm{e}^{-\mathrm{i}k\cdot x} + a^\dagger(\boldsymbol{k})\mathrm{e}^{\mathrm{i}k\cdot x}) , \tag{7.B}$$

where $a(\boldsymbol{k})$ and $b(\boldsymbol{k})$ are *annihilation operators*; $a^\dagger(\boldsymbol{k})$ and $b^\dagger(\boldsymbol{k})$ *creation operators* and $a(\boldsymbol{k}) = b(\boldsymbol{k})$ is valid for a real scalar field. Real scalar fields are associated to neutral particles, while complex fields describe charged particles.

- The fields canonically conjugate to ϕ and $\phi^\dagger$ are

$$\pi = \frac{\partial \mathcal{L}}{\partial \dot{\phi}} = \dot{\phi}^\dagger , \quad \pi^\dagger = \frac{\partial \mathcal{L}}{\partial \dot{\phi}^\dagger} = \dot{\phi} .$$

Equal–time commutation relations take the following form:

$$\begin{aligned} [\phi(\boldsymbol{x},t), \pi(\boldsymbol{y},t)] = [\phi^\dagger(\boldsymbol{x},t), \pi^\dagger(\boldsymbol{y},t)] = \mathrm{i}\delta^{(3)}(\boldsymbol{x}-\boldsymbol{y}) , \\ [\phi(\boldsymbol{x},t), \phi(\boldsymbol{y},t)] = [\phi(\boldsymbol{x},t), \phi^\dagger(\boldsymbol{y},t)] = [\pi(\boldsymbol{x},t), \pi(\boldsymbol{y},t)] = 0 , \\ [\pi(\boldsymbol{x},t), \pi^\dagger(\boldsymbol{y},t)] = [\phi(\boldsymbol{x},t), \pi^\dagger(\boldsymbol{y},t)] = 0 . \end{aligned} \tag{7.C}$$

From (7.C) we obtain:

$$\begin{aligned} [a(\boldsymbol{k}), a^\dagger(\boldsymbol{q})] = [b(\boldsymbol{k}), b^\dagger(\boldsymbol{q})] = \delta^{(3)}(\boldsymbol{k}-\boldsymbol{q}) , \\ [a(\boldsymbol{k}), a(\boldsymbol{q})] = [a^\dagger(\boldsymbol{k}), a^\dagger(\boldsymbol{q})] = [a(\boldsymbol{k}), b^\dagger(\boldsymbol{q})] = [a^\dagger(\boldsymbol{k}), b^\dagger(\boldsymbol{q})] = 0 , \\ [b(\boldsymbol{k}), b(\boldsymbol{q})] = [b^\dagger(\boldsymbol{k}), b^\dagger(\boldsymbol{q})] = [a(\boldsymbol{k}), b(\boldsymbol{q})] = [a^\dagger(\boldsymbol{k}), b(\boldsymbol{q})] = 0 . \end{aligned} \tag{7.D}$$

- The vacuum $|0\rangle$ is defined by $a(\boldsymbol{k})|0\rangle = 0, \quad b(\boldsymbol{k})|0\rangle = 0$, for all $\boldsymbol{k}$. A state $a^\dagger(\boldsymbol{k})|0\rangle$ describes scalar particle with momentum $\boldsymbol{k}$, $b^\dagger(\boldsymbol{k})|0\rangle$ an antiparticle with momentum $\boldsymbol{k}$. Many–particle states are obtained by acting repeatedly with creation operators on the vacuum state.

- In normal ordering, denoted by : :, the creation operators stand to the left of all the annihilation operators. For example:

$$: a_1 a_2 a_3^\dagger a_4 a_5^\dagger := a_3^\dagger a_5^\dagger a_1 a_2 a_4 \; .$$

- The Hamiltonian, linear momentum and angular momentum of a scalar field are

$$H = \frac{1}{2}\int \mathrm{d}^3\boldsymbol{x}[(\partial_0\phi)^2 + (\nabla\phi)^2 + m^2\phi^2] \; ,$$

$$\boldsymbol{P} = -\int \mathrm{d}^3\boldsymbol{x}\partial_0\phi\nabla\phi \; ,$$

$$M^{\mu\nu} = \int \mathrm{d}^3\boldsymbol{x}(x^\mu T^{0\nu} - x^\nu T^{0\mu}) \; .$$

- *The Feynman propagator* of a complex field is defined by

$$\mathrm{i}\Delta_{\mathrm{F}}(x-y) = \langle 0| T(\phi(x)\phi^\dagger(y)) |0\rangle \; . \tag{7.E}$$

Time ordering is defined by

$$T\left(\phi(x)\phi^\dagger(y)\right) = \theta(x_0 - y_0)\phi(x)\phi^\dagger(y) + \theta(y_0 - x_0)\phi^\dagger(y)\phi(x) \; .$$

- The transformation rules for a scalar field under Poincaré transformations are given in Problem 7.20. Problems 7.21, 7.22 and 7.23 present the transformations of a scalar field under discrete transformations.

7.1. Starting from the canonical commutators

$$[\phi(\boldsymbol{x},t), \dot\phi(\boldsymbol{y},t)] = \mathrm{i}\delta^{(3)}(\boldsymbol{x}-\boldsymbol{y}) \; ,$$

$$[\phi(\boldsymbol{x},t), \phi(\boldsymbol{y},t)] = [\dot\phi(\boldsymbol{x},t), \dot\phi(\boldsymbol{y},t)] = 0 \; ,$$

derive the following commutation relations for creation and annihilation operators:

$$[a(\boldsymbol{k}), a^\dagger(\boldsymbol{q})] = \delta^{(3)}(\boldsymbol{k}-\boldsymbol{q}) \; ,$$

$$[a(\boldsymbol{k}), a(\boldsymbol{q})] = [a^\dagger(\boldsymbol{k}), a^\dagger(\boldsymbol{q})] = 0 \; .$$

7.2. At $t = 0$, a real scalar field and its time derivative are given by

$$\phi(t=0,\boldsymbol{x}) = 0, \quad \dot\phi(t=0,\boldsymbol{x}) = c \; ,$$

where c is a constant. Find the scalar field $\phi(t,\boldsymbol{x})$ at an arbitrary moment $t > 0$.

7.3. Calculate the energy $:H:$, momentum $:\boldsymbol{P}:$ and charge $:Q:$ of a complex scalar field. Compare these results to the results obtained in Problems 2.2, 2.3 and 2.4.

7.4. Prove that the modes

$$u_k = \frac{1}{\sqrt{2(2\pi)^3\omega_k}} \mathrm{e}^{-\mathrm{i}\omega_k t + \mathrm{i}\boldsymbol{k}\cdot\boldsymbol{x}} ,$$

are orthonormal with respect to the scalar product

$$\langle f|g\rangle = -\mathrm{i}\int \mathrm{d}^3x[f(x)\partial_0 g^*(x) - g^*(x)\partial_0 f(x)] .$$

7.5. Show that the vacuum expectation value of the scalar field Hamiltonian is given by

$$\langle 0|\, H\, |0\rangle = -\frac{1}{4}\pi m^4 \delta^{(3)}(0)\Gamma(-2) .$$

As one can see, this expression is the product of two divergent terms. Note that normal ordering gets rid of this c–number divergent term.

7.6. Calculate the following commutators: (Assume that the scalar field is a real one except for case (d))

(a) $[P^\mu, \phi(x)]$,
(b) $[P^\mu, F(\phi(x), \pi(x))]$, where F is an arbitrary polynomial function of fields and momenta,
(c) $[H, a^\dagger(\boldsymbol{k})a(\boldsymbol{q})]$,
(d) $[Q, P^\mu]$,
(e) $[N, H]$, where $N = \int \mathrm{d}^3\boldsymbol{k} a^\dagger(\boldsymbol{k})a(\boldsymbol{k})$ is the particle number operator,
(f) $\int \mathrm{d}^3\boldsymbol{x}[H, \phi(x)]\mathrm{e}^{-\mathrm{i}\boldsymbol{p}\cdot\boldsymbol{x}}$.

7.7. Prove that $\mathrm{e}^{\mathrm{i}Q}\phi(x)\mathrm{e}^{-\mathrm{i}Q} = \mathrm{e}^{-\mathrm{i}q}\phi(x)$.

7.8. The angular momentum of a scalar field $M_{\mu\nu}$, is obtained in Problem 5.18. Instead of the classical field, use the corresponding operator. Prove the following relations:

(a) $[M_{\mu\nu}, \phi(x)] = -\mathrm{i}(x_\mu\partial_\nu - x_\nu\partial_\mu)\phi(x)$,
(b) $[M_{\mu\nu}, P_\lambda] = \mathrm{i}(g_{\lambda\nu}P_\mu - g_{\lambda\mu}P_\nu)$,
(c) $[M_{\mu\nu}, M_{\rho\sigma}] = \mathrm{i}(g_{\mu\sigma}M_{\nu\rho} + g_{\nu\rho}M_{\mu\sigma} - g_{\mu\rho}M_{\nu\sigma} - g_{\nu\sigma}M_{\mu\rho})$.

7.9. Prove that $\phi_{\boldsymbol{k}}(x) = \langle \boldsymbol{k}|\phi(x)|0\rangle$ satisfies the Klein–Gordon equation.

7.10. Calculate the charges $Q^a = \int \mathrm{d}^3\boldsymbol{x} j_0^a(x)$, where j_0^a are zero components of the Noether currents for the symmetries defined in Problems 5.12 and 5.15.

(a) Prove that in both cases the charges satisfy the commutation relations of the SU(2) algebra.

(b) Calculate

$$[Q^a, \phi_i], \quad [Q^a, \phi_i^\dagger], \quad (i = 1, 2) \ ,$$

for the symmetry defined in Problem 5.12 and

$$[Q_k, \phi_i], \quad (i = 1, 2, 3) \ ,$$

for the symmetry defined in 5.15.

7.11. In Problem 5.19, it is shown that the action of a free massless scalar field is invariant under dilatations.

(a) Calculate the conserved charge $D = \int \mathrm{d}^3\boldsymbol{x} j^0$.
(b) Prove that relations $\rho[D, \phi(x)] = \mathrm{i}\delta_0\phi(x)$ and $\rho[D, \pi(x)] = \mathrm{i}\delta_0\pi(x)$ hold.
(c) Calculate the commutator $[D, F(\phi, \pi)]$, where F is an arbitrary analytic function.
(d) Prove that $[D, P^\mu] = \mathrm{i}P^\mu$.

7.12. If, instead of the field $\phi(x)$, we define the smeared field

$$\phi_f(\boldsymbol{x}, t) = \int \mathrm{d}^3\boldsymbol{y} \phi(t, \boldsymbol{y}) f(\boldsymbol{x} - \boldsymbol{y}) \ ,$$

where f is given by

$$f(\boldsymbol{x}) = \frac{1}{(a^2\pi)^{3/2}} \mathrm{e}^{-\boldsymbol{x}^2/a^2} \ ,$$

calculate the vacuum expectation value $\langle 0| \phi_f(t, \boldsymbol{x})\phi_f(t, \boldsymbol{x}) |0\rangle$. Find the result in the limit of vanishing mass.

7.13. The creation and annihilation operators of the free bosonic string α_m^μ $(0 < m \in Z)$, and α_m^μ $(0 > m \in Z)$, satisfy the commutation relations

$$[\alpha_m^\mu, \alpha_n^\nu] = -m\delta_{m+n,0} g^{\mu\nu} \ .$$

Show that the operators $L_m = -\frac{1}{2}\sum \alpha_{m-n}^\mu \alpha_{n\mu}$ satisfy

$$[L_m, L_n] = (m - n)L_{m+n} \ .$$

The operators L_m form the classical Virasora algebra. Upon normal ordering of the L_m's one can obtain the full algebra (with central charge):

$$[L_m, L_n] = (m - n)L_{m+n} + \frac{D-2}{12}(m^3 - m)\delta_{m+n,0} \ .$$

D is number of scalar fields.

7.14. Calculate the vacuum expectation value

$$\langle 0| \{\phi(x), \phi(y)\} |0\rangle \ ,$$

where { , } is the anticommutator. Assume that the scalar field is massless. Prove that the obtained expression satisfies the Klein–Gordon equation.

7.15. Calculate

$$\langle 0|\,\phi(x_1)\phi(x_2)\phi(x_3)\phi(x_4)\,|0\rangle$$

for a free scalar field.

7.16. Find

$$\langle 0|\,\phi(x)\phi(y)\,|0\rangle$$

in two dimensions, for a massless scalar field.

7.17. Prove the relation

$$(\Box_x + m^2)\,\langle 0|\,T(\phi(x)\phi(y))\,|0\rangle = -\mathrm{i}\delta^{(4)}(x-y)\ .$$

7.18. The Lagrangian density of a spinless Schrödinger field ψ, is given by

$$\mathcal{L} = \mathrm{i}\psi^\dagger \frac{\partial \psi}{\partial t} - \frac{1}{2m}\nabla\psi^\dagger \cdot \nabla\psi - V(\boldsymbol{r})\psi^\dagger\psi\ .$$

(a) Find the equations of motion.
(b) Express the free fields ψ and $\psi^\dagger$ in terms of creation and annihilation operators and find commutation relations between them.
(c) Calculate the Green function

$$G(x_0, \boldsymbol{x}, y_0, \boldsymbol{y}) = -\mathrm{i}\,\langle 0|\,\psi(x_0, \boldsymbol{x})\psi^\dagger(y_0, \boldsymbol{y})\,|0\rangle\,\theta(x_0 - y_0)$$

and prove that it satisfies the equation

$$\left(\mathrm{i}\frac{\partial}{\partial t} + \frac{1}{2m}\triangle\right) G(t, \boldsymbol{x}, 0, 0) = \delta(t)\delta^{(3)}(\boldsymbol{x})\ .$$

(d) Calculate the Green function for one–dimensional particle in the potential

$$V = \begin{cases} 0, & x > 0 \\ \infty, & x < 0 \end{cases}.$$

(e) Show that the free Schrödinger equation is invariant under Galilean transformations, which contain:
- spatial translations $\psi'(t, \boldsymbol{r} + \boldsymbol{\epsilon}) = \psi(t, \boldsymbol{r})$,
- time translations $\psi'(t + \delta, \boldsymbol{r}) = \psi(t, \boldsymbol{r})$,
- spatial rotations $\psi'(t, \boldsymbol{r} + \boldsymbol{\theta} \times \boldsymbol{r}) = \psi(t, \boldsymbol{r})$,
- "boost" $\psi'(t, \boldsymbol{r} - \boldsymbol{v}t) = \mathrm{e}^{-im\boldsymbol{v}\cdot\boldsymbol{r} + imv^2 t/2}\psi(t, \boldsymbol{r})$.

Without the phase factor in the last transformation rule the Schrödinger equation will not be invariant, unless $m = 0$. Consequently this representation of the Galilean group is projective.
(f) Find the conserved quantities associated with these transformations and commutations relations between them, i.e. the Galilean algebra.

7.19. Let

$$f(x) = \int \frac{\mathrm{d}^3\boldsymbol{p}}{2\omega_p} \tilde{f}(p)\mathrm{e}^{-\mathrm{i}p\cdot x},$$

be a classical function which satisfies the Klein–Gordon equation. Introduce the operators

$$a = C \int \frac{\mathrm{d}^3\boldsymbol{p}}{\sqrt{2\omega_p}} \tilde{f}^*(p) a(\boldsymbol{p}) ,$$

$$a^\dagger = C \int \frac{\mathrm{d}^3\boldsymbol{p}}{\sqrt{2\omega_p}} \tilde{f}(p) a^\dagger(\boldsymbol{p}) ,$$

where $a(\boldsymbol{p})$ and $a^\dagger(\boldsymbol{p})$ are annihilation and creation operators for scalar field, and C is a constant given by

$$C = \frac{1}{\sqrt{\int \frac{\mathrm{d}^3\boldsymbol{p}}{2\omega_p} |\tilde{f}(p)|^2}} .$$

A coherent state is defined by

$$|z\rangle = \mathrm{e}^{-|z|^2/2} \mathrm{e}^{z a^\dagger} |0\rangle ,$$

where z is a complex number.

(a) Calculate the following commutators:

$$[a(\boldsymbol{p}), a^\dagger], \quad [a(\boldsymbol{p}), a] .$$

(b) Prove the relation

$$[a(\boldsymbol{p}), (a^\dagger)^n] = C \frac{n\tilde{f}(p)}{\sqrt{2\omega_p}} (a^\dagger)^{n-1} .$$

(c) Show that the coherent state is an eigenstate of the operator $a(\boldsymbol{p})$.
(d) Calculate the standard deviation of a scalar field in the coherent state

$$\sqrt{\langle z| : \phi^2(x) : |z\rangle - (\langle z| \phi(x) |z\rangle)^2} .$$

(e) Find the expectation value of the Hamiltonian in the coherent state, $\langle z| H |z\rangle$.

7.20. Under the Poincaré transformation, $x \to x' = \Lambda x + a$, the real scalar field transforms as follows:

$$U(\Lambda, a)\phi(x)U^{-1}(\Lambda, a) = \phi(\Lambda x + a) ,$$

where $U(\Lambda, a)$ is a representation of the Poincaré group in space of the fields.

(a) Prove the following transformation rules for creation and annihilation operators:

$$U(\Lambda,a)a(\boldsymbol{k})U^{-1}(\Lambda,a) = \sqrt{\frac{\omega_{k'}}{\omega_k}}\exp(-\mathrm{i}\Lambda^\mu{}_\nu k^\nu a_\mu)a(\Lambda k)\ ,$$

$$U(\Lambda,a)a^\dagger(\boldsymbol{k})U^{-1}(\Lambda,a) = \sqrt{\frac{\omega_{k'}}{\omega_k}}\exp(\mathrm{i}\Lambda^\mu{}_\nu k^\nu a_\mu)a^\dagger(\Lambda k)\ .$$

(b) Prove that the transformation rule of the n–particle state $|k_1,\dots,k_n\rangle$ is given by

$$U(\Lambda,a)\,|k_1,\ k_2,\dots,k_n\rangle = \sqrt{\frac{\omega_{k_1'}\cdots\omega_{k_n'}}{\omega_{k_1}\cdots\omega_{k_n}}}\mathrm{e}^{\mathrm{i}a_\mu\Lambda^\mu{}_\nu(k_1^\nu+\dots+k_n^\nu)}\,|\Lambda k_1,\dots,\Lambda k_n\rangle\ .$$

(c) Prove that the momentum operator, P^μ of a scalar field is a vector under Lorentz transformations:

$$U(\Lambda,0)P^\mu U^{-1}(\Lambda,0) = \Lambda_\nu{}^\mu P^\nu\ .$$

(d) Prove that the commutator $[\phi(x),\phi(y)]$ is invariant with respect to Lorentz transformations.

7.21. The parity operator of a scalar field is given by

$$P = \exp\left[-\mathrm{i}\frac{\pi}{2}\int \mathrm{d}^3k\,\left(a^\dagger(\boldsymbol{k})a(\boldsymbol{k}) - \eta_p a^\dagger(\boldsymbol{k})a(-\boldsymbol{k})\right)\right]\ ,$$

where $\eta_p = \pm 1$ is the intrinsic parity of the field.

(a) Prove that P commutes with the Hamiltonian.
(b) Prove the relation $PM_{ij}P^{-1} = M_{ij}$, where M_{ij} is the angular momentum for scalar field.

7.22. Under time reversal, the scalar field is transformed according to

$$\tau\phi(x)\tau^{-1} = \eta\phi(-t,\ \boldsymbol{x})\ ,$$

where τ is an antiunitary operator, while η is a phase.

(a) Prove the relations:

$$\tau a(\boldsymbol{k})\tau^{-1} = \eta a(-\boldsymbol{k})\ ,$$
$$\tau a^\dagger(\boldsymbol{k})\tau^{-1} = \eta^* a^\dagger(-\boldsymbol{k})\ .$$

(b) Derive the transformation rules for the Hamiltonian and momentum under the time reversal.

7.23. Charge conjugation for the charged scalar field is defined by

$$C\phi(x)C^{-1} = \eta_c\phi^\dagger(x)\ ,$$

where η_c is a phase factor. Prove that

$$CQC^{-1} = -Q\ ,$$

where Q is the charge operator.

Canonical quantization of the Dirac field

- The operators of a Dirac field are:

$$\psi(x) = \frac{1}{(2\pi)^{\frac{3}{2}}} \sum_{r=1}^{2} \int d^3\boldsymbol{p} \sqrt{\frac{m}{E_{\boldsymbol{p}}}} \left(u_r(\boldsymbol{p}) c_r(\boldsymbol{p}) e^{-ip\cdot x} + v_r(\boldsymbol{p}) d_r^\dagger(\boldsymbol{p}) e^{ip\cdot x} \right) , \quad (8.A)$$

$$\bar{\psi}(x) = \frac{1}{(2\pi)^{\frac{3}{2}}} \sum_{r=1}^{2} \int d^3\boldsymbol{p} \sqrt{\frac{m}{E_{\boldsymbol{p}}}} \left(\bar{u}_r(\boldsymbol{p}) c_r^\dagger(\boldsymbol{p}) e^{ip\cdot x} + \bar{v}_r(\boldsymbol{p}) d_r(\boldsymbol{p}) e^{-ip\cdot x} \right) . \quad (8.B)$$

The operators $c_r^\dagger(\boldsymbol{p})$ and $d_r^\dagger(\boldsymbol{p})$ are *creation operators*, while $c_r(\boldsymbol{p}), d_r(\boldsymbol{p})$ are *annihilation operators.*

- From the Dirac Lagrangian density,

$$\mathcal{L} = \bar{\psi}(i\gamma^\mu \partial_\mu - m)\psi ,$$

one obtains the expressions for the conjugate momenta:

$$\pi_\psi = \frac{\partial \mathcal{L}}{\partial \dot{\psi}} = i\psi^\dagger , \ \pi_{\bar{\psi}} = \frac{\partial \mathcal{L}}{\partial \dot{\bar{\psi}}} = 0 .$$

Particles of spin 1/2 obey Fermi-Dirac statistics. We impose the canonical equal-time anticommutation relations:

$$\{\psi_a(t, \boldsymbol{x}), \psi_b^\dagger(t, \boldsymbol{y})\} = \delta_{ab}\delta^{(3)}(\boldsymbol{x} - \boldsymbol{y}) , \quad (8.C)$$

$$\{\psi_a(t, \boldsymbol{x}), \psi_b(t, \boldsymbol{y})\} = \{\psi_a^\dagger(t, \boldsymbol{x}), \psi_b^\dagger(t, \boldsymbol{y})\} = 0 . \quad (8.D)$$

From this we obtain the corresponding anticommutation relations between creation and annihilation operators:

$$\{c_r(\boldsymbol{p}), c_s^\dagger(\boldsymbol{q})\} = \{d_r(\boldsymbol{p}), d_s^\dagger(\boldsymbol{q})\} = \delta_{rs}\delta^{(3)}(\boldsymbol{p} - \boldsymbol{q}) . \quad (8.E)$$

All other anticommutators are zero.

- *The Fock space* of states is obtained as usual, by acting with creation operators on the vacuum $|0\rangle$. The states $c^\dagger(\boldsymbol{p},r)|0\rangle$, and $d^\dagger(\boldsymbol{p},r)|0\rangle$ are the electron and positron one–particle states, respectively with defined momentum and polarization.
- *Normal ordering* is defined as in the case scalar field but now the anticommutation relations (8.E) have to be taken into account, e.g.

$$:c(\boldsymbol{q})c^\dagger(\boldsymbol{p}):=-c^\dagger(\boldsymbol{p})c(\boldsymbol{q})\,,$$

$$:c(\boldsymbol{q})c(\boldsymbol{k})c^\dagger(\boldsymbol{p}):=c^\dagger(\boldsymbol{p})c(\boldsymbol{q})c(\boldsymbol{k})\,.$$

- The Hamiltonian, momentum and angular moment of the Dirac field are:

$$H=\int \mathrm{d}^3\boldsymbol{x}\bar{\psi}[-\mathrm{i}\gamma\nabla+m]\psi\,,$$

$$\boldsymbol{P}=-\mathrm{i}\int \mathrm{d}^3\boldsymbol{x}\psi^\dagger\nabla\psi\,,$$

$$M_{\mu\nu}=\int \mathrm{d}^3\boldsymbol{x}\psi^\dagger(\mathrm{i}(x_\mu\partial_\nu-x_\nu\partial_\mu)+\frac{1}{2}\sigma_{\mu\nu})\psi\,.$$

- *The Feynman propagator* is given by

$$\mathrm{i}S_F(x-y)=\langle 0|\,T\left(\psi(x)\bar{\psi}(y)\right)|0\rangle\,. \tag{8.F}$$

  Time ordering is defined by

$$T\left(\psi(x)\bar{\psi}(y)\right)=\theta(x_0-y_0)\psi(x)\bar{\psi}(y)-\theta(y_0-x_0)\bar{\psi}(y)\psi(x)\,.$$

- Under *the Lorentz transformation,* $x'=\Lambda x$ the operator $\psi(x)$ transforms according to:

$$U(\Lambda)\psi(x)U^{-1}(\Lambda)=S^{-1}(\Lambda)\psi(\Lambda x)\,. \tag{8.G}$$

  Here $U(\Lambda)$ is a unitary operator in spinor representation which generates the Lorentz transformation.
- *Parity,* $t'=t,\ \boldsymbol{x}'=-\boldsymbol{x}$ changes the Dirac field as follows

$$P\psi(t,\boldsymbol{x})P^{-1}=\gamma_0\psi(t,-\boldsymbol{x})\,, \tag{8.H}$$

  where P is the appropriate unitary operator.
- *Time reversal,* $t'=-t,\ \boldsymbol{x}'=\boldsymbol{x}$ is represented by an antiunitary operator. The transformation law is given by

$$\tau\psi(t,\boldsymbol{x})\tau^{-1}=T\psi(-t,\boldsymbol{x})\,. \tag{8.I}$$

  Properties of the matrix T, are given in Chapter 4. One should not forget that time reversal includes complex conjugation:

$$\tau(c\ldots)\tau^{-1}=c^*\tau\ldots\tau^{-1}\,.$$

- The operator $\mathcal{C}$ generates *charge conjugation* in the space of spinors:

$$\mathcal{C}\psi_a(x)\mathcal{C}^{-1} = (C\gamma_0^T)_{ab}\psi_b^\dagger(x) . \tag{8.J}$$

Properties of the matrix C are given in Chapter 4. The charge conjugation transforms a particle into an antiparticle and vice–versa.

- In this chapter we will very often use the identities:

$$[AB, C] = A[B, C] + [A, C]B ,$$
$$[AB, C] = A\{B, C\} - \{A, C\}B . \tag{8.K}$$

8.1. Starting from the anticommutation relations (8.E) show that:

$$iS(x-y) = \{\psi(x), \bar{\psi}(y)\} = i(i\gamma_\mu\partial^\mu + m)\Delta(x-y)$$
$$\{\psi(x), \psi(y)\} = 0 ,$$

where the function $\Delta(x-y)$ is to be determined. Prove that for $x_0 = y_0$ the function $iS(x-y)$ becomes $\gamma_0\delta^{(3)}(\boldsymbol{x}-\boldsymbol{y})$, i.e. the equal-time anticommutation relations for the Dirac field is obtained.

8.2. Express the following quantities in terms of creation and annihilation operators:

(a) charge $Q = -e\int d^3\boldsymbol{x} : \psi^+\psi : ,$
(b) energy $H = \int d^3\boldsymbol{x}[: \bar{\psi}(-i\gamma^i\partial_i + m)\psi :] ,$
(c) momentum $\boldsymbol{P} = -i\int d^3\boldsymbol{x} : \psi^\dagger\nabla\psi : .$

8.3. (a) Show that $i[H,\ \psi(x)] = \frac{\partial}{\partial t}\psi(x)$. Comment on this result.
(b) If the Dirac field is quantized according to the Bose-Einstein rather than Fermi-Dirac statistics, what would be the energy of the field?

8.4. Calculate $[H,\ c_r^\dagger(\boldsymbol{p})c_r(\boldsymbol{p})]$.

8.5. Starting from the transformation law for the classical Dirac field under Lorentz transformations show that the generators of these transformations are given by

$$M_{\mu\nu} = i(x_\mu\partial_\nu - x_\nu\partial_\mu) + \frac{1}{2}\sigma_{\mu\nu} .$$

8.6. The angular momentum of the Dirac field is

$$M_{\mu\nu} = \int d^3\boldsymbol{x}\psi^\dagger(x)\left[i(x_\mu\partial_\nu - x_\nu\partial_\mu) + \frac{1}{2}\sigma_{\mu\nu}\right]\psi(x) .$$

(a) Prove that

$$[M_{\mu\nu}, \psi(x)] = -\mathrm{i}(x_\mu \partial_\nu - x_\nu \partial_\mu)\psi(x) - \frac{1}{2}\sigma_{\mu\nu}\psi(x) ,$$

and comment on this result.

(b) Also, prove

$$[M_{\mu\nu}, P_\rho] = \mathrm{i}(g_{\nu\rho}P_\mu - g_{\mu\rho}P_\nu) ,$$

where P_μ is the four-vector of momentum.

8.7. Show that the helicity of the Dirac field is given by

$$S_p = \frac{1}{2}\sum_r \int \mathrm{d}^3\boldsymbol{p}(-1)^{r+1}[c_r^\dagger(\boldsymbol{p})c_r(\boldsymbol{p}) + d_r^\dagger(\boldsymbol{p})d_r(\boldsymbol{p})] .$$

8.8. Let $|\boldsymbol{p}_1, r_1; \boldsymbol{p}_2, r_2\rangle = c_{r_1}^\dagger(\boldsymbol{p}_1)c_{r_2}^\dagger(\boldsymbol{p}_2)\,|0\rangle$ be a two-particle state. Find the energy, charge and helicity of this state. Here $r_{1,2}$ are helicities of one-particle states.

8.9. Prove that the charges found in Problem 5.13 satisfy the commutation relation:

$$[Q^a, Q^b] = \mathrm{i}\epsilon^{abc}Q^c .$$

8.10. Find conserved charges for the symmetry in Problem 5.17 and calculate the commutators:

(a) $[Q^a, Q^b]$,
(b) $[Q^b, \pi^a(x)]$, $[Q^b, \psi_i(x)]$, $[Q^b, \bar{\psi}_i(x)]$.

8.11. In Problem 5.20 we showed that the action for a massless Dirac field is invariant under dilatations. Find the conserved charge $D = \int \mathrm{d}^3\boldsymbol{x} j^0$ for this symmetry and show that the relation

$$[D, P^\mu] = \mathrm{i}P^\mu ,$$

is satisfied.

8.12. Let the Lagrangian density be given by

$$\mathcal{L} = \mathrm{i}\bar{\psi}\gamma^\mu\partial_\mu\psi - gx^2\bar{\psi}\psi ,$$

where g is a constant.

(a) Derive the expression for the energy–momentum tensor $T_{\mu\nu}$. Find its divergence, $\partial_\mu T^{\mu\nu}$. Comment on this result.
(b) Calculate the commutator $[P^0(t), P^i(t)]$.
(c) Find the four divergence of the angular momentum operator $M^{\mu\alpha\beta}$.

8.13. Consider the current commutator $[J_\mu(x), J_\nu(y)]$ where $J_\mu = \bar{\psi}\gamma_\mu\psi$.

(a) Prove that the commutator given above is Lorentz covariant.
(b) Show that the commutator is equal to zero for space–like interval, i.e. for $(x-y)^2 < 0$.

8.14. Calculate $\langle 0| \bar{\psi}(x_1)\psi(x_2)\psi(x_3)\bar{\psi}(x_4) |0\rangle$. The result should be expressed in terms of vacuum expectation value of two fields.

8.15. Prove that $: \bar{\psi}\gamma^\mu\psi := \frac{1}{2}[\bar{\psi}, \gamma^\mu\psi]$.

8.16. Prove that $\langle 0| T(\bar{\psi}(x)\Gamma\psi(y)) |0\rangle$ is equal to zero for $\Gamma = \{\gamma_5, \gamma_5\gamma_\mu\}$, while for $\Gamma = \gamma_\mu\gamma_\nu$ one gets the result $-4\mathrm{i}mg_{\mu\nu}\Delta_F(y-x)$.

8.17. The Dirac spinor in terms of two Weyl spinors φ and χ is of the form

$$\psi = \begin{pmatrix} \varphi \\ -\mathrm{i}\sigma_2\chi^* \end{pmatrix} .$$

(a) Show that the Majorana spinor equals

$$\psi_M = \begin{pmatrix} \chi \\ -\mathrm{i}\sigma_2\chi^* \end{pmatrix} .$$

(b) Prove the identities:

$$\begin{aligned}
\bar{\psi}_M\phi_M &= \bar{\phi}_M\psi_M \, , \\
\bar{\psi}_M\gamma^\mu\phi_M &= -\bar{\phi}_M\gamma^\mu\psi_M \, , \\
\bar{\psi}_M\gamma_5\phi_M &= \bar{\phi}_M\gamma_5\psi_M \, , \\
\bar{\psi}_M\gamma^\mu\gamma_5\phi_M &= \bar{\phi}_M\gamma^\mu\gamma_5\psi_M \, , \\
\bar{\psi}_M\sigma_{\mu\nu}\phi_M &= -\bar{\phi}_M\sigma_{\mu\nu}\psi_M \, .
\end{aligned}$$

(c) Express the Majorana field operator, $\psi_M = \frac{1}{\sqrt{2}}(\psi + \psi_c)$ using creation and annihilation operators of a Dirac field. Introduce creation and annihilation operators for Majorana spinors and find corresponding anticomutation relations.
(d) Rewrite the QED Lagrangian density using Majorana spinors.

8.18. Find the transformation laws of the quantities $V_\mu(x) = \bar{\psi}(x)\gamma_\mu\psi(x)$ and $A_\mu(x) = \bar{\psi}(x)\gamma_5\partial_\mu\psi(x)$ under Lorentz and discrete transformations.

8.19. Show that the Lagrangian density

$$\mathcal{L} = \mathrm{i}\bar{\psi}(x)\gamma^\mu\partial_\mu\psi(x) + m\bar{\psi}(x)\psi(x) \, ,$$

is invariant under the Lorentz and discrete transformations.

8.20. Show that the quantity $T_{\mu\nu}(x) = \bar{\psi}(x)\sigma_{\mu\nu}\psi(x)$ transforms as a tensor under Lorentz transformations. Find its transformation rules under discrete symmetries.

9

Canonical quantization of the electromagnetic field

- The Lagrangian density of the electromagnetic field in the presence of an exterior current j_μ is
$$\mathcal{L} = -\frac{1}{4}F_{\mu\nu}F^{\mu\nu} - j^\mu A_\mu \ .$$
From this expression we derive the equations of motion to be:
$$\partial_\mu F^{\mu\nu} = j^\nu \Rightarrow (\delta^\nu_\mu \Box - \partial_\mu \partial^\nu)A^\mu = j^\nu \ . \tag{9.A}$$
It is easy to see that the field strength $F_{\mu\nu}$ satisfies the identity:
$$\partial_\mu F_{\nu\rho} + \partial_\nu F_{\rho\mu} + \partial_\rho F_{\mu\nu} = 0 \ . \tag{9.B}$$
Equations (9.A-B) are *the Maxwell equations*; (9.B) is the so–called, Bianchi identity and is a kinematical condition.
- Electrodynamics is invariant under *the gauge transformation*
$$A^\mu \to A^\mu + \partial^\mu \Lambda(x) \ ,$$
where $\Lambda(x)$ is an arbitrary function. The gauge symmetry can be fixed by imposing a "gauge condition". The following choices are often convenient:
$$\begin{aligned} \text{Lorentz gauge} \ \ & \partial_\mu A^\mu = 0 \ , \\ \text{Coulomb gauge} \ \ & \nabla \cdot \boldsymbol{A} = 0 \ , \\ \text{Time gauge} \ \ & A_0 = 0 \ , \\ \text{Axial gauge} \ \ & A_3 = 0 \ . \end{aligned}$$
- The general solution of the vacuum Maxwell equations ($j^\mu = 0$) takes the form:
$$A^\mu(x) = \sum_{\lambda=0}^{3} \frac{1}{(2\pi)^{\frac{3}{2}}} \int \frac{\mathrm{d}^3\boldsymbol{k}}{\sqrt{2\omega_k}} \left(a_\lambda(\boldsymbol{k})\epsilon^\mu_\lambda(\boldsymbol{k})\mathrm{e}^{-ik\cdot x} + a^\dagger_\lambda(\boldsymbol{k})\epsilon^\mu_\lambda(\boldsymbol{k})\mathrm{e}^{ik\cdot x} \right) \ , \tag{9.C}$$
where $\omega_k = |\boldsymbol{k}|$,$\epsilon^\mu_\lambda(\boldsymbol{k})$ are polarization vectors. The transverse polarization vectors which satisfy $\boldsymbol{\epsilon}(\boldsymbol{k}) \cdot \boldsymbol{k} = 0$ we denote by $\epsilon^\mu_1(\boldsymbol{k})$ and $\epsilon^\mu_2(\boldsymbol{k})$. The scalar

polarization vector is $\epsilon_0^\mu = n^\mu$, where n^μ is a unit time–like vector. We can choose $n^\mu = (1,0,0,0)$. The longitudinal polarization vector, $\epsilon_3^\mu(\boldsymbol{k})$ is given by

$$\epsilon_3^\mu(\boldsymbol{k}) = \frac{k^\mu - (n\cdot k)n^\mu}{(n\cdot k)} \ .$$

Due to gauge symmetry only two polarizations are independent. The polarization vectors satisfy the orthonormality relations:

$$g_{\mu\nu}\epsilon_\lambda^\mu(\boldsymbol{k})\epsilon_{\lambda'}^\nu(\boldsymbol{k}) = -\delta_{\lambda\lambda'} \ .$$

In (9.C) we assumed the polarization vectors to be real valued.

- The polarization vectors satisfy the following *completeness relations*:

$$\sum_{\lambda=0}^{3} g_{\lambda\lambda}\epsilon_\lambda^\mu(\boldsymbol{k})\epsilon_\lambda^\nu(\boldsymbol{k}) = g^{\mu\nu} \ . \tag{9.D}$$

From (9.D) follows that the sum over transverse photons is

$$\sum_{\lambda=1}^{2} \epsilon_\lambda^i(\boldsymbol{k})\epsilon_\lambda^j(\boldsymbol{k}) = -g^{ij} - \frac{k^ik^j}{(k\cdot n)^2} + \frac{k^in^j + k^jn^i}{k\cdot n} \ . \tag{9.E}$$

- In the Lorentz gauge *the equal-time commutation relations* are:

$$\begin{aligned}
[A^\mu(t,\boldsymbol{x}), \pi^\nu(t,\boldsymbol{y})] &= \mathrm{i}g^{\mu\nu}\delta^{(3)}(\boldsymbol{x}-\boldsymbol{y}) \ , \\
[A^\mu(t,\boldsymbol{x}), A^\nu(t,\boldsymbol{y})] &= 0 \ , \\
[\pi^\mu(t,\boldsymbol{x}), \pi^\nu(t,\boldsymbol{y})] &= 0 \ .
\end{aligned} \tag{9.F}$$

where $\pi^\nu = -\dot{A}^\nu$. Creation and annihilation operators of the photon field satisfy the following commutation relations:

$$\begin{aligned}
[a_\lambda(\boldsymbol{k}), a_{\lambda'}^\dagger(\boldsymbol{q})] &= -g_{\lambda\lambda'}\delta^{(3)}(\boldsymbol{k}-\boldsymbol{q}) \ , \\
[a_\lambda(\boldsymbol{k}), a_{\lambda'}(\boldsymbol{q})] &= 0 \ , \\
[a_\lambda^\dagger(\boldsymbol{k}), a_{\lambda'}^\dagger(\boldsymbol{q})] &= 0 \ .
\end{aligned} \tag{9.G}$$

The physical states, $|\Phi\rangle$ satisfy the operator condition

$$\partial^\mu A_\mu^{(+)}|\Phi\rangle = 0.$$

This is the Gupta–Bleuler method of quantization.

- In the Coulomb gauge we have

$$\boldsymbol{A}(x) = \sum_{\lambda=1}^{2}\frac{1}{(2\pi)^{\frac{3}{2}}}\int\frac{\mathrm{d}^3\boldsymbol{k}}{\sqrt{2\omega_k}}\left(a_\lambda(\boldsymbol{k})\boldsymbol{\epsilon}_\lambda(\boldsymbol{k})\mathrm{e}^{-ik\cdot x} + a_\lambda^\dagger(\boldsymbol{k})\boldsymbol{\epsilon}_\lambda(\boldsymbol{k})\mathrm{e}^{ik\cdot x}\right) \ , \tag{9.H}$$

while $A^0 = 0$. *The equal-time commutation relations* are:

$$\begin{aligned}
[A^i(t,\boldsymbol{x}), \pi^j(t,\boldsymbol{y})] &= -\mathrm{i}\delta^{(3)}_{\perp ij}(\boldsymbol{x}-\boldsymbol{y}) \ , \\
[A^i(t,\boldsymbol{x}), A^j(t,\boldsymbol{y})] &= 0 \ , \\
[\pi^i(t,\boldsymbol{x}), \pi^j(t,\boldsymbol{y})] &= 0 \ ,
\end{aligned} \tag{9.I}$$

where $\boldsymbol{\pi} = \boldsymbol{E}$ and $\delta^{(3)}_{\perp ij}(\boldsymbol{x}-\boldsymbol{y})$ is the transversal delta function given by

$$\delta^{(3)}_{\perp ij}(\boldsymbol{x}-\boldsymbol{y}) = \frac{1}{(2\pi)^3}\int \mathrm{d}^3\boldsymbol{k}\mathrm{e}^{\mathrm{i}\boldsymbol{k}\cdot(\boldsymbol{x}-\boldsymbol{y})}\left(\delta_{ij} - \frac{k_i k_j}{\boldsymbol{k}^2}\right) .$$

Creation and annihilation operators obey

$$\begin{aligned}
[a_\lambda(\boldsymbol{k}), a^\dagger_{\lambda'}(\boldsymbol{q})] &= \delta_{\lambda\lambda'}\delta^{(3)}(\boldsymbol{k}-\boldsymbol{q}) \ , \\
[a_\lambda(\boldsymbol{k}), a_{\lambda'}(\boldsymbol{q})] &= 0 \ , \\
[a^\dagger_\lambda(\boldsymbol{k}), a^\dagger_{\lambda'}(\boldsymbol{q})] &= 0 \ .
\end{aligned} \tag{9.J}$$

- *The Feynman propagator* for the electromagnetic field is given by

$$\mathrm{i}D_F^{\mu\nu}(x-y) = \langle 0|\, T(A^\mu(x)A^\nu(y))\, |0\rangle \ . \tag{9.K}$$

9.1. Starting from the commutation relations (9.G) prove that

$$[A^\mu(t,\boldsymbol{x}), \dot{A}^\nu(t,\boldsymbol{y})] = -\mathrm{i}g^{\mu\nu}\delta^{(3)}(\boldsymbol{x}-\boldsymbol{y}) \ .$$

9.2. Find the commutator

$$\mathrm{i}D^{\mu\nu}(x-y) = [A^\mu(x), A^\nu(y)] \ ,$$

in the Lorentz gauge.

9.3. Calculate the commutators between components of the electric and the magnetic fields:

$$\begin{aligned}
&[E^i(x), E^j(y)] \ , \\
&[B^i(x), B^j(y)] \ , \\
&[E^i(x), B^j(y)] \ .
\end{aligned}$$

Also calculate the previous commutators for equal times, $x^0 = y^0$.

9.4. Prove that $[P^\mu, A^\nu] = -\mathrm{i}\partial^\mu A^\nu$.

9.5. Determine the helicity of photons described by polarization vectors $\epsilon_+^\mu(k\boldsymbol{e}_z) = 2^{-1/2}(0,1,\mathrm{i},0)^\mathrm{T}$ and $\epsilon_-^\mu(k\boldsymbol{e}_z) = 2^{-1/2}(0,1,-\mathrm{i},0)^\mathrm{T}$.

9.6. A photon linearly polarized along the x–axis is moving along the z–direction with momentum $\boldsymbol{k}$. Determine the polarization of the photon for observer S' moving in the x–direction with velocity $\boldsymbol{v}$.

9.7. The arbitrary state not containing transversal photons has the form

$$|\Phi\rangle = \sum_n C_n |\Phi_n\rangle \ ,$$

where C_n are constants and

$$|\Phi_n\rangle = \int \mathrm{d}^3\boldsymbol{k}_1 \dots \mathrm{d}^3\boldsymbol{k}_n f(\boldsymbol{k}_1, \dots, \boldsymbol{k}_n) \prod_{i=1}^{n} (a_0^\dagger(\boldsymbol{k}_i) - a_3^\dagger(\boldsymbol{k}_i)) |0\rangle \ ,$$

where $f(\boldsymbol{k}_1, \dots, \boldsymbol{k}_n)$ are arbitrary functions. The state $|\Phi_0\rangle$ is a vacuum.

(a) Prove that $\langle\Phi_n|\Phi_n\rangle = \delta_{n,0}$.
(b) Show that $\langle\Phi| A^\mu(x) |\Phi\rangle$ is a pure gauge.

9.8. Let

$$P^{\mu\nu} = g^{\mu\nu} - \frac{k^\mu \bar{k}^\nu + k^\nu \bar{k}^\mu}{k \cdot \bar{k}} \ ,$$

and

$$P_\perp^{\mu\nu} = \frac{k^\mu \bar{k}^\nu + k^\nu \bar{k}^\mu}{k \cdot \bar{k}} \ ,$$

where $\bar{k}^\mu = (k^0, -\boldsymbol{k})$.
Calculate: $P^{\mu\nu}P_{\nu\sigma}$, $P_\perp^{\mu\nu}P_{\nu\sigma}^\perp$, $P^{\mu\nu} + P_\perp^{\mu\nu}$, $g^{\mu\nu}P_{\mu\nu}$, $g^{\mu\nu}P_{\mu\nu}^\perp$, $P^\mu{}_\nu P_\perp^{\nu\sigma}$, if $k^2 = 0$.

9.9. The angular momentum of the photon field is defined by $J^l = \frac{1}{2}\epsilon^{lij}M^{ij}$, where M^{ij} was found in Problem 5.18.

(a) Express $\boldsymbol{J}$ in terms of the potentials in the Coulomb gauge.
(b) Express the spin part of the angular momentum in terms of $a_\lambda(\boldsymbol{k})$, $a_\lambda^\dagger(\boldsymbol{k})$ and diagonalize it.
(c) Show that the states

$$a_\pm^\dagger(\boldsymbol{q})|0\rangle = \frac{1}{\sqrt{2}}(a_1^\dagger(\boldsymbol{q}) \pm \mathrm{i}a_2^\dagger(\boldsymbol{q}))|0\rangle \ ,$$

are the eigenstates of the helicity operator with the eigenvalues ± 1.
(d) Calculate the commutator $[J^l, A^m(\boldsymbol{y}, t)]$.

9.10. Calculate:

(a) $\langle 0| \{E^i(x), B^j(y)\} |0\rangle$,
(b) $\langle 0| \{B^i(x), B^j(y)\} |0\rangle$,

(c) $\langle 0| \{E^i(x), E^j(y)\} |0\rangle$.

9.11. Consider the quantization of the electromagnetic field in space between two parallel square plates located at $z = 0$ and $z = a$. The plates are squares with size of length L. They are perfect conductors.

(a) Find the general solution for the electromagnetic potential inside this capacitor.
(b) Quantize the electromagnetic field using canonical quantization.
(c) Find the Hamiltonian H and show that the vacuum energy is

$$E = \frac{1}{2}L^2 \int \frac{d^2k}{(2\pi)^2} \left[2\sum_{n=1}^{\infty} \sqrt{k_1^2 + k_2^2 + \left(\frac{n\pi}{a}\right)^2} + \sqrt{k_1^2 + k_2^2} \right] . \tag{9.1}$$

(d) Define the quantity

$$\epsilon = \frac{E - E_0}{L^2} ,$$

which is the difference between the vacuum energies per unit area in the presence and in the absent of plates. This quantity is divergent and can be regularized introducing the function

$$f(k) = \begin{cases} 1, & k < \Lambda \\ 0, & k > \Lambda \end{cases} ,$$

into the integral; Λ is a cutoff parameter. Calculate ϵ and show that there is an attractive force between the plates. This is *the Casimir effect.*
(e) The energy per unit area, E/L^2 can be regularized in a different way. Calculate integral

$$I = \int d^2k \frac{1}{(k^2 + m^2)^\alpha} ,$$

for $\operatorname{Re}\alpha > 0$, and then analitically continue this integral to $\operatorname{Re}\alpha \leq 0$. Show that

$$E/L^2 = -\frac{\pi^2}{6a^3} \sum_{n=1}^{\infty} n^3 .$$

Regularize the sum in the previous expression using the Rieman ζ–function

$$\zeta(s) = \sum_{n=1}^{\infty} n^{-s} .$$

Calculate the energy and the force per unit area.

10

Processes in the lowest order of perturbation theory

- *The Wick's theorem* states

$$T(ABC\dots YZ) =: \{ABC\dots YZ + \text{"all contractions"}\} : . \tag{10.A}$$

In the case of fermions we have to take care about anticommutation relations, i.e. every time when we interchange neighboring fermionic operators a minus sign appears.

- *The S–matrix* is given by

$$S = \sum_{n=0}^{\infty} \frac{(-\mathrm{i})^n}{n!} \int \dots \int \mathrm{d}^4x_1 \dots \mathrm{d}^4x_n \, T\left(\mathcal{H}_\mathrm{I}(x_1) \cdots \mathcal{H}_\mathrm{I}(x_n)\right) , \tag{10.B}$$

where $\mathcal{H}_\mathrm{I}$ is the Hamiltonian density of interaction in the interaction picture.

- *S*–matrix elements have the general form

$$S_\mathrm{fi} = (2\pi)^4 \delta^{(4)}(p_\mathrm{f} - p_\mathrm{i}) \mathrm{i}\mathcal{M} \prod_b \frac{1}{\sqrt{2VE}} \prod_f \sqrt{\frac{m}{VE}} , \tag{10.C}$$

where p_i and p_f are the initial and the final momenta, respectively; $\mathrm{i}\mathcal{M}$ is the Feynman amplitude for the process, which will be determined using Feynman diagrams. The delta function in (10.C) is a consequence of the conservation of energy and momentum in the process. Normalization factors also appear in the expression (10.C) and they are different for bosonic and fermionic particles. In this Chapter we will use so–called box normalization.

- *The differential cross section* for the scattering of two particles into N final particles is

$$\mathrm{d}\sigma = \frac{|S_\mathrm{fi}|^2}{T} \frac{1}{|\boldsymbol{J}_\mathrm{in}|} \prod_{i=1}^{N} \frac{V\mathrm{d}^3\boldsymbol{p}_\mathrm{i}}{(2\pi)^3} , \tag{10.D}$$

where $\boldsymbol{J}_\mathrm{in}$ is the flux of initial particles:

$$|\boldsymbol{J}_\mathrm{in}| = \frac{v_\mathrm{rel}}{V} .$$

The relative velocity v_{rel} is given by

$$v_{\text{rel}} = \frac{|\boldsymbol{p}_1|}{E_1} \ ,$$

in the laboratory frame of reference (particle 2 is at rest), while in the center–of–mass frame we have

$$v_{\text{rel}} = |\boldsymbol{p}_1| \frac{E_1 + E_2}{E_1 E_2} \ ,$$

$\boldsymbol{p}_1$ is the momentum of particle 1, and $E_{1,2}$ are energies of particles. In expression (10.D), $V \mathrm{d}^3\boldsymbol{p}/(2\pi)^3$ is the volume element of phase space.

- *Feynman rules for QED*:
  - Vertex:

$$= \mathrm{i}e\gamma^{\mu}$$

  - Photon and lepton propagators:

$\mathrm{i}D_{F\mu\nu} =$ 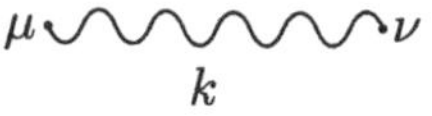

$= -\dfrac{\mathrm{i}g_{\mu\nu}}{k^2 + \mathrm{i}\epsilon},$

$\mathrm{i}S_F(p) =$ 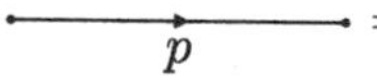

$= \dfrac{\mathrm{i}}{\not p - m + \mathrm{i}\epsilon}.$

  - External lines:

a) leptons (e.g. electron): 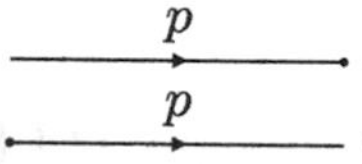

$= u(p,s)$ final
$= \bar{u}(p,s)$ initial

b) antileptons (e.g. positron): 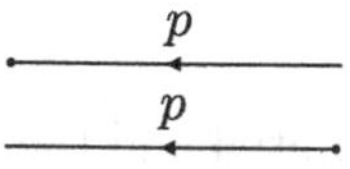

$= v(p,s)$ final
$= \bar{v}(p,s)$ initial

c) photons:

$= \varepsilon_{\mu}(k,\lambda)$ final

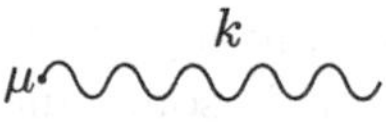

$= \varepsilon^*_{\mu}(k,\lambda)$ initial

  - Spinor factor are written from the left to the right along each of the fermionic lines. The order of writing is important, because it is a question of matrix multiplication of the corresponding factors.
  - For all loops with momentum k, we must integrate over the momentum: $\int \mathrm{d}^4k/(2\pi)^4$. This corresponds to the addition of quantum mechanical amplitudes.
  - For fermion loops we have to take the trace and multiply it by the factor -1.

- If two diagrams differ for an odd number of fermionic interchanges, then they must differ by a relative minus sign.

10.1. For the process

$$\mathrm{A}(E_1, \boldsymbol{p}_1) + \mathrm{B}(E_2, \boldsymbol{p}_2) \to \mathrm{C}(E_1', \boldsymbol{p}_1') + \mathrm{D}(E_2', \boldsymbol{p}_2')$$

prove that the differential cross section in the center of mass frame is given by

$$\left(\frac{\mathrm{d}\sigma}{\mathrm{d}\Omega}\right)_{\mathrm{cm}} = \frac{1}{4\pi^2(E_1+E_2)^2}\frac{|\boldsymbol{p}_1'|}{|\boldsymbol{p}_1|} m_{\mathrm{A}} m_{\mathrm{B}} m_{\mathrm{C}} m_{\mathrm{D}} |\mathcal{M}|^2 ,$$

where $\mathrm{i}\mathcal{M}$ is the Feynman amplitude. Assume that all particles in the process are fermions.

10.2. Consider the following integral:

$$I = \int \frac{\mathrm{d}^3\boldsymbol{p}}{2E_p}\frac{d^3\boldsymbol{q}}{2E_q}\delta^{(3)}(\boldsymbol{p}+\boldsymbol{q}-\boldsymbol{P})\delta(E_p+E_q-P^0) ,$$

where $E_p^2 = \boldsymbol{p}^2 + m^2$ and $E_q^2 = \boldsymbol{q}^2 + m'^2$. Show that the integral I is Lorentz invariant. Calculate it in the frame where $\boldsymbol{P} = 0$.

10.3. If

$$\mathrm{i}\mathcal{M} = \bar{u}(\boldsymbol{p}, r)\gamma_\mu(1-\gamma_5)u(\boldsymbol{q}, s)\epsilon^\mu(\boldsymbol{k}, \lambda) ,$$

calculate the sum

$$\sum_{\lambda=1}^{2}\sum_{r,s=1}^{2} |\mathcal{M}|^2 .$$

10.4. Using the Wick theorem evaluate:

(a) $\langle 0|\, T(\phi^4(x)\phi^4(y))\, |0\rangle$,
(b) $T(:\phi^4(x): \, :\phi^4(y):)$,
(c) $\langle 0|\, T(\bar{\psi}(x)\psi(x)\bar{\psi}(y)\psi(y))\, |0\rangle$.

10.5. In ϕ^4 theory the interaction Lagrangian density is $\mathcal{L}_{int} = -\frac{\lambda}{4!}\phi^4$. Using the Wick theorem determine the symmetry factor S, for the following diagrams:

(a)

(b)

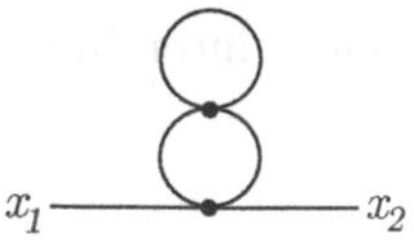

(c)

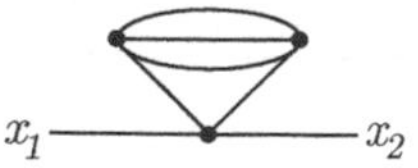

Also, check the results using the formula [6]:

$$S = g \prod_{n=2,3,..} 2^{\beta}(n!)^{\alpha_n} \ ,$$

where g is the number of possible permutations of vertices which leave unchanged the diagram with fixed external lines, α_n is the number of vertex pairs connected by n identical lines, and β is the number of lines connecting a vertex with itself.

10.6. In ϕ^3 theory calculate

$$\frac{1}{2}\left(\frac{-\mathrm{i}\lambda}{3!}\right)^2 \int \mathrm{d}^4y_1\mathrm{d}^4y_2 \, \langle 0| \, T(\phi(x_1)\phi(x_2)\phi^3(y_1)\phi^3(y_2)) \, |0\rangle \ .$$

10.7. For the QED processes :

(a) $\mu^-\mu^+ \to e^-e^+$,
(b) $e^-\mu^+ \to e^-\mu^+$,

write the expressions for amplitudes using Feynman rules. Calculate $\langle |\mathcal{M}|^2 \rangle$ averaging over all initial polarization states and summing over the final polarization states of particles. Calculate the differential cross sections in center–of–mass system in an ultrarelativistic limit.

10.8. Show that the Feynman amplitude for the Compton scattering is a gauge invariant quantity.

10.9. Find the differential cross section for the scattering of an electron in the external electromagnetic field (a, g, k are constants)

(a) $A^{\mu}(x) = (a\mathrm{e}^{-k^2\boldsymbol{x}^2}, \ 0, \ 0, \ 0)$,
(b) $A^{\mu}(x) = (0, \ 0, \ 0, \ \frac{g}{r}\mathrm{e}^{-r/a})$.

The initial electron is moving along z–axis.

10.10. Calculate the cross section per unit volume for the creation of electron–positron pairs by the electromagnetic potential

$$A^{\mu} = (0, \ 0, \ a\mathrm{e}^{-\mathrm{i}\omega t}, \ 0) \ ,$$

where ω and a are constants.

10.11. Find the differential cross section for the scattering of an electron in the external potential

$$A^{\mu} = (0,\ 0,\ 0,\ ae^{-k^2\boldsymbol{x}^2})\ ,$$

for a theory which is the same as QED except the fact that the vertex $\mathrm{i}e\gamma_{\mu}$ is replaced by $\mathrm{i}e\gamma_{\mu}(1-\gamma_5)$. The initial electron is moving along z–axis.

10.12. Find the differential cross section for the scattering of a positron in the external potential

$$A^{\mu} = (\frac{g}{r},\ 0,\ 0,\ 0)\ ,$$

where g is a constant. The S–matrix element is given by

$$S_{\mathrm{fi}} = \mathrm{i}e \int \mathrm{d}^4x \bar{\psi}_{\mathrm{f}}(x)\partial_{\mu}\psi_{\mathrm{i}}(x)A^{\mu}(x)\ .$$

10.13. Calculate the cross section for the scattering of an electron with positive helicity in the electromagnetic potential

$$A^{\mu} = (a\delta^{(3)}(\boldsymbol{x}),\ 0,\ 0,\ 0)\ ,$$

where a is a constant.

10.14. Calculate the differential cross section for scattering of e^- and a muon μ^+

$$e^-\mu^+ \to e^-\mu^+\ ,$$

in the center–of–mass system. Assume that initial particles have negative helicity, while the spin states of final particles are arbitrary.

10.15. Consider the theory of interaction of a spinor and scalar field:

$$\mathcal{L} = \frac{1}{2}(\partial\phi)^2 - \frac{M^2}{2}\phi^2 + \bar{\psi}(\mathrm{i}\gamma_{\mu}\partial^{\mu} - m)\psi - g\bar{\psi}\gamma_5\psi\phi\ .$$

Calculate the cross section for the scattering of two fermions in the lowest order.

10.16. Write the expressions for the Feynman amplitudes for diagrams given in the figure.

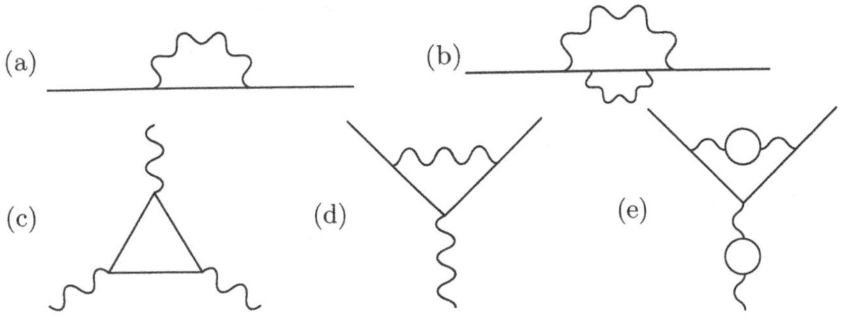

(f) 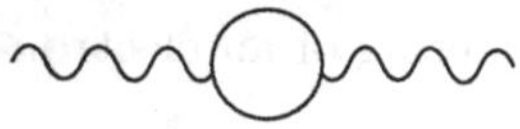

(g) 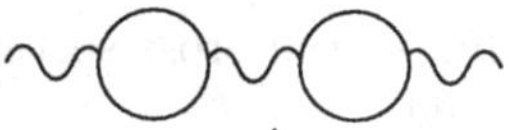

(h) 

(i) 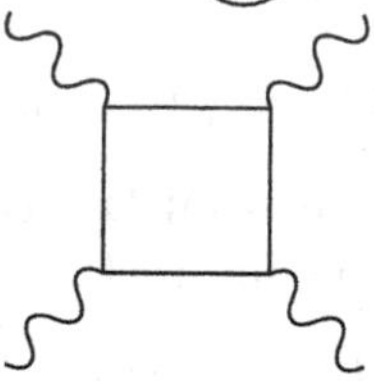

11

Renormalization and regularization

- Table of D-dimensional integrals in Minkowski spacetime:

$$\int \mathrm{d}^D k \frac{1}{(k^2 + 2p\cdot k - m^2 + \mathrm{i}\epsilon)^n} = \frac{\mathrm{i}(-1)^n \pi^{\frac{D}{2}}}{\Gamma(n)(m^2+p^2)^{n-\frac{D}{2}}} \Gamma(n - \frac{D}{2}), \quad (11.A)$$

$$\int \mathrm{d}^D k \frac{k^\mu}{(k^2 + 2p\cdot k - m^2 + \mathrm{i}\epsilon)^n} = \frac{-\mathrm{i}(-1)^n \pi^{\frac{D}{2}}}{\Gamma(n)(m^2+p^2)^{n-\frac{D}{2}}} p^\mu \Gamma(n - \frac{D}{2}), \quad (11.B)$$

$$\int \mathrm{d}^D k \frac{k^\mu k^\nu}{(k^2 + 2p\cdot k - m^2 + \mathrm{i}\epsilon)^n} = \frac{\mathrm{i}(-1)^n \pi^{\frac{D}{2}}}{\Gamma(n)(m^2+p^2)^{n-\frac{D}{2}}} \Big[p^\mu p^\nu \Gamma(n - \frac{D}{2})$$
$$- \frac{1}{2} g^{\mu\nu}(p^2+m^2)\Gamma(n - \frac{D}{2} - 1)\Big], \quad (11.C)$$

$$\int \mathrm{d}^D k \frac{k^\mu k^\nu k^\rho}{(k^2 + 2p\cdot k - m^2 + \mathrm{i}\epsilon)^n} = \frac{-\mathrm{i}(-1)^n \pi^{\frac{D}{2}}}{\Gamma(n)(m^2+p^2)^{n-\frac{D}{2}}} \Big[p^\mu p^\nu p^\rho \Gamma(n - \frac{D}{2})$$
$$- \frac{1}{2}(g^{\mu\nu}p^\rho + g^{\mu\rho}p^\nu + g^{\nu\rho}p^\mu)(p^2+m^2)\Gamma(n - \frac{D}{2} - 1)\Big], \quad (11.D)$$

$$\int \mathrm{d}^D k \frac{k^\mu k^\nu k^\rho k^\sigma}{(k^2 + 2p\cdot k - m^2 + \mathrm{i}\epsilon)^n} = \frac{\mathrm{i}(-1)^n \pi^{D/2}}{\Gamma(n)(m^2+p^2)^{n-\frac{D}{2}}} \Big[p^\mu p^\nu p^\rho p^\sigma \Gamma(n - \frac{D}{2})$$
$$- \frac{1}{2}(g^{\mu\nu}p^\rho p^\sigma + g^{\mu\rho}p^\nu p^\sigma + g^{\mu\sigma}p^\nu p^\rho + g^{\nu\rho}p^\mu p^\sigma + g^{\nu\sigma}p^\rho p^\mu + g^{\rho\sigma}p^\mu p^\nu)$$
$$\times (p^2+m^2)\Gamma(n - \frac{D}{2} - 1)$$
$$+ \frac{1}{4}(g_{\mu\nu}g_{\rho\sigma} + g_{\mu\rho}g_{\nu\sigma} + g_{\mu\sigma}g_{\rho\nu})(p^2+m^2)^2 \Gamma\left(n - \frac{D}{2} - 2\right)\Big]. \quad (11.E)$$

- The gamma–function obeys

$$\Gamma(-n+\epsilon)=\frac{(-1)^n}{n!}\left(\frac{1}{\epsilon}+\psi(n+1)+o(\epsilon)\right)\,, \tag{11.F}$$

where $n \in N$ and

$$\psi(n+1)=1+\frac{1}{2}+\ldots+\frac{1}{n}-\gamma\,.$$

The $\gamma = 0,5772$ is the Euler–Mascheroni constant.

- The general expression for *Feynman parametrization* is given in Problem 11.1. The most frequently used parameterizations are:

$$\frac{1}{AB}=\int_0^1 \mathrm{d}x\frac{1}{[xA+(1-x)B]^2}\,, \tag{11.G}$$

$$\frac{1}{ABC}=2\int_0^1 \mathrm{d}x\int_0^{1-x}\mathrm{d}z\frac{1}{[A+(B-A)x+(C-A)z]^3}\,. \tag{11.H}$$

- *Cutkosky rule* for computing discontinuity of any Feynman diagram contains the following steps:
  1. Cut through the diagram in all possible ways such that the cut propagators can be put on–shell.
  2. For each cut, make the replacement

  $$\frac{1}{p^2-m^2}\to(-2i\pi)\delta^{(4)}(p^2-m^2)\theta(p^0)\,.$$

  3. Sum the contributions of all possible cuts.

11.1. Prove the following formula (the Feynman parametrization)

$$\frac{1}{A_1\ldots A_n}=(n-1)!\int_0^1\cdots\int_0^1 \mathrm{d}x_1\ldots\mathrm{d}x_n\frac{\delta(x_1+\ldots+x_n-1)}{(x_1A_1+\ldots+x_nA_n)^n}\,.$$

11.2. Show that expression (11.A) holds.

11.3. Prove the formula (11.F).

11.4. Regularize the integral

$$I=\int \mathrm{d}^4k\frac{1}{k^2}\frac{1}{(k+p)^2-m^2}\,,$$

using Pauli–Villars regularization.

11.5. Compute

$$I_{\alpha\beta\mu\nu\rho\sigma} = \int \mathrm{d}^D k \frac{k_\alpha k_\beta k_\mu k_\nu k_\rho k_\sigma}{(k^2)^n} \,.$$

Also, find the divergent part of the previous integral for $n = 5$. Apply the dimensional regularization.

11.6. Consider the interacting theory of two scalar fields ϕ and χ:

$$\mathcal{L} = \frac{1}{2}(\partial\phi)^2 - \frac{1}{2}m^2\phi^2 + \frac{1}{2}(\partial\chi)^2 - \frac{1}{2}M^2\chi^2 - g\phi^2\chi \,.$$

(a) Find the self–energy of the χ particle, $-\mathrm{i}\Pi(p^2)$.
(b) Calculate the decay rate of the χ particle into two ϕ particles.
(c) Prove that

$$\mathrm{Im}\, \Pi(M^2) = -M\Gamma.$$

11.7. Consider the theory

$$\mathcal{L} = \frac{1}{2}(\partial_\mu\phi)^2 - \frac{m^2}{2}\phi^2 - \frac{g}{3!}\phi^3 - \frac{\lambda}{4!}\phi^4 \,.$$

Find the expression for the self–energy and the mass shift δm.

11.8. The Lagrangian density is given by

$$\mathcal{L} = \frac{1}{2}(\partial_\mu\sigma)^2 + \frac{1}{2}(\partial_\mu\pi)^2 - \frac{m^2}{2}\sigma^2 - \lambda v\sigma^3 - \lambda v\sigma\pi^2 - \frac{\lambda}{4}(\sigma^2 + \pi^2)^2 \,,$$

where σ and π are scalar fields, and $v^2 = \frac{m^2}{2\lambda}$ is constant. Classically, π field is massless. Show that it also remains massless when the one–loop corrections are included.

11.9. Find the divergent part of the diagram

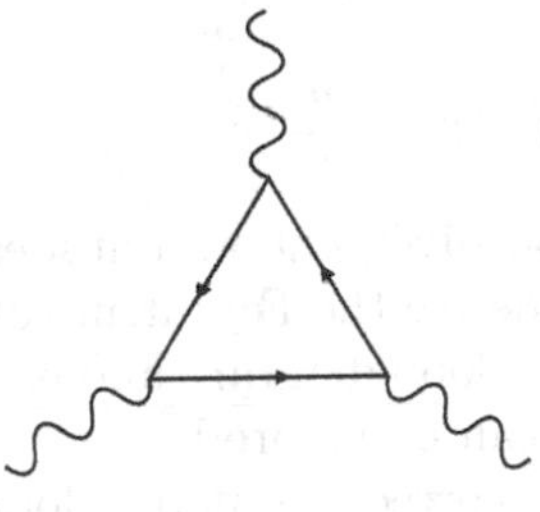

Prove that this diagram cancels with the diagram of the reverse orientation inside the fermion loop.

11.10. The polarization of vacuum in QED has form

$$-\mathrm{i}\Pi_{\mu\nu}(q) = -\mathrm{i}(q_\mu q_\nu - q^2 g_{\mu\nu})\Pi(q^2) \,.$$

Prove the following expression:

$$\mathrm{Im}\, \Pi(q^2) = -\frac{e^2}{12\pi}\left(1 + \frac{2m^2}{q^2}\right)\sqrt{1 - \frac{4m^2}{q^2}}\,\theta\left(1 - \frac{4m^2}{q^2}\right) \,.$$

11.11. In scalar electrodynamics two diagrams give contribution to the polarization of vacuum. Using dimensional regularization derive the following expression for the divergent part of the vacuum polarization:

$$\frac{ie^2}{24\pi^2}\frac{1}{\epsilon}(p_\mu p_\nu - p^2 g_{\mu\nu}) \, .$$

11.12. The Lagrangian density for the pseudoscalar Yukawa theory is given by

$$\mathcal{L} = \frac{1}{2}(\partial\phi)^2 - \frac{m^2}{2}\phi^2 + \bar{\psi}(i\gamma_\mu\partial^\mu - M)\psi - ig\bar{\psi}\gamma_5\psi\phi - \frac{\lambda}{4!}\phi^4 \, .$$

(a) Find the superficial degree of divergence for this theory and the corresponding divergent amplitudes. Write the bare Lagrangian density as a sum of the initial Lagrangian density and counterterms. Write out the Feynman rules in the renormalized theory.
(b) Find the self–energy of the spinor field at one–loop and determine the corresponding counterterms.
(c) Find the self–energy of the scalar field at one–loop and determine the corresponding counterterms.
(d) Calculate the one–loop vertex correction $\phi\bar{\psi}\psi$ and δg.
(e) Calculate the one–loop vertex correction ϕ^4 and $\delta\lambda$.

11.13. Consider massless two–dimensional QED, the so–called Schwinger model.

(a) Calculate the vacuum polarization at one–loop.
(b) Find the full photon propagator and read off the mass of the photon.

11.14. Consider ϕ^3 theory in six–dimensional spacetime, with the Lagrangian density given by

$$\mathcal{L} = \frac{1}{2}(\partial\phi)^2 - \frac{m^2}{2}\phi^2 - \frac{g}{3!}\phi^3 - h\phi \, .$$

(a) Determine the superficial divergent amplitudes. Write the renormalized Lagrangian density and derive the Feynman rules.
(b) Calculate the tadpole one–loop diagram and explain why the contribution of the tadpole diagrams can be ignored.
(c) Calculate the propagator correction at one–loop order and determine δZ and δm. Use the minimal subtraction (MS) scheme.
(d) Calculate the vertex correction and find δg.
(e) Derive the relations $m_0 = m_0(m, g, \epsilon)$ and $g_0 = g_0(m, g, \epsilon)$.

Part II

Solutions

1

Lorentz and Poincaré symmetries

1.1 The square of the length of a four–vector, $\boldsymbol{x}$ is $x^2 = g_{\mu\nu}x^\mu x^\nu$. By substituting $x'^\mu = \Lambda^\mu{}_\rho x^\rho$ into the condition $x'^2 = x^2$ one obtains:

$$g_{\mu\nu}\Lambda^\mu{}_\rho \Lambda^\nu{}_\sigma x^\rho x^\sigma = g_{\rho\sigma}x^\rho x^\sigma \; . \tag{1.1}$$

Since (1.1) is valid for any vector $\boldsymbol{x} \in M_4$, we get $\Lambda^\mu{}_\rho g_{\mu\nu}\Lambda^\nu{}_\sigma = g_{\rho\sigma}$. The previous condition can be rewritten in the following form

$$(\Lambda^T)_\rho{}^\mu g_{\mu\nu}\Lambda^\nu{}_\sigma = g_{\rho\sigma} \Rightarrow \Lambda^T g\Lambda = g \; , \tag{1.2}$$

and we have obtained the requested expression.

Now, we shall show that the Lorentz transformations form a group. If Λ_1 and Λ_2 are Lorentz transformations then their product, $\Lambda_1\Lambda_2$ is Lorentz transformation because it satisfies the condition (1.2):

$$(\Lambda_1\Lambda_2)^T g(\Lambda_1\Lambda_2) = \Lambda_2^T(\Lambda_1^T g\Lambda_1)\Lambda_2 = \Lambda_2^T g\Lambda_2 = g \; .$$

Thus, we have shown the closure axiom. Multiplication of matrices is generally an associative operation, so this property is valid for Lorentz matrices Λ. Identity matrix I satisfies the condition (1.2) and it is the unit element of the group. Taking determinant of the expression (1.2) we obtain $\det\Lambda = \pm 1$. Since $\det\Lambda \neq 0$ the inverse element Λ^{-1} exists for every Lorentz matrix. From (1.2) we see that the inverse element is given by $\Lambda^{-1} = g^{-1}\Lambda^T g$. In the component notation the previous relation takes the following form:

$$(\Lambda^{-1})^\mu{}_\nu = g^{\mu\rho}\Lambda^\sigma{}_\rho g_{\sigma\nu} = \Lambda_\nu{}^\mu \; .$$

1.2 By substituting infinitesimal form of the Lorentz transformation into the formula (1.2), one gets:

$$(\delta^\mu_\rho + \omega^\mu{}_\rho)g_{\mu\nu}(\delta^\nu_\sigma + \omega^\nu{}_\sigma) + o(\omega^2) = g_{\rho\sigma} \; ,$$

$$g_{\rho\sigma} + \omega^{\mu}{}_{\rho} g_{\mu\nu} \delta^{\nu}_{\sigma} + \omega^{\nu}{}_{\sigma} g_{\mu\nu} \delta^{\mu}_{\rho} + o(\omega^2) = g_{\rho\sigma} \ .$$

from which follows that

$$\omega_{\rho\sigma} + \omega_{\sigma\rho} = 0 \Rightarrow \omega_{\rho\sigma} = -\omega_{\sigma\rho} \ .$$

Since the parameters of the Lorentz group $\omega_{\mu\nu}$ are antisymmetric only six of them are independent, so the Lorentz group is six–parameters group. Moreover the Lorentz group is a Lie group.

1.3 Given relation is in agreement with definitions of the ϵ symbol and determinant.

1.4 From (1.2) follows that $\delta^{\sigma}_{\rho} = \delta^{\nu}_{\mu} \Lambda^{\mu}{}_{\rho} \Lambda_{\nu}{}^{\sigma}$, so we conclude that $\delta'^{\sigma}_{\rho} = \delta^{\sigma}_{\rho}$. In the same way we have

$$\epsilon'_{\mu\nu\rho\sigma} = \Lambda_{\mu}{}^{\alpha} \Lambda_{\nu}{}^{\beta} \Lambda_{\rho}{}^{\gamma} \Lambda_{\sigma}{}^{\delta} \epsilon_{\alpha\beta\gamma\delta} = \det(\Lambda^{-1}) \epsilon_{\mu\nu\rho\sigma} = \epsilon_{\mu\nu\rho\sigma} \ ,$$

since $\det\Lambda^{-1} = 1$ for the proper orthochronous Lorentz transformations. Thus, Levi-Civita symbol is defined independently of the inertial frame. Note that the components $\epsilon_{\mu\nu\rho\sigma}$ are obtained by applying the antisymmetric tensor ϵ on basis vectors $\boldsymbol{e}_0, \dots, \boldsymbol{e}_3$:

$$\epsilon(\boldsymbol{e}_{\mu}, \boldsymbol{e}_{\nu}, \boldsymbol{e}_{\rho}, \boldsymbol{e}_{\sigma}) = \epsilon_{\mu\nu\rho\sigma} \ .$$

The ϵ tensor can be written in the form

$$\epsilon = \boldsymbol{\theta}^0 \wedge \boldsymbol{\theta}^1 \wedge \boldsymbol{\theta}^2 \wedge \boldsymbol{\theta}^3 \ ,$$

where $\boldsymbol{\theta}^{\mu}$ are basic one-forms.

1.5 The results are given below

$$\begin{aligned}
\epsilon^{\mu\nu\rho\sigma} \epsilon_{\mu\beta\gamma\delta} &= -\delta^{\nu}_{\beta} \delta^{\rho}_{\gamma} \delta^{\sigma}_{\delta} + \delta^{\nu}_{\gamma} \delta^{\rho}_{\beta} \delta^{\sigma}_{\delta} + \delta^{\nu}_{\beta} \delta^{\rho}_{\delta} \delta^{\sigma}_{\gamma} - \delta^{\nu}_{\gamma} \delta^{\rho}_{\delta} \delta^{\sigma}_{\beta} - \delta^{\nu}_{\delta} \delta^{\rho}_{\beta} \delta^{\sigma}_{\gamma} + \delta^{\nu}_{\delta} \delta^{\rho}_{\gamma} \delta^{\sigma}_{\beta} \ , \\
\epsilon^{\mu\nu\rho\sigma} \epsilon_{\mu\nu\gamma\delta} &= -2(\delta^{\rho}_{\gamma} \delta^{\sigma}_{\delta} - \delta^{\rho}_{\delta} \delta^{\sigma}_{\gamma}) \ , \\
\epsilon^{\mu\nu\rho\sigma} \epsilon_{\mu\nu\rho\delta} &= -6 \delta^{\sigma}_{\delta} \ , \\
\epsilon^{\mu\nu\rho\sigma} \epsilon_{\mu\nu\rho\sigma} &= -24 \ .
\end{aligned}$$

1.6

(a) The matrix X is

$$X = \begin{pmatrix} x^0 - x^3 & -x^1 + ix^2 \\ -x^1 - ix^2 & x^0 + x^3 \end{pmatrix} \ ,$$

so $\det X = (x^0)^2 - (\boldsymbol{x})^2 = x^2$. It is not difficult to see that from the transformation law, $X' = SXS^{\dagger}$, follows that

$$\det X' = \det S \det X \det S^{\dagger} = \det X \ ,$$

which means that $x'^2 = x^2$.

(b) Multiplying the expression $X = x_\mu \sigma^\mu$ by $\bar{\sigma}^\nu$ and taking trace we obtain the requested relation. The matrices σ^μ satisfy the following orthogonality relation $\mathrm{tr}[\bar{\sigma}^\mu \sigma^\nu] = 2g^{\mu\nu}$.

1.7 The result follows from

$$x'^\mu = \frac{1}{2}\mathrm{tr}(\bar{\sigma}^\mu X') = \frac{1}{2}x^\nu \mathrm{tr}(\bar{\sigma}^\mu S \sigma_\nu S^\dagger) = \Lambda^\mu{}_\nu x^\nu \ .$$

1.8 An arbitrary Lorentz transformation, which is connected with the unit element, can be written in the form $U(\omega) = \exp\left(-\frac{\mathrm{i}}{2}M_{\mu\nu}\omega^{\mu\nu}\right)$, where $M_{\mu\nu}$ are generators. There are three (independent) rotations and three (also independent) boosts. Rotation around $z-$axis for angle θ_3 is represented by matrix

$$\Lambda(\theta_3) = \begin{pmatrix} 1 & 0 & 0 & 0 \\ 0 & \cos\theta_3 & \sin\theta_3 & 0 \\ 0 & -\sin\theta_3 & \cos\theta_3 & 0 \\ 0 & 0 & 0 & 1 \end{pmatrix} \approx I + \begin{pmatrix} 0 & 0 & 0 & 0 \\ 0 & 0 & \theta_3 & 0 \\ 0 & -\theta_3 & 0 & 0 \\ 0 & 0 & 0 & 1 \end{pmatrix} .$$

From the previous expression we conclude that $\omega^1{}_2 = -\omega_{12} = \theta_3$. The generator of this transformation is

$$M_{12} = \mathrm{i}\frac{\mathrm{d}\Lambda(\theta_3)}{\mathrm{d}\omega^{12}}\bigg|_{\omega_{12}=0} = -\,\mathrm{i}\frac{\mathrm{d}\Lambda(\theta_3)}{\mathrm{d}\theta_3}\bigg|_{\theta_3=0} = \mathrm{i}\begin{pmatrix} 0 & 0 & 0 & 0 \\ 0 & 0 & -1 & 0 \\ 0 & 1 & 0 & 0 \\ 0 & 0 & 0 & 0 \end{pmatrix} . \qquad (1.3)$$

In the same way we obtain the other two generators:

$$M_{13} = \mathrm{i}\begin{pmatrix} 0 & 0 & 0 & 0 \\ 0 & 0 & 0 & -1 \\ 0 & 0 & 0 & 0 \\ 0 & 1 & 0 & 0 \end{pmatrix}, \quad M_{23} = \mathrm{i}\begin{pmatrix} 0 & 0 & 0 & 0 \\ 0 & 0 & 0 & 0 \\ 0 & 0 & 0 & -1 \\ 0 & 0 & 1 & 0 \end{pmatrix} . \qquad (1.4)$$

In this case the relation between the parameters ω_{ij} and the angles of rotations θ_i around x_i-axis is $\theta_i = -\frac{1}{2}\epsilon_{ijk}\omega_{jk}$.

The matrix of the boost along $x-$axis is

$$\Lambda(\varphi_1) = \begin{pmatrix} \mathrm{ch}\varphi_1 & -\mathrm{sh}\varphi_1 & 0 & 0 \\ -\mathrm{sh}\varphi_1 & \mathrm{sh}\varphi_1 & 0 & 0 \\ 0 & 0 & 1 & 0 \\ 0 & 0 & 0 & 1 \end{pmatrix} \approx I + \begin{pmatrix} 0 & -\varphi_1 & 0 & 0 \\ -\varphi_1 & 0 & 0 & 0 \\ 0 & 0 & 0 & 0 \\ 0 & 0 & 0 & 0 \end{pmatrix} ,$$

where $\omega^0{}_1 = -\varphi_1 = -\mathrm{arc\,th}\, v_1$. The corresponding generator is

$$M_{01} = \mathrm{i}\,\frac{\mathrm{d}\Lambda(\varphi_1)}{\mathrm{d}\omega^{01}}\bigg|_{\varphi_1=0} = \mathrm{i}\,\frac{\mathrm{d}\Lambda(\varphi_1)}{\mathrm{d}\varphi_1}\bigg|_{\varphi_1=0} = -\mathrm{i}\begin{pmatrix} 0 & 1 & 0 & 0 \\ 1 & 0 & 0 & 0 \\ 0 & 0 & 0 & 0 \\ 0 & 0 & 0 & 0 \end{pmatrix} . \qquad (1.5)$$

The other two generators are

$$M_{03} = -\mathrm{i}\begin{pmatrix} 0 & 0 & 0 & 1 \\ 0 & 0 & 0 & 0 \\ 0 & 0 & 0 & 0 \\ 1 & 0 & 0 & 0 \end{pmatrix}, \quad M_{02} = -\mathrm{i}\begin{pmatrix} 0 & 0 & 1 & 0 \\ 0 & 0 & 0 & 0 \\ 1 & 0 & 0 & 0 \\ 0 & 0 & 0 & 0 \end{pmatrix}. \tag{1.6}$$

The boost parameters (rapidity) are $\omega_{oi} = -\varphi_i = -\operatorname{arc\,th}(v_i)$, where v_i is the velocity of the inertial frame moving along the x_i–axis.

1.10 The multiplication rule is

$$(\Lambda_1, a_1)(\Lambda_2, a_2) = (\Lambda_1\Lambda_2, \Lambda_1 a_2 + a_1)\ .$$

Unit element is $(I, 0)$, while the inverse is $(\Lambda, a)^{-1} = (\Lambda^{-1}, -\Lambda^{-1}a)$.

1.11

(a) Since this relation is valid in the defining representation then it is also valid in any arbitrary representation. By using this relation one gets:

$$U^{-1}(\Lambda, 0)(1 + \mathrm{i}\epsilon^{\mu}P_{\mu})U(\Lambda, 0) = 1 + \mathrm{i}(\Lambda^{-1})^{\mu}{}_{\nu}\epsilon^{\nu}P_{\mu}\ . \tag{1.7}$$

From the expression (1.7) we obtain

$$U^{-1}(\Lambda, 0)P_{\mu}U(\Lambda, 0) = (\Lambda^{-1})^{\nu}{}_{\mu}P_{\nu}\ . \tag{1.8}$$

The formula (1.8) is transformation law of the momentum P_{μ} under Lorentz transformations; the momentum is a four–vector. By substituting

$$U(\omega, 0) = \exp\left(-\frac{\mathrm{i}}{2}M_{\mu\nu}\omega^{\mu\nu}\right) = 1 - \frac{\mathrm{i}}{2}M_{\mu\nu}\omega^{\mu\nu} + o(\omega^2)$$

into (1.8) we get

$$(1 + \frac{\mathrm{i}}{2}M_{\rho\sigma}\omega^{\rho\sigma})P_{\mu}(1 - \frac{\mathrm{i}}{2}M_{\rho\sigma}\omega^{\rho\sigma}) = (\delta^{\alpha}_{\mu} - \omega^{\alpha}{}_{\mu})P_{\alpha}\ , \tag{1.9}$$

and then

$$\mathrm{i}\omega^{\rho\sigma}(M_{\rho\sigma}P_{\mu} - P_{\mu}M_{\rho\sigma}) = -\omega^{\rho\sigma}(g_{\mu\sigma}P_{\rho} - g_{\mu\rho}P_{\sigma})\ . \tag{1.10}$$

We had to antisymmetrize the right hand side of Equation (1.10) in order to eliminate antisymmetric parameters $\omega^{\rho\sigma}$. Finally, we obtain

$$[M_{\rho\sigma}, P_{\mu}] = \mathrm{i}(g_{\mu\sigma}P_{\rho} - g_{\mu\rho}P_{\sigma})\ . \tag{1.11}$$

(b) If we take an infinitesimal transformation $\Lambda' = I + \omega'$ then

$$(\Lambda^{-1}\Lambda'\Lambda)^{\mu}{}_{\nu} = \delta^{\mu}_{\nu} + (\Lambda^{-1})^{\mu}{}_{\rho}\Lambda^{\sigma}{}_{\nu}\omega'^{\rho}{}_{\sigma}\ , \tag{1.12}$$

so that

$$U^{-1}(\Lambda,0)(1-\frac{\mathrm{i}}{2}\omega'^{\rho\sigma}M_{\rho\sigma})U(\Lambda,0)=1-\frac{\mathrm{i}}{2}M_{\mu\nu}(\Lambda^{-1})^{\mu\rho}\Lambda^{\sigma\nu}\omega'_{\rho\sigma}\,. \quad (1.13)$$

From the last expression follows

$$U^{-1}(\Lambda,0)M_{\rho\sigma}U(\Lambda,0)=(\Lambda^{-1})^{\mu}{}_{\rho}(\Lambda^{-1})^{\nu}{}_{\sigma}M_{\mu\nu}\,. \quad (1.14)$$

The last equation is the transformation law of the second rank tensor. For an infinitesimal Lorentz transformation $\Lambda^{\mu}{}_{\nu}=\delta^{\mu}_{\nu}+\omega^{\mu}{}_{\nu}$ from Equation (1.14) follows

$$\frac{\mathrm{i}}{2}\omega^{\mu\nu}[M_{\mu\nu},M_{\rho\sigma}]=\frac{1}{2}\omega^{\mu\nu}(g_{\sigma\mu}M_{\rho\nu}-g_{\rho\nu}M_{\mu\sigma}-g_{\sigma\nu}M_{\rho\mu}+g_{\rho\mu}M_{\nu\sigma})\,,$$

or

$$[M_{\mu\nu},M_{\rho\sigma}]=\mathrm{i}(g_{\sigma\mu}M_{\nu\rho}+g_{\rho\nu}M_{\mu\sigma}-g_{\rho\mu}M_{\nu\sigma}-g_{\sigma\nu}M_{\mu\rho})\,. \quad (1.15)$$

(c) It is easy to prove that

$$[P_\mu,P_\nu]=0\,. \quad (1.16)$$

The relations (1.11), (1.15) and (1.16) are the commutation relations of the Poincaré algebra.

1.12 In the given representation the generator of the rotation around z–axis is

$$M_{12}=\mathrm{i}\begin{pmatrix}0&0&0&0&0\\0&0&-1&0&0\\0&1&0&0&0\\0&0&0&0&0\\0&0&0&0&0\end{pmatrix}.$$

The time translation generator has the form

$$T_0=-\mathrm{i}\begin{pmatrix}0&0&0&0&1\\0&0&0&0&0\\0&0&0&0&0\\0&0&0&0&0\\0&0&0&0&0\end{pmatrix}.$$

The other generators have similar structure and they can be computed easily. The relations (1.11), (1.15) and (1.16) are fulfilled.

1.13 Under the Poincaré transformation

$$x'=\Lambda x+a\approx x+\delta x\,,$$

a classical scalar field transforms as follows

$$\phi'(x+\delta x)=\phi(x)\,.$$

From the last relation we have

$$\phi'(x) = \phi(x - \delta x) = \phi(x) - \delta x^\mu \partial_\mu \phi \ . \tag{1.17}$$

Form variation of a scalar field is given by

$$\delta_0 \phi = \phi'(x) - \phi(x) = -\delta x^\mu \partial_\mu \phi \ . \tag{1.18}$$

For the Lorentz transformation $\delta x^\mu = \omega^\mu{}_\nu x^\nu$, and therefore

$$\delta_0 \phi = -\omega^{\mu\nu} x_\nu \partial_\mu \phi = -\frac{1}{2}\omega^{\mu\nu}(x_\nu \partial_\mu - x_\mu \partial_\nu)\phi \ . \tag{1.19}$$

On the other hand

$$\delta_0 \phi = -\frac{\mathrm{i}}{2}\omega^{\mu\nu} M_{\mu\nu}\phi \ . \tag{1.20}$$

By comparing two previous results we get that Lorentz's generators are

$$M_{\mu\nu} = \mathrm{i}(x_\mu \partial_\nu - x_\nu \partial_\mu) \ . \tag{1.21}$$

For translations $\delta x^\mu = \epsilon^\mu$ and

$$\delta_0 \phi = -\epsilon^\mu \partial_\mu \phi = \mathrm{i}\epsilon^\mu P_\mu \phi \ . \tag{1.22}$$

Hence

$$P_\mu = \mathrm{i}\partial_\mu \ . \tag{1.23}$$

Since

$$[x_\mu \partial_\nu, x_\rho \partial_\sigma] = g_{\nu\rho} x_\mu \partial_\sigma - g_{\sigma\mu} x_\rho \partial_\nu \ , \tag{1.24}$$

and

$$[x_\mu \partial_\nu, \partial_\rho] = -g_{\rho\mu}\partial_\nu \tag{1.25}$$

we get the commutation relations of the Poincaré algebra:

$$\begin{aligned}
[P_\mu, P_\nu] &= 0 \\
[M_{\rho\sigma}, P_\mu] &= \mathrm{i}(g_{\mu\sigma}P_\rho - g_{\mu\rho}P_\sigma) \\
[M_{\mu\nu}, M_{\rho\sigma}] &= \mathrm{i}(g_{\sigma\mu}M_{\nu\rho} + g_{\rho\nu}M_{\mu\sigma} - g_{\rho\mu}M_{\nu\sigma} - g_{\sigma\nu}M_{\mu\rho}) \ .
\end{aligned}$$

1.14

(a) $W_\mu P^\mu = \frac{1}{2}\epsilon_{\mu\nu\rho\sigma}M^{\nu\rho}P^\sigma P^\mu = 0$, as $P^\sigma P^\mu$ is a symmetric tensor with respect to indices σ and μ. Using the same argument, we obtain $[W_\mu, P_\nu] = 0$.

(b) Using the result of Problem 1.11 we obtain

$$\begin{aligned}
W^2 &= \frac{1}{4}\epsilon_{\mu\nu\rho\sigma}\epsilon^{\mu\alpha\beta\gamma}M^{\nu\rho}P^\sigma M_{\alpha\beta}P_\gamma \\
&= \frac{1}{4}\epsilon_{\mu\nu\rho\sigma}\epsilon^{\mu\alpha\beta\gamma}M^{\nu\rho}\left(M_{\alpha\beta}P^\sigma - \mathrm{i}\delta^\sigma_\beta P_\alpha + \mathrm{i}\delta^\sigma_\alpha P_\beta\right)P_\gamma \\
&= \frac{1}{4}\epsilon_{\mu\nu\rho\sigma}\epsilon^{\mu\alpha\beta\gamma}M^{\nu\rho}M_{\alpha\beta}P^\sigma P_\gamma \ .
\end{aligned} \tag{1.26}$$

The contraction of two ϵ symbols in the last line of (1.26) has been calculated in 1.5 so that:

$$
\begin{aligned}
W^2 &= -\frac{1}{4}(\delta^\alpha_\nu\delta^\beta_\rho\delta^\gamma_\sigma + \delta^\beta_\nu\delta^\gamma_\rho\delta^\alpha_\sigma + \delta^\gamma_\nu\delta^\alpha_\rho\delta^\beta_\sigma - \delta^\beta_\nu\delta^\alpha_\rho\delta^\gamma_\sigma - \delta^\alpha_\nu\delta^\gamma_\rho\delta^\beta_\sigma - \delta^\gamma_\nu\delta^\beta_\rho\delta^\alpha_\sigma) \\
&\times M^{\nu\rho}M_{\alpha\beta}P^\sigma P_\gamma \\
&= -\frac{1}{4}\left(2M^{\nu\rho}M_{\nu\rho}P^2 - M^{\nu\rho}M_{\nu\sigma}P^\sigma P_\rho + M^{\nu\rho}M_{\sigma\nu}P^\sigma P_\rho + \right. \\
&\left. + M^{\nu\rho}M_{\rho\sigma}P^\sigma P_\nu - M^{\nu\rho}M_{\sigma\rho}P^\sigma P_\nu\right) \\
&= -\frac{1}{2}M^{\nu\rho}M_{\nu\rho}P^2 + M^{\nu\rho}M_{\nu\sigma}P^\sigma P_\rho \; . \qquad (1.27)
\end{aligned}
$$

(c) Using the previous result we have

$$[W^2, M_{\rho\sigma}] = -\frac{1}{2}[M^{\mu\nu}M_{\mu\nu}P^2, M_{\rho\sigma}] + [M_{\mu\alpha}M^{\nu\alpha}P^\mu P_\nu, M_{\rho\sigma}] \; . \qquad (1.28)$$

The first commutator in (1.28) we denote by A, while the second one by B. Using (1.15) we obtain that $A = 0$; this result is obvious since the P^2 and $M_{\mu\nu}M^{\mu\nu}$ are Lorentz scalars. The commutator B is

$$
\begin{aligned}
B = M_{\mu\alpha}M^{\nu\alpha}&\left(P^\mu[P_\nu, M_{\rho\sigma}] + [P^\mu, M_{\rho\sigma}]P_\nu\right) + \\
&+ M_{\mu\alpha}[M^{\nu\alpha}, M_{\rho\sigma}]P^\mu P_\nu + [M_{\mu\alpha}, M_{\rho\sigma}]M^{\nu\alpha}P^\mu P_\nu \; . \qquad (1.29)
\end{aligned}
$$

Using the commutation relations (1.11) and (1.15) we get $B = 0$. Therefore, we have

$$[W^2, M_{\rho\sigma}] = 0 \; .$$

1.15 By using the result of Problem 1.14 (b) and $P^\mu\,|p^\mu, s, \sigma\rangle = p^\mu\,|p^\mu, s, \sigma\rangle$ we get

$$
\begin{aligned}
W^2\,|\boldsymbol{p} = 0, m, s, \sigma\rangle &= -m^2\left(\frac{1}{2}M^{\mu\nu}M_{\mu\nu} - M_{0i}M^{0i}\right)|\boldsymbol{p} = 0, m, s, \sigma\rangle \\
&= -\frac{1}{2}M_{ij}M^{ij}m^2\,|\boldsymbol{p} = 0, m, s, \sigma\rangle \\
&= -m^2\left((M_{12})^2 + (M_{13})^2 + (M_{23})^2\right)|\boldsymbol{p} = 0, m, s, \sigma\rangle \\
&= -m^2\boldsymbol{J}^2\,|\boldsymbol{p} = 0, m, s, \sigma\rangle \\
&= -m^2 s(s+1)\,|\boldsymbol{p} = 0, m, s, \sigma\rangle \; ,
\end{aligned}
$$

because $J_i = \frac{1}{2}\epsilon_{ijk}M_{jk}$ are the components of the angular momentum tensor.

1.16

(a) Under Lorentz transformations W_μ transforms according to:

$$U^{-1}(\Lambda)W_\sigma U(\Lambda) = \Lambda_\sigma{}^\alpha W_\alpha \; . \qquad (1.30)$$

From Equation (1.30) we have

$$\frac{\mathrm{i}}{2}[M_{\mu\nu}, W_\sigma]\omega^{\mu\nu} = \omega^{\mu\nu} g_{\sigma\mu} W_\nu = \frac{1}{2}(g_{\sigma\mu}W_\nu - g_{\sigma\nu}W_\mu)\omega^{\mu\nu} .$$

From the previous expression we easily obtain the requested result.

(b) Using the result of the previous part we have

$$\begin{aligned}[][W_\mu, W_\nu] &= \frac{1}{2}\epsilon_{\mu\alpha\beta\gamma}[M^{\alpha\beta}P^\gamma, W_\nu] \\ &= \frac{1}{2}\epsilon_{\mu\alpha\beta\gamma}\left(M^{\alpha\beta}[P^\gamma, W_\nu] + [M^{\alpha\beta}, W_\nu]P^\gamma\right) \\ &= \mathrm{i}\epsilon_{\mu\alpha\nu\gamma}W^\alpha P^\gamma .\end{aligned}$$

1.17

(a) Applying the result of Problem 1.16 (a) we get

$$[W_\mu,\ M^2] = -2\mathrm{i}(W^\alpha M_{\alpha\mu} + M_{\alpha\mu}W^\alpha) .$$

(b) $[M_{\mu\nu}, W^\mu W^\nu] = 0$. Take care that $\delta^\mu_\mu = 4$.

(c) Using the formula (1.11) we obtain $[M^2, P_\mu] = 2\mathrm{i}(P^\alpha M_{\alpha\mu} + M_{\alpha\mu}P^\alpha)$. This result and the result in the first part of this Problem are similar, since W_μ and P_μ are both four–vectors.

(d) $[\epsilon^{\mu\nu\rho\sigma}M_{\mu\nu}M_{\rho\sigma}, M_{\alpha\beta}] = 0$.

1.18 In the case of massive particles, $m^2 > 0$ since the Lorentz transformations, $\Lambda^\mu{}_\nu = \delta^\mu_\nu + \omega^\mu{}_\nu$ leave p^μ invariant (i.e. $\Lambda^\mu{}_\nu p^\nu = p^\mu$) the following relation is satisfied:

$$\begin{pmatrix} 0 & \omega_{01} & \omega_{02} & \omega_{03} \\ \omega_{01} & 0 & -\omega_{12} & -\omega_{13} \\ \omega_{02} & \omega_{12} & 0 & -\omega_{23} \\ \omega_{03} & \omega_{13} & \omega_{23} & 0 \end{pmatrix} \begin{pmatrix} m \\ 0 \\ 0 \\ 0 \end{pmatrix} = \begin{pmatrix} 0 \\ 0 \\ 0 \\ 0 \end{pmatrix} .$$

From here follows

$$\omega_{01} = \omega_{02} = \omega_{03} = 0,\ \omega_{ij} \neq 0 .$$

The corresponding generators are M^{12}, M^{13} and M^{23} and they are generators of the spatial rotations. Therefore, for massive particles little group is SO(3). The little group for the quantum mechanical Lorentz group, i.e. SL(2, C) group, is $\overline{\mathrm{SO}(3)} = \mathrm{SU}(2)$.

For massless particles we have

$$\begin{pmatrix} 0 & \omega_{01} & \omega_{02} & \omega_{03} \\ \omega_{01} & 0 & -\omega_{12} & -\omega_{13} \\ \omega_{02} & \omega_{12} & 0 & -\omega_{23} \\ \omega_{03} & \omega_{13} & \omega_{23} & 0 \end{pmatrix} \begin{pmatrix} k \\ 0 \\ 0 \\ k \end{pmatrix} = \begin{pmatrix} 0 \\ 0 \\ 0 \\ 0 \end{pmatrix} ,$$

which gives $\omega_{03} = 0$, $\omega_{01} = \omega_{13}$, $\omega_{02} = \omega_{23}$ while the parameter ω_{12} is arbitrary. It corresponds to the rotation around z–axis. The generator of this transformation is M_{12}. From the conditions derived above follows that there are two independent generators $M^{01}+M^{13}$ and $-(M^{02}+M^{23})$. Note that $W_1=(M^{02}+M^{23})k$, $W_2=-(M^{01}+M^{13})k$ as well as $W_0=-M^{12}k$. Then, using Problem 1.16 (b) we obtain

$$[W_1,W_2]=0, \quad [W_0/k,W_1]=-\mathrm{i}W_2, \quad [W_0/k,W_2]=\mathrm{i}W_1 \ .$$

These commutation relations define E(2) algebra. Thus, for massless particles little group is euclidian group E(2) in two dimensions.

1.19 It is easy to prove that Lorentz transformations, dilatations and SCT form a group. It is the conformal group, $C(1,3)$. An arbitrary element of this group is

$$U(\omega,\epsilon,\rho,c)=\mathrm{e}^{\mathrm{i}(P_\mu\epsilon^\mu-\frac{1}{2}M_{\mu\nu}\omega^{\mu\nu}+\rho D+c_\mu K^\mu)} \ ,$$

where D is generator of dilatation, and K^μ are four generators for SCT . Conformal group has 15 parameters. The commutation relations of the algebra can be evaluated from multiplication rules of the group. Let (Λ,a,ρ,c) denote group element. If we start from

$$(\Lambda^{-1},0,0,0)(I,0,0,c)(\Lambda,0,0,0)=(I,0,0,\Lambda^{-1}c)$$

for infinitesimal SCT we obtain

$$U^{-1}(\Lambda)K_\rho U(\Lambda)=(\Lambda^{-1})^\mu{}_\rho K_\mu \ .$$

For infinitesimal Lorentz transformations we get:

$$[M_{\mu\nu},K_\rho]=\mathrm{i}(g_{\nu\rho}K_\mu-g_{\mu\rho}K_\nu). \tag{1.31}$$

From $U^{-1}(\Lambda,0,0,0)U(I,0,\rho,0)U(\Lambda,0,0,0)=U(I,0,\rho,0)$, follows

$$[M_{\mu\nu},D]=0 \ . \tag{1.32}$$

Starting from

$$\begin{aligned}(I,0,\rho,0)^{-1}(I,0,0,c)(I,0,\rho,0)x^\mu &= (I,0,\rho,0)^{-1}(I,0,0,c)\mathrm{e}^{-\rho}x^\mu \\ &= (I,0,\rho,0)^{-1}\frac{\mathrm{e}^{-\rho}x^\mu+c^\mu\mathrm{e}^{-2\rho}x^2}{1+2(c\cdot x)\mathrm{e}^{-\rho}+c^2\mathrm{e}^{-2\rho}x^2} \\ &= \frac{x^\mu+c^\mu\mathrm{e}^{-\rho}x^2}{1+2(c\cdot x)\mathrm{e}^{-\rho}+c^2\mathrm{e}^{-2\rho}x^2} \\ &= (I,0,0,\mathrm{e}^{-\rho}c)x^\mu \ ,\end{aligned}$$

we obtain

$$\mathrm{e}^{-\mathrm{i}\rho D}(1+iK^\mu c_\mu)\mathrm{e}^{\mathrm{i}\rho D}=1+\mathrm{i}K^\mu\mathrm{e}^{-\rho}c_\mu \ ,$$

for infinitesimal SCT. From the last expression follows

$$\mathrm{e}^{-\mathrm{i}\rho D}K^\mu\mathrm{e}^{\mathrm{i}\rho D}=\mathrm{e}^{-\rho}K^\mu \ .$$

This is the transformation law of SCT generators under dilatation. For infinitesimal dilatations we get:

$$[D, K^{\mu}] = -\mathrm{i}K^{\mu} \ . \tag{1.33}$$

Similar procedure gives us the following commutators:

$$[P_{\mu}, D] = -\mathrm{i}P_{\mu} \ , \tag{1.34}$$

$$[D, D] = 0, \tag{1.35}$$

$$[K_{\mu}, K_{\nu}] = 0, \tag{1.36}$$

$$[P_{\mu}, K_{\nu}] = 2\mathrm{i}(g_{\mu\nu}D + M_{\mu\nu}). \tag{1.37}$$

Equations (1.31)–(1.37) together with (1.11), (1.15) and (1.16) are commutation relations of the conformal algebra.

2

The Klein–Gordon equation

2.1 A particular solution of the Klein–Gordon equation

$$(\Box + m^2)\phi(x) = 0 \,, \tag{2.1}$$

is plane wave,

$$\mathrm{e}^{-\mathrm{i}k\cdot x} = \mathrm{e}^{-\mathrm{i}Et+\mathrm{i}\boldsymbol{k}\cdot\boldsymbol{x}} \,, \tag{2.2}$$

where E and $\boldsymbol{k}$ are energy and momentum respectively. We see that from

$$\mathrm{i}\frac{\partial}{\partial t}\mathrm{e}^{-\mathrm{i}k\cdot x} = E\mathrm{e}^{-\mathrm{i}k\cdot x} \,,$$

and

$$-\mathrm{i}\nabla\mathrm{e}^{-\mathrm{i}k\cdot x} = \boldsymbol{k}\mathrm{e}^{-\mathrm{i}k\cdot x} \,.$$

By inserting the solution (2.2) into (2.1) we obtain $k^2 = m^2$ i.e. $E = \pm\sqrt{\boldsymbol{k}^2 + m^2} = \pm\omega_k$. Therefore, the plane wave (2.2) is a solution of the Klein–Gordon equation if the previous relation is satisfied.

For momentum $\boldsymbol{k}$ there are two independent solutions $\mathrm{e}^{-\mathrm{i}\omega_k t+\mathrm{i}\boldsymbol{k}\cdot\boldsymbol{x}}$ and $\mathrm{e}^{+\mathrm{i}\omega_k t+\mathrm{i}\boldsymbol{k}\cdot\boldsymbol{x}}$. The general solution of (2.1) is

$$\phi(x) = \frac{1}{(2\pi)^{3/2}}\int\frac{\mathrm{d}^3\boldsymbol{k}}{\sqrt{2\omega_k}}\left(a(\boldsymbol{k})\mathrm{e}^{-\mathrm{i}(\omega_k t-\boldsymbol{k}\cdot\boldsymbol{x})} + b^\dagger(-\boldsymbol{k})\mathrm{e}^{\mathrm{i}(\omega_k t+\boldsymbol{k}\cdot\boldsymbol{x})}\right) \,, \tag{2.3}$$

where $a(\boldsymbol{k})$ and $b^\dagger(\boldsymbol{k})$ are complex coefficients. In the second term in (2.3) we make the following change $\boldsymbol{k} \to -\boldsymbol{k}$. Then (2.3) becomes

$$\phi(x) = \frac{1}{(2\pi)^{3/2}}\int\frac{\mathrm{d}^3\boldsymbol{k}}{\sqrt{2\omega_k}}\left(a(\boldsymbol{k})\mathrm{e}^{-\mathrm{i}k\cdot x} + b^\dagger(\boldsymbol{k})\mathrm{e}^{\mathrm{i}k\cdot x}\right) \,, \tag{2.4}$$

where $k^\mu = (\omega_k, \boldsymbol{k})$. If $\phi(x)$ is a real field then $a(\boldsymbol{k}) = b(\boldsymbol{k})$.

2.2 Using (2.4) we get

$$
\begin{aligned}
Q &= \mathrm{i}q \int \mathrm{d}^3\boldsymbol{x} \left(\phi^* \frac{\partial \phi}{\partial t} - \phi \frac{\partial \phi^*}{\partial t} \right) \\
&= \mathrm{i}\frac{q}{2(2\pi)^3} \int \frac{\mathrm{d}^3\boldsymbol{x}\mathrm{d}^3\boldsymbol{k}\mathrm{d}^3\boldsymbol{k'}}{\sqrt{\omega_k \omega_{k'}}} \left[\left(a^\dagger(\boldsymbol{k})\mathrm{e}^{\mathrm{i}k\cdot x} + b(\boldsymbol{k})\mathrm{e}^{-\mathrm{i}k\cdot x}\right) \right. \\
&\times \left(-\mathrm{i}\omega_{k'} a(\boldsymbol{k'})\mathrm{e}^{-\mathrm{i}k'\cdot x} + \mathrm{i}\omega_{k'} b^\dagger(\boldsymbol{k'})\mathrm{e}^{\mathrm{i}k'\cdot x}\right) - \left(a(\boldsymbol{k})\mathrm{e}^{-\mathrm{i}k\cdot x} + b^\dagger(\boldsymbol{k})\mathrm{e}^{\mathrm{i}k\cdot x}\right) \\
&\times \left. \left(\mathrm{i}\omega_{k'} a^\dagger(\boldsymbol{k'})\mathrm{e}^{\mathrm{i}k'\cdot x} - \mathrm{i}\omega_{k'} b(\boldsymbol{k'})\mathrm{e}^{-\mathrm{i}k'\cdot x}\right) \right] . \qquad (2.5)
\end{aligned}
$$

By integrating over $\boldsymbol{x}$ in (2.5), we obtain

$$
\begin{aligned}
Q = -\frac{q}{2} \int \mathrm{d}^3\boldsymbol{k}\mathrm{d}^3\boldsymbol{k'} \sqrt{\frac{\omega_{k'}}{\omega_k}} & \left(-a^\dagger(\boldsymbol{k})a(\boldsymbol{k'})\mathrm{e}^{\mathrm{i}(\omega_k - \omega_{k'})t}\delta^{(3)}(\boldsymbol{k}-\boldsymbol{k'}) \right. \\
&+ a^\dagger(\boldsymbol{k})b^\dagger(\boldsymbol{k'})\mathrm{e}^{\mathrm{i}(\omega_k+\omega_{k'})t}\delta^{(3)}(\boldsymbol{k}+\boldsymbol{k'}) - b(\boldsymbol{k})a(\boldsymbol{k'})\mathrm{e}^{-\mathrm{i}(\omega_k+\omega_{k'})t}\delta^{(3)}(\boldsymbol{k}+\boldsymbol{k'}) \\
&+ \left. b^\dagger(\boldsymbol{k})b(\boldsymbol{k'})\mathrm{e}^{-\mathrm{i}(\omega_k-\omega_{k'})t}\delta^{(3)}(\boldsymbol{k}-\boldsymbol{k'}) + \text{c.c.} \right) . \qquad (2.6)
\end{aligned}
$$

where c.c. denotes complex conjugation. If in expression (2.6) we integrate over the momentum $\boldsymbol{k'}$ we obtain

$$
Q = \frac{q}{2} \int \mathrm{d}^3\boldsymbol{k} \left[a^\dagger(\boldsymbol{k})a(\boldsymbol{k}) + a(\boldsymbol{k})a^\dagger(\boldsymbol{k}) - b^\dagger(\boldsymbol{k})b(\boldsymbol{k}) - b(\boldsymbol{k})b^\dagger(\boldsymbol{k}) \right] . \qquad (2.7)
$$

In the result (2.7) we do not take care about ordering of $a(\boldsymbol{k}), a^\dagger(\boldsymbol{k})$ and $b(\boldsymbol{k}), b^\dagger(\boldsymbol{k})$ since they are complex numbers. This will be different in Chapter 7 where $a(\boldsymbol{k})$ and $b^\dagger(\boldsymbol{k})$ are going to be operators.

2.3 If we first integrate over $\boldsymbol{x}$ we get

$$
\begin{aligned}
H = -\frac{1}{4} \int \frac{\mathrm{d}^3\boldsymbol{k}\mathrm{d}^3\boldsymbol{k'}}{\sqrt{\omega_k\omega_{k'}}} & \left(a(\boldsymbol{k})a(\boldsymbol{k'})(\omega_k\omega_{k'} + \boldsymbol{k}\cdot\boldsymbol{k'} - m^2)\mathrm{e}^{-\mathrm{i}(\omega_k+\omega_{k'})t}\delta^{(3)}(\boldsymbol{k}+\boldsymbol{k'}) \right. \\
&+ a^\dagger(\boldsymbol{k})a^\dagger(\boldsymbol{k'})(\omega_k\omega_{k'} + \boldsymbol{k}\cdot\boldsymbol{k'} - m^2)\mathrm{e}^{\mathrm{i}(\omega_k+\omega_{k'})t}\delta^{(3)}(\boldsymbol{k}+\boldsymbol{k'}) \\
&- a(\boldsymbol{k})a^\dagger(\boldsymbol{k'})(\omega_k\omega_{k'} + \boldsymbol{k}\cdot\boldsymbol{k'} + m^2)\mathrm{e}^{-\mathrm{i}(\omega_k-\omega_{k'})t}\delta^{(3)}(\boldsymbol{k}-\boldsymbol{k'}) \\
&- \left. a^\dagger(\boldsymbol{k})a(\boldsymbol{k'})(\omega_k\omega_{k'} + \boldsymbol{k}\cdot\boldsymbol{k'} + m^2)\mathrm{e}^{\mathrm{i}(\omega_k-\omega_{k'})t}\delta^{(3)}(\boldsymbol{k}-\boldsymbol{k'}) \right) . \qquad (2.8)
\end{aligned}
$$

Performing integration over momentum $\boldsymbol{k'}$, and using the relation $\boldsymbol{k}^2 + m^2 = \omega_k^2$, we obtain

$$
H = \frac{1}{2} \int \mathrm{d}^3\boldsymbol{k}\,\omega_k \left(a^\dagger(\boldsymbol{k})a(\boldsymbol{k}) + a(\boldsymbol{k})a^\dagger(\boldsymbol{k}) \right) . \qquad (2.9)
$$

2.4 Solution of this problem is very similar to the solutions of the previous two. The result is

$$
\boldsymbol{P} = \int \mathrm{d}^3\boldsymbol{k}\,\boldsymbol{k}\,a^\dagger(\boldsymbol{k})a(\boldsymbol{k}) .
$$

2.5 The four–divergence of the current j^μ is

$$\partial_\mu j^\mu = -\frac{\mathrm{i}}{2}(\partial_\mu \phi \partial^\mu \phi^* + \phi \Box \phi^* - \partial_\mu \phi \partial^\mu \phi^* - \phi^* \Box \phi) .$$

Using the equations of motion we obtain the requested result $\partial_\mu j^\mu = 0$.

2.6 It is easy to see that

$$\partial_\mu j^\mu = -\frac{\mathrm{i}}{2}(\partial_\mu \phi \partial^\mu \phi^* + \phi \Box \phi^* - \partial_\mu \phi \partial^\mu \phi^* - \phi^* \Box \phi) - \\ - q(\phi A^\mu \partial_\mu \phi^* + \phi \phi^* \partial_\mu A^\mu + \phi^* A^\mu \partial_\mu \phi) . \tag{2.10}$$

The equations of motion are

$$\left[\Box - \mathrm{i}q(\partial_\mu A^\mu + 2A^\mu \partial_\mu - \mathrm{i}q A_\mu A^\mu) + m^2\right] \phi^*(x) = 0 , \tag{2.11}$$

$$\left[\Box + \mathrm{i}q(\partial_\mu A^\mu + 2A^\mu \partial_\mu + \mathrm{i}q A_\mu A^\mu) + m^2\right] \phi(x) = 0 . \tag{2.12}$$

If we multiply Equation (2.11) by ϕ and Equation (2.12) by ϕ^* and then subtract obtained equations we get

$$\phi \Box \phi^* - \phi^* \Box \phi - 2\mathrm{i}q(\phi \phi^* \partial_\mu A^\mu + A^\mu \phi^* \partial_\mu \phi + A^\mu \phi \partial_\mu \phi^*) = 0 .$$

Combining the previous expression and (2.10), one easily obtains

$$\partial_\mu j^\mu = 0 .$$

2.7 The equation of motion for a scalar particle in a electromagnetic field is

$$\left[(\partial_\mu + \mathrm{i}q A_\mu)(\partial^\mu + \mathrm{i}q A^\mu) + m^2\right] \phi(x) = 0 . \tag{2.13}$$

In the region $r < a$ Equation (2.13) becomes

$$\left[\left(\frac{\partial}{\partial t} - \mathrm{i}V\right)\left(\frac{\partial}{\partial t} - \mathrm{i}V\right) - \Delta + m^2\right] \phi(x) = 0 . \tag{2.14}$$

For stationary states $\phi(x) = \mathrm{e}^{-\mathrm{i}Et} F(\boldsymbol{r})$ one gets

$$\left[-(E+V)^2 - \Delta + m^2\right] F(\boldsymbol{r}) = 0 . \tag{2.15}$$

If we assume that a solution of the previous equation is given by

$$F = \frac{f(r)}{r} Q(\theta, \varphi) ,$$

then from (2.15) we get the following two equations:

$$\frac{\mathrm{d}^2 f}{\mathrm{d}r^2} + \left[(E+V)^2 - m^2\right] f = \frac{l(l+1)}{r^2} f , \tag{2.16}$$

$$\frac{1}{\sin\theta} \frac{\partial}{\partial \theta}\left(\sin\theta \frac{\partial Q}{\partial \theta}\right) + \frac{1}{\sin^2\theta} \frac{\partial^2 Q}{\partial \varphi^2} = -l(l+1) Q . \tag{2.17}$$

The particular solutions of (2.17) are spherical harmonics, Y_{lm}. In the case $l = 0$, the corresponding spherical harmonic Y_{00} is a constant. The solution of (2.16) is

$$f = A\sin(qr) + B\cos(qr) , \tag{2.18}$$

where

$$q^2 = [(E+V)^2 - m^2] > 0 . \tag{2.19}$$

Constant B has to be zero since function $f(r)/r$ should not be singular in the $r \to 0$ limit. In the region $r > a$ ($A^0 = 0$) the solution is given by

$$f = C\mathrm{e}^{-kr} + D\mathrm{e}^{kr} , \tag{2.20}$$

where $k^2 = m^2 - E^2$. But, the constant D has to be zero since the wave function has to be finite in the large r limit. Therefore, the wave function is

$$\phi_< = A\frac{\sin qr}{r} , \quad r < a \tag{2.21}$$

$$\phi_> = C\frac{\mathrm{e}^{-kr}}{r} , \quad r > a . \tag{2.22}$$

At $r = a$ we should apply the continuity conditions: $\phi_<(a) = \phi_>(a)$ and $\phi'_<(a) = \phi'_>(a)$ for the wave function and its first derivative. These boundary conditions give:

$$A\sin(qa) - C\mathrm{e}^{-ka} = 0 , \tag{2.23}$$

$$Aq\cos(qa) + Ck\mathrm{e}^{-ka} = 0 . \tag{2.24}$$

The homogenous system (2.23–2.24) has non–trivial solutions if and only if its determinant is equal to zero. Finally, we obtain the condition

$$\frac{\tan(qa)}{q} = -\frac{1}{k} . \tag{2.25}$$

The dispersion relation (2.25) will be analyzed graphically in the case $V < 2m$. Solid line in Fig. 2.1 is function $\tan(qa)/q$ while dashed line is

$$f(q) = -\frac{1}{k} = -\frac{1}{\sqrt{2V\sqrt{q^2+m^2} - V^2 - q^2}} .$$

There is only one bound state (in case $V < 2m$) if the condition

$$\frac{\pi}{2a} < \sqrt{V(V+2m)} \le \frac{3\pi}{2a} .$$

is satisfied.

2.8 The wave equation is

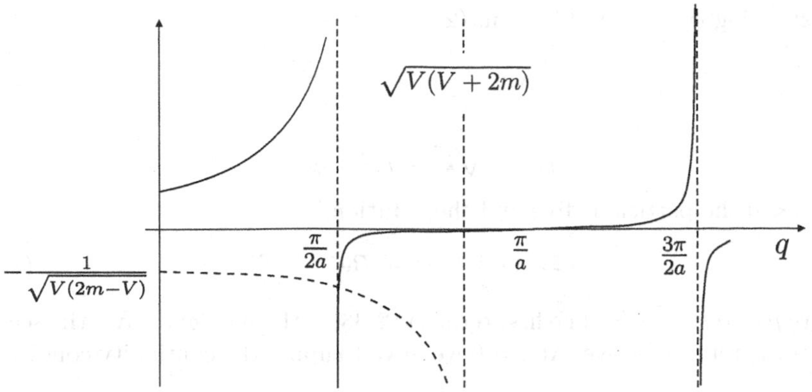

Fig. 2.1. Graphical solution of the dispersion relation (2.25) for $V < 2m$

$$\left[\frac{\partial^2}{\partial t^2} - \left(\frac{\partial}{\partial x} + \mathrm{i}qBy\right)^2 - \frac{\partial^2}{\partial y^2} - \frac{\partial^2}{\partial z^2} + m^2\right]\phi(x) = 0\,. \tag{2.26}$$

It is easy to see that the operators $\hat{p}_x = -\mathrm{i}\frac{\partial}{\partial x}$ and $\hat{p}_z = -\mathrm{i}\frac{\partial}{\partial z}$ commute with the Hamiltonian, so we can assume that the solution of (2.26) has the following form

$$\phi = \mathrm{e}^{-\mathrm{i}(Et - k_x x - k_z z)}\varphi(y)\,. \tag{2.27}$$

From (2.26) and (2.27) we get

$$\left(\frac{\mathrm{d}^2}{\mathrm{d}y^2} - (k_x + qBy)^2 + E^2 - k_z^2 - m^2\right)\varphi(y) = 0\,. \tag{2.28}$$

Introducing the new variable $\xi = k_x + qBy$, Equation (2.28) takes the same form as the Schrödinger equation for the oscillator

$$\left(\frac{\mathrm{d}^2}{\mathrm{d}\xi^2} - \frac{1}{(qB)^2}\xi^2 + \frac{E^2 - k_z^2 - m^2}{(qB)^2}\right)\tilde{\varphi}(\xi) = 0\,.$$

Then the energy levels are

$$E_n = \sqrt{m^2 + k_z^2 + (2n+1)qB}\,, \quad n = 0, 1, 2, \ldots\,.$$

Eigenfunctions are

$$\phi_n(x) = (q\pi B)^{-1/4}\frac{1}{\sqrt{2^n n!}}\mathrm{e}^{-\mathrm{i}E_n t + ik_x x + ik_z z}\mathrm{e}^{-(k_x + qBy)^2/2qB}H_n\Big(\frac{k_x + qBy}{\sqrt{qB}}\Big)\,, \tag{2.29}$$

where H_n are the Hermite polynomials.

2.9 In the region $z > 0$ the equation of motion is

$$\left[\Box - q^2U_0^2 + 2\mathrm{i}qU_0\frac{\partial}{\partial t} + m^2\right]\phi_{II}(x) = 0\,. \tag{2.30}$$

Substituting $\phi_{II} = C\mathrm{e}^{-\mathrm{i}Et+\mathrm{i}kz}$ in (2.30), we get

$$k = \pm K = \pm\sqrt{(E - qU_0)^2 - m^2}\ , \tag{2.31}$$

or

$$E = \pm\sqrt{k^2 + m^2} + qU_0\ . \tag{2.32}$$

For $z < 0$ the particle is free and the solution is

$$\phi_I = A\mathrm{e}^{-\mathrm{i}Et+\mathrm{i}pz} + B\mathrm{e}^{-\mathrm{i}Et-\mathrm{i}pz}\ , \tag{2.33}$$

where $p = \sqrt{E^2 - m^2}$. The first term in (2.33) is the incident wave, the second one is the reflected wave. At $z = 0$ we have to apply the continuity conditions:

$$\phi_I(0) = \phi_{II}(0),\ \ \phi'_I(0) = \phi'_{II}(0)\ .$$

They give

$$A = \frac{1}{2}\left(1 + \frac{k}{p}\right)C,\quad B = \frac{1}{2}\left(1 - \frac{k}{p}\right)C\ . \tag{2.34}$$

We will separately discuss three different possibilities:
Case 1: $E > m + qU_0$.
For this value of energy the sign in the expressions (2.31) and (2.32) is plus. The formula for the current has been given in Problem 2.5. The reflection coefficient is

$$R = \frac{-(j_r)_z}{(j_{in})_z} = \frac{|B|^2}{|A|^2} = \left|\frac{p-K}{p+K}\right|^2\ ,$$

while the transmission coefficient is $T = 1 - R$.
Case 2: $E < -m + qU_0$.
In this case the momentum is negative, $k = -K$. The reflection coefficient is different comparing to the previous case:

$$R = \left|\frac{p+K}{p-K}\right|^2\ ,\ \ T = 1 - R\ .$$

As we immediately see the reflection coefficient is larger than 1: the potential is strong enough to create particle–antiparticle pairs. The antiparticles are moving to the right producing a negative charge current and therefore we obtain negative transmission coefficient. This is the *Klein paradox.*
Case 3: $|E - qU_0| < m$.
We leave to the reader to show that in this case $R = 1,\ T = 0$.

2.10 For $z < 0$ and $z > 0$ a wave function satisfies the free Klein–Gordon equation, while in the region $0 < z < a$ the equation is

$$\left[\Box - q^2U_0^2 + 2\mathrm{i}qU_0\frac{\partial}{\partial t} + m^2\right]\phi_{II}(x) = 0\ .$$

The solution is given by:

$$\begin{aligned}\phi_I &= A\mathrm{e}^{-\mathrm{i}Et+\mathrm{i}pz} + B\mathrm{e}^{-\mathrm{i}Et-\mathrm{i}pz} \ , \\ \phi_{II} &= C\mathrm{e}^{-\mathrm{i}Et+\mathrm{i}kz} + D\mathrm{e}^{-\mathrm{i}Et-\mathrm{i}kz} \\ \phi_{III} &= F\mathrm{e}^{-\mathrm{i}Et+\mathrm{i}pz} \ , \end{aligned} \tag{2.35}$$

where $k = \sqrt{(E - qU_0)^2 - m^2}$ and $p = \sqrt{E^2 - m^2}$. From the continuity conditions follows:

$$\begin{aligned} A + B &= C + D \ , \\ A - B &= \frac{k}{p}(C - D) \ , \\ C\mathrm{e}^{\mathrm{i}ka} + D\mathrm{e}^{-\mathrm{i}ka} &= F\mathrm{e}^{\mathrm{i}pa} \ , \\ C\mathrm{e}^{\mathrm{i}ka} - D\mathrm{e}^{-\mathrm{i}ka} &= \frac{p}{k}F\mathrm{e}^{\mathrm{i}pa} \ . \end{aligned} \tag{2.36}$$

Thus, one gets:

$$T = \left|\frac{F}{A}\right|^2 = \frac{16}{|2 + \frac{p}{k} + \frac{k}{p} + (2 - \frac{p}{k} - \frac{k}{p})\mathrm{e}^{2\mathrm{i}ka}|^2} \ .$$

If $(E - qU_0)^2 - m^2 < 0$ the momentum k becomes imaginary, i.e.

$$k = \mathrm{i}\kappa = \mathrm{i}\sqrt{m^2 - (E - qU_0)^2} \ .$$

It is easy to show that the transmission coefficient is equal to one if $E = \frac{E}{2}$.

2.11 The Klein–Gordon equation for a particle in the Coulomb potential is

$$\left[\left(\frac{\partial}{\partial t} - \mathrm{i}e\frac{Ze}{r}\right)^2 - \Delta + m^2\right]\phi(x) = 0 \ . \tag{2.37}$$

By substituting $\phi = \mathrm{e}^{-\mathrm{i}Et}R(r)Y(\theta, \varphi)$ in (2.37) and using (2.17) we obtain:

$$-\frac{1}{2m}\frac{1}{r}\frac{\mathrm{d}^2}{\mathrm{d}r^2}(rR) + \frac{l(l+1) - Z^2e^4}{2mr^2}R - \frac{Ze^2E}{mr}R = \frac{E^2 - m^2}{2m}R \ .$$

This equation has the same form as the Schrödinger equation for hydrogen atom. By comparing these equations we get

$$E_{n,l} = m\frac{1}{\sqrt{1 + Z^2e^4(n - l - \frac{1}{2}) + \sqrt{(l + \frac{1}{2})^2 - Z^2e^4}}} \ .$$

In the nonrelativistic limit the result is

$$E_n - m = -\frac{mZ^2e^4}{2n^2} - Z^3e^6\frac{m}{n^3}\left(\frac{1}{2l+1} - \frac{3}{8n}\right) \ .$$

2.12 The Klein–Gordon equation in the Schrödinger form is

$$\mathrm{i}\frac{\partial}{\partial t}\begin{pmatrix}\theta\\ \chi\end{pmatrix}=H\begin{pmatrix}\theta\\ \chi\end{pmatrix}, \tag{2.38}$$

where the Hamiltonian is given by

$$H=\left[-\frac{\Delta}{2m}\begin{pmatrix}1 & 1\\ -1 & -1\end{pmatrix}+m\begin{pmatrix}1 & 0\\ 0 & -1\end{pmatrix}\right].$$

2.13 The eigenequation, $H\phi=E\phi$ in the momentum representation takes the following form

$$\begin{pmatrix}\frac{\boldsymbol{p}^2}{2m}+m & \frac{\boldsymbol{p}^2}{2m}\\ -\frac{\boldsymbol{p}^2}{2m} & -\frac{\boldsymbol{p}^2}{2m}-m\end{pmatrix}\begin{pmatrix}\theta_0\\ \chi_0\end{pmatrix}=E\begin{pmatrix}\theta_0\\ \chi_0\end{pmatrix}. \tag{2.39}$$

The eigenvalues of the Hamiltonian are evaluated easily and they are $E=\pm\omega_p=\pm\sqrt{\boldsymbol{p}^2+m^2}$.

In order to find nonrelativistic limit we suppose that the solution has the following form

$$\begin{pmatrix}\theta\\ \chi\end{pmatrix}=\begin{pmatrix}\theta_0\\ \chi_0\end{pmatrix}\mathrm{e}^{-\mathrm{i}(m+T)t}, \tag{2.40}$$

where T is the kinetic energy of the particle. From (2.38) we get

$$\begin{pmatrix}-\frac{\triangle}{2m}+m & -\frac{\triangle}{2m}\\ \frac{\triangle}{2m} & \frac{\triangle}{2m}-m\end{pmatrix}\begin{pmatrix}\theta_0\\ \chi_0\end{pmatrix}=(m+T)\begin{pmatrix}\theta_0\\ \chi_0\end{pmatrix}, \tag{2.41}$$

i.e.

$$\begin{aligned}\left(-\frac{\triangle}{2m}+m\right)\theta_0-\frac{\triangle}{2m}\chi_0&=(m+T)\theta_0,\\ \frac{\triangle}{2m}\theta_0+\left(\frac{\triangle}{2m}-m\right)\chi_0&=(T+m)\chi_0.\end{aligned} \tag{2.42}$$

From the second equation in (2.42) we obtain

$$\chi_0\approx\frac{\triangle}{4m^2}\theta_0, \tag{2.43}$$

in nonrelativistic limit. Using this the first equation in (2.42) becomes

$$T\theta_0=\left(-\frac{\triangle}{2m}-\frac{\triangle^2}{8m^3}\right)\theta_0. \tag{2.44}$$

Also, from (2.43) we see that $\chi_0\ll\theta_0$ and χ is so called small component. From the expression (2.44) follows that first relativistic correction of nonrelativistic Hamiltonian is $-\nabla^4/8m^3$.

2.14 Velocity operator is

$$\boldsymbol{v}=\frac{\partial H}{\partial \boldsymbol{p}}=\frac{\boldsymbol{p}}{m}\begin{pmatrix}1 & 1\\ -1 & -1\end{pmatrix}.$$

The eigenvalue of the velocity operator is zero.

2.15 Show that $<\psi,H\chi>=<H\psi,\chi>$. The average value is $\langle\boldsymbol{v}\rangle=\frac{\boldsymbol{p}}{m}$.

3

The γ–matrices

3.1

(a) In the Dirac representation of γ–matrices we have

$$(\gamma^0)^\dagger = \begin{pmatrix} I & 0 \\ 0 & -I \end{pmatrix}^\dagger = \begin{pmatrix} I & 0 \\ 0 & -I \end{pmatrix} = \gamma^0\gamma^0\gamma^0 = \gamma^0 \,,$$

$$(\gamma^i)^\dagger = \begin{pmatrix} 0 & \sigma_i \\ -\sigma_i & 0 \end{pmatrix}^\dagger = -\begin{pmatrix} 0 & \sigma_i \\ -\sigma_i & 0 \end{pmatrix} = -\gamma^0\gamma^0\gamma^i = \gamma^0\gamma^i\gamma^0 \,,$$

where we used the facts that $(\gamma^0)^2 = 1$, γ^0 and γ^i anticommute, and the Pauli matrices are hermitian. This relation is true in any representation of γ–matrices which is obtained by a unitary transformation from the Dirac representation.

(b) Using the previous result we find

$$\begin{aligned}
\sigma^\dagger_{\mu\nu} &= -\frac{\mathrm{i}}{2}(\gamma_\mu\gamma_\nu - \gamma_\nu\gamma_\mu)^\dagger \\
&= -\frac{\mathrm{i}}{2}(\gamma^\dagger_\nu\gamma^\dagger_\mu - \gamma^\dagger_\mu\gamma^\dagger_\nu) \\
&= -\frac{\mathrm{i}}{2}\gamma_0(\gamma_\nu\gamma_\mu - \gamma_\mu\gamma_\nu)\gamma_0 \\
&= \gamma_0\sigma_{\mu\nu}\gamma_0 \,.
\end{aligned}$$

3.2

(a) Taking the adjoint of γ_5 we obtain

$$\begin{aligned}
\gamma_5^\dagger &= \mathrm{i}\gamma_3^\dagger\gamma_2^\dagger\gamma_1^\dagger\gamma_0^\dagger \\
&= \mathrm{i}\gamma_0\gamma_3\gamma_0\gamma_0\gamma_2\gamma_0\gamma_0\gamma_1\gamma_0\gamma_0\gamma_0\gamma_0 \\
&= \mathrm{i}\gamma_0\gamma_3\gamma_2\gamma_1 \\
&= -\mathrm{i}\gamma_0\gamma_1\gamma_2\gamma_3 = \gamma_5 \,.
\end{aligned}$$

The property $\gamma_5^{-1} = \gamma_5$ can be proven by using $\gamma_0^{-1} = \gamma_0$ and $\gamma_i^{-1} = -\gamma_i = \gamma^i$. Both of these relations follow from anticommutation relations $\{\gamma_\mu, \gamma^\nu\} = 2\delta_\mu^\nu$.

(b) Using the definition of the ϵ symbol we find

$$-\frac{\mathrm{i}}{4!}\epsilon_{\mu\nu\rho\sigma}\gamma^\mu\gamma^\nu\gamma^\rho\gamma^\sigma = \frac{\mathrm{i}}{4!}(\gamma^0\gamma^1\gamma^2\gamma^3 - \gamma^0\gamma^1\gamma^3\gamma^2 + \ldots + \gamma^3\gamma^2\gamma^1\gamma^0)$$
$$= \mathrm{i}\gamma^0\gamma^1\gamma^2\gamma^3 = \gamma_5.$$

(c) This is a consequence of (a) result.

(d) In a similar manner, we have:

$$(\gamma_5\gamma_\mu)^\dagger = \gamma_\mu^\dagger\gamma_5^\dagger = \gamma_0\gamma_\mu\gamma^0\gamma_5 = \gamma^0\gamma_5\gamma_\mu\gamma^0 .$$

3.3

(a) For $\mu = 0$ we have

$$\begin{aligned}\{\gamma_5, \gamma^0\} &= \gamma_5\gamma^0 + \gamma^0\gamma_5 \\ &= -\mathrm{i}\gamma_0\gamma_1\gamma_2\gamma_3\gamma_0 - \mathrm{i}\gamma_0\gamma_0\gamma_1\gamma_2\gamma_3 \\ &= \mathrm{i}\gamma_1\gamma_2\gamma_3 - \mathrm{i}\gamma_1\gamma_2\gamma_3 = 0 ,\end{aligned} \tag{3.1}$$

and similarly for other three cases.

(b) By a straightforward calculation one gets:

$$\begin{aligned}[\sigma_{\mu\nu}, \gamma_5] &= \frac{\mathrm{i}}{2}[\gamma_\mu\gamma_\nu - \gamma_\nu\gamma_\mu, \gamma_5] \\ &= \frac{\mathrm{i}}{2}(\gamma_\mu\{\gamma_\nu, \gamma_5\} - \{\gamma_\mu, \gamma_5\}\gamma_\nu - \gamma_\nu\{\gamma_\mu, \gamma_5\} + \{\gamma_\mu, \gamma_5\}\gamma_\nu) \\ &= 0\end{aligned}$$

since $\{\gamma_\mu, \gamma_5\} = 0$.

3.4 $\not{a}\not{a} = a^\mu a^\nu\gamma_\mu\gamma_\nu = \frac{1}{2}a^\mu a^\nu(\gamma_\mu\gamma_\nu + \gamma_\nu\gamma_\mu) = g^{\mu\nu}a_\mu a_\nu = a^2$

3.5

(a) From the relation $\{\gamma_\mu, \gamma^\mu\} = 2\gamma_\mu\gamma^\mu = 2\delta_\mu^\mu = 8$ it follows that $\gamma_\mu\gamma^\mu = 4$.

(b) $\gamma_\mu\gamma_\nu\gamma^\mu = (2g_{\mu\nu} - \gamma_\nu\gamma_\mu)\gamma^\mu = 2\gamma_\nu - 4\gamma_\nu = -2\gamma_\nu$.

(c) $\gamma_\mu\gamma_\alpha\gamma^\beta\gamma^\mu = (2g_{\mu\alpha} - \gamma_\alpha\gamma_\mu)\gamma^\beta\gamma^\mu = 2\gamma^\beta\gamma_\alpha + 2\gamma_\alpha\gamma^\beta = 4\delta_\alpha^\beta$, where we used the second part of this problem and (3.A).

(d) By commuting γ_μ and γ^α and making use of the previous result, one gets:

$$\begin{aligned}\gamma_\mu\gamma^\alpha\gamma^\beta\gamma^\gamma\gamma^\mu &= (2\delta_\mu^\alpha - \gamma^\alpha\gamma_\mu)\gamma^\beta\gamma^\gamma\gamma^\mu \\ &= 2\gamma^\beta\gamma^\gamma\gamma^\alpha - 4\gamma^\alpha g^{\beta\gamma} \\ &= -2(2g^{\beta\gamma} - \gamma^\beta\gamma^\gamma)\gamma^\alpha \\ &= -2\gamma^\gamma\gamma^\beta\gamma^\alpha .\end{aligned}$$

(e) By using the definition $\sigma_{\mu\nu}$–matrices, one obtains:

$$\sigma_{\mu\nu}\sigma^{\mu\nu} = -\frac{1}{4}(\gamma^\mu\gamma^\nu\gamma_\mu\gamma_\nu - \gamma^\mu\gamma^\nu\gamma_\nu\gamma_\mu - \gamma^\nu\gamma^\mu\gamma_\mu\gamma_\nu + \gamma^\nu\gamma^\mu\gamma_\nu\gamma_\mu) \,.$$

By using parts (a) and (b) of this problem, one gets $\sigma_{\mu\nu}\sigma^{\mu\nu} = 12$.

(f) Use Problem 3.3 and parts (a) and (b) of this problem.

(g) By direct calculation, one finds

$$\begin{aligned}\sigma_{\alpha\beta}\gamma_\mu\sigma^{\alpha\beta} &= -\frac{1}{4}(\gamma^\alpha\gamma^\beta\gamma_\mu\gamma_\alpha\gamma_\beta - \gamma^\alpha\gamma^\beta\gamma_\mu\gamma_\beta\gamma_\alpha \\ &\quad -\gamma^\beta\gamma^\alpha\gamma_\mu\gamma_\alpha\gamma_\beta + \gamma^\beta\gamma^\alpha\gamma_\mu\gamma_\beta\gamma_\alpha) \\ &= -\frac{1}{4}(4\delta^\beta_\mu\gamma_\beta - 4\gamma_\mu - 4\gamma_\mu + 4g_{\mu\beta}\gamma^\beta) = 0 \,.\end{aligned}$$

(h)

$$\begin{aligned}\sigma^{\alpha\beta}\sigma^{\mu\nu}\sigma_{\alpha\beta} &= -\frac{\mathrm{i}}{8}(\gamma^\alpha\gamma^\beta\gamma^\mu\gamma^\nu\gamma_\alpha\gamma_\beta - \gamma^\alpha\gamma^\beta\gamma^\mu\gamma^\nu\gamma_\beta\gamma_\alpha \\ &\quad -\gamma^\alpha\gamma^\beta\gamma^\nu\gamma^\mu\gamma_\alpha\gamma_\beta + \gamma^\alpha\gamma^\beta\gamma^\nu\gamma^\mu\gamma_\beta\gamma_\alpha - \gamma^\beta\gamma^\alpha\gamma^\mu\gamma^\nu\gamma_\alpha\gamma_\beta \\ &\quad +\gamma^\beta\gamma^\alpha\gamma^\mu\gamma^\nu\gamma_\beta\gamma_\alpha + \gamma^\beta\gamma^\alpha\gamma^\nu\gamma^\mu\gamma_\alpha\gamma_\beta - \gamma^\beta\gamma^\alpha\gamma^\nu\gamma^\mu\gamma_\beta\gamma_\alpha) \\ &= -\frac{\mathrm{i}}{8}(-8\gamma^\nu\gamma^\mu - 16g^{\mu\nu} + 8\gamma^\mu\gamma^\nu \\ &\quad +16g^{\mu\nu} - 16g^{\mu\nu} - 8\gamma^\nu\gamma^\mu + 16g^{\mu\nu} + 8\gamma^\mu\gamma^\nu) \\ &= -2\mathrm{i}(\gamma^\mu\gamma^\nu - \gamma^\nu\gamma^\mu) = -4\sigma^{\mu\nu} \,.\end{aligned}$$

(i) Use part (g) of this problem.

(j) $\sigma_{\mu\nu}\gamma_5\sigma^{\mu\nu} = \frac{\mathrm{i}}{2}(\gamma_\mu\gamma_\nu - \gamma_\nu\gamma_\mu)\gamma_5\sigma^{\mu\nu} = \gamma_5\sigma_{\mu\nu}\sigma^{\mu\nu} = 12\gamma_5$.

3.6

(a) By using the trace property $\mathrm{tr}(A_1A_2\dots A_n) = \mathrm{tr}(A_2A_3\dots A_nA_1)$, Problem 3.3(a), and $(\gamma_5)^2 = 1$, it follows that

$$\begin{aligned}\mathrm{tr}(\gamma_\mu) &= \mathrm{tr}(\gamma_\mu\gamma_5\gamma_5) \\ &= \mathrm{tr}(\gamma_5\gamma_\mu\gamma_5) \\ &= -\mathrm{tr}((\gamma_5)^2\gamma_\mu) \\ &= -\mathrm{tr}(\gamma_\mu) \,.\end{aligned}$$

From the previous expression we get $\mathrm{tr}(\gamma_\mu) = 0$.

(b) Taking trace of the relation $\{\gamma_\mu, \gamma_\nu\} = 2g_{\mu\nu}$, we easy obtain the requested result.

(c) By applying the basic anticommutation relation (3.A), one gets:

$$\begin{aligned}\mathrm{tr}(\gamma_\mu\gamma_\nu\gamma_\rho\gamma_\sigma) &= \mathrm{tr}\left[(2g_{\mu\nu} - \gamma_\nu\gamma_\mu)\gamma_\rho\gamma_\sigma\right] \\ &= 2g_{\mu\nu}\mathrm{tr}(\gamma_\rho\gamma_\sigma) - \mathrm{tr}[\gamma_\nu(2g_{\mu\rho} - \gamma_\rho\gamma_\mu)\gamma_\sigma] \\ &= 2g_{\mu\nu}\mathrm{tr}(\gamma_\rho\gamma_\sigma) - 2g_{\mu\rho}\mathrm{tr}(\gamma_\nu\gamma_\sigma) + 2g_{\mu\sigma}\mathrm{tr}(\gamma_\nu\gamma_\rho) \\ &\quad - \mathrm{tr}(\gamma_\nu\gamma_\rho\gamma_\sigma\gamma_\mu) \,.\end{aligned}$$

From the previous part of this problem and relation $\mathrm{tr}(\gamma_\mu\gamma_\nu\gamma_\rho\gamma_\sigma) = \mathrm{tr}(\gamma_\nu\gamma_\rho\gamma_\sigma\gamma_\mu)$, one easily obtains the requested result.

(d) $\mathrm{tr}\gamma_5 = \mathrm{tr}(\gamma_5\gamma_0\gamma_0) = -\mathrm{tr}(\gamma_0\gamma_5\gamma_0)$, where we used Problem 3.3 (a). Further, from the trace property and $(\gamma_0)^2 = 1$ it follows that:

$$\mathrm{tr}\gamma_5 = -\mathrm{tr}(\gamma_0\gamma_0\gamma_5) = -\mathrm{tr}\gamma_5 \ ,$$

which implies $\mathrm{tr}\gamma_5 = 0$.

(e) Since $\gamma_\alpha\gamma^\alpha = 4$, we have

$$\begin{aligned}
\mathrm{tr}(\gamma_5\gamma_\mu\gamma_\nu) &= \frac{1}{4}\mathrm{tr}(\gamma_5\gamma^\alpha\gamma_\alpha\gamma_\mu\gamma_\nu) \\
&= \frac{1}{4}\mathrm{tr}(\gamma_\alpha\gamma_\mu\gamma_\nu\gamma_5\gamma^\alpha) \\
&= -\frac{1}{4}\mathrm{tr}(\gamma_5\gamma_\alpha\gamma_\mu\gamma_\nu\gamma^\alpha) \\
&= -g_{\mu\nu}\mathrm{tr}(\gamma_5) = 0 \ .
\end{aligned}$$

In the previous calculation we used the trace property and Problem 3.5 (c).

(f) The quantity $\mathrm{tr}(\gamma_5\gamma_\mu\gamma_\nu\gamma_\rho\gamma_\sigma)$ is an antisymmetric tensor with respect to the indexes $(\mu,\ \nu,\ \rho,\ \sigma)$. Thus, it must be proportional to the Levi-Civita tensor. The constant of proportionality can be determined by substituting $\mu = 0,\ \nu = 1,\ \rho = 2$ and $\sigma = 3$.

(g) From $(\gamma_5)^2 = 1$, $\{\gamma_5, \gamma_\mu\} = 0$ and the trace property follows:

$$\begin{aligned}
\mathrm{tr}(\not{a}_1 \dots \not{a}_{2n+1}) &= \mathrm{tr}(\gamma_5\gamma_5\not{a}_1\cdots\not{a}_{2n+1}) \\
&= (-1)^{2n+1}\mathrm{tr}(\gamma_5\not{a}_1\cdots\not{a}_{2n+1}\gamma_5) \\
&= -\mathrm{tr}(\gamma_5\gamma_5\not{a}_1\cdots\not{a}_{2n+1}) \\
&= -\mathrm{tr}(\not{a}_1 ... \not{a}_{2n+1}) \ .
\end{aligned}$$

Hence, $\mathrm{tr}(\not{a}_1 \dots \not{a}_{2n+1}) = 0$.

(h) $\mathrm{tr}(\not{a}_1\cdots\not{a}_{2n}) = \mathrm{tr}(C\not{a}_1C^{-1}C\cdots C^{-1}C\not{a}_{2n}C^{-1})$, where the matrix C satisfies the relation $C\gamma_\mu C^{-1} = -\gamma_\mu^T$. Thus,

$$\mathrm{tr}(\not{a}_1\cdots\not{a}_{2n}) = (-1)^{2n}\mathrm{tr}(\not{a}_1^T\cdots\not{a}_{2n}^T) = \mathrm{tr}(\not{a}_{2n}\cdots\not{a}_1) \ .$$

(i) $\mathrm{tr}(\gamma_5\gamma_\mu) = -i\mathrm{tr}(\gamma_0\gamma_1\gamma_2\gamma_3\gamma_\mu) = 0$, since it is the trace of odd number of γ–matrices.

3.7

$$\begin{aligned}
&\mathrm{tr}(\not{a}_1\not{a}_2\cdots\not{a}_6) = \\
&4\,\{(a_1\cdot a_2)\,[(a_3\cdot a_4)(a_5\cdot a_6) - (a_3\cdot a_5)(a_4\cdot a_6) + (a_3\cdot a_6)(a_4\cdot a_5)] \\
&-(a_1\cdot a_3)\,[(a_2\cdot a_4)(a_5\cdot a_6) - (a_2\cdot a_5)(a_4\cdot a_6) + (a_2\cdot a_6)(a_4\cdot a_5)] \\
&+(a_1\cdot a_4)\,[(a_2\cdot a_3)(a_5\cdot a_6) - (a_2\cdot a_5)(a_3\cdot a_6) + (a_2\cdot a_6)(a_3\cdot a_5)] \\
&-(a_1\cdot a_5)\,[(a_2\cdot a_3)(a_4\cdot a_6) - (a_2\cdot a_4)(a_3\cdot a_6) + (a_2\cdot a_6)(a_3\cdot a_4)] \\
&+(a_1\cdot a_6)\,[(a_2\cdot a_3)(a_4\cdot a_5) - (a_2\cdot a_4)(a_3\cdot a_5) + (a_2\cdot a_5)(a_3\cdot a_4)]\} \ .
\end{aligned}$$

3.8 $4\left[p_\mu q_\nu - (p\cdot q)g_{\mu\nu} + p_\nu q_\mu + \mathrm{i}\epsilon_{\alpha\mu\beta\nu}p^\alpha q^\beta - m^2 g_{\mu\nu}\right]$.

3.9 $-2\not{p} - 2\gamma_5\not{p} - 4m - 4m\gamma_5$.

3.10 Expanding the exponential function in series, we find

$$e^{\gamma_5 \not{a}} = 1 + (\gamma_5\not{a}) + \frac{1}{2}(\gamma_5\not{a})^2 + \frac{1}{3!}(\gamma_5\not{a})^3 + \cdots . \tag{3.2}$$

By substituting $(\gamma_5\not{a})^2 = -a^2$, $(\gamma_5\not{a})^3 = -a^2(\gamma_5\not{a}), \ldots$ into expression (3.2), we get

$$\begin{aligned} e^{\gamma_5\not{a}} &= (1 - \frac{a^2}{2!} + \frac{a^4}{4!} + \cdots) + (\gamma_5\not{a})(1 - \frac{a^2}{3!} + \frac{a^4}{5!} - \cdots) \\ &= \cos(\sqrt{a^2}) + \frac{1}{\sqrt{a^2}}\sin(\sqrt{a^2})\gamma_5\not{a} \ , \end{aligned}$$

where $a^2 = a_\mu a^\mu$.

3.11 The fact that the product of any two Γ–matrices is again a Γ matrix (modulo $\pm 1, \pm\mathrm{i}$) can be proved directly. For example, $\gamma_5\sigma_{01} = -\mathrm{i}\sigma_{23}$.
Now, we shall prove that Γ–matrices are linearly independent. Multiplying the relation $\sum_a c_a \Gamma^a = 0$ by $\Gamma_b = (\Gamma^b)^{-1}$, we obtain

$$c_b \Gamma^b \Gamma_b + \sum_{a\neq b} c_a \Gamma^a \Gamma_b = 0 \ ,$$

where the b–term is separated. Using the ordering lemma, the last expression becomes

$$c_b I + \sum_{d,\Gamma^d \neq I} c_d \eta \Gamma^d = 0 \ , \tag{3.3}$$

where $\eta \in \{\pm 1, \pm \mathrm{i}\}$. After taking trace of (3.3) and using the fact that

$$\mathrm{tr}(\Gamma^a) = \begin{cases} 0, & \Gamma^a \neq I \\ 4, & \Gamma^a = I \end{cases} \ ,$$

one obtains $c_b = 0$ $(\forall b)$. This means that Γ–matrices are linearly independent one.

3.12 Multiplying the equation $A = \sum_a c_a \Gamma^a$ by Γ_b from the right and separating the b–term in the sum, we have

$$A\Gamma_b = c_b \Gamma^b \Gamma_b + \sum_{a\neq b} c_a \Gamma^a \Gamma_b = c_b I + \sum_{d,\Gamma^d\neq I} c_d \eta \Gamma^d .$$

Taking the trace of previous relation we obtain the requesting relation.

3.13 The coefficients can be calculated by using the formula obtained in the previous problem.

(a) From the traces (which were actually calculated in Problem 3.6):

$$\begin{aligned}
\mathrm{tr}(\gamma_\mu\gamma_\nu\gamma_\rho) &= 0 \ ,\\
\mathrm{tr}(\gamma_\mu\gamma_\nu\gamma_\rho\gamma_\sigma) &= 4(g_{\mu\nu}g_{\rho\sigma} - g_{\mu\rho}g_{\nu\sigma} + g_{\mu\sigma}g_{\nu\rho}) \ ,\\
\mathrm{tr}(\gamma_\mu\gamma_\nu\gamma_\rho\gamma_\sigma\gamma_5) &= -4\mathrm{i}\epsilon_{\mu\nu\rho\sigma} \ ,\\
\mathrm{tr}(\gamma_\mu\gamma_\nu\gamma_\rho\gamma_5) &= \mathrm{tr}(\gamma_\mu\gamma_\nu\gamma_\rho\sigma_{\alpha\beta}) = 0 \ ,
\end{aligned}$$

follows $\gamma_\mu\gamma_\nu\gamma_\rho = (g_{\mu\nu}g_{\rho\sigma} - g_{\mu\rho}g_{\sigma\nu} + g_{\mu\sigma}g_{\rho\nu})\gamma^\sigma + \mathrm{i}\epsilon_{\sigma\mu\nu\rho}\gamma_5\gamma^\sigma$.
(b) $\gamma_5\gamma_\mu\gamma_\nu = g_{\mu\nu}\gamma_5 + \frac{1}{2}\epsilon^{\alpha\beta}{}_{\mu\nu}\sigma_{\alpha\beta}$,
(c) $\sigma_{\mu\nu}\gamma_\rho\gamma_5 = \epsilon_{\alpha\mu\nu\rho}\gamma^\alpha - \mathrm{i}g_{\nu\rho}\gamma_5\gamma_\mu + \mathrm{i}g_{\mu\rho}\gamma_5\gamma_\nu$.

3.14 From Problem 3.13 (a), it follows that $\{\gamma_\mu, \sigma_{\nu\rho}\} = -2\epsilon_{\alpha\mu\nu\rho}\gamma^5\gamma^\alpha$.

3.15 By applying the result of Problem 3.13 (a) the trace can be transformed as follows

$$\begin{aligned}
\mathrm{tr}(\gamma_\mu\gamma_\nu\gamma_\rho\gamma_\sigma\gamma_\alpha\gamma_\beta\gamma_5) &= (g_{\mu\nu}g_{\rho\delta} - g_{\mu\rho}g_{\nu\delta} + g_{\mu\delta}g_{\rho\nu})\mathrm{tr}(\gamma^\delta\gamma_\sigma\gamma_\alpha\gamma_\beta\gamma_5)\\
&+ \mathrm{i}\epsilon_{\delta\mu\nu\rho}\mathrm{tr}(\gamma^\delta\gamma_\sigma\gamma_\alpha\gamma_\beta) \ .
\end{aligned}$$

Using 3.6 (c), (f), we get

$$\begin{aligned}
\mathrm{tr}(\gamma_\mu\gamma_\nu\gamma_\rho\gamma_\sigma\gamma_\alpha\gamma_\beta\gamma_5) &= 4\mathrm{i}(-g_{\mu\nu}\epsilon_{\rho\sigma\alpha\beta} + g_{\mu\rho}\epsilon_{\nu\sigma\alpha\beta}\\
&\quad - g_{\rho\nu}\epsilon_{\mu\sigma\alpha\beta} + g_{\alpha\beta}\epsilon_{\sigma\mu\nu\rho} - g_{\sigma\beta}\epsilon_{\alpha\mu\nu\rho} + g_{\sigma\alpha}\epsilon_{\beta\mu\nu\rho}) \ .
\end{aligned}$$

3.16 Use the solution of Problem 3.13 (b).

3.17 Applying the formulae

$$[A, BC] = [A, B]C + B[A, C] \ ,$$

and

$$[AB, C] = A\{B, C\} - \{A, C\}B \ ,$$

as well as the anticommutation relations (3.A), we obtain

$$\begin{aligned}
[\gamma_\mu\gamma_\nu, \gamma_\rho\gamma_\sigma] &= \gamma_\mu\{\gamma_\nu, \gamma_\rho\}\gamma_\sigma - \{\gamma_\mu, \gamma_\rho\}\gamma_\nu\gamma_\sigma\\
&+ \gamma_\rho\gamma_\mu\{\gamma_\nu, \gamma_\sigma\} - \gamma_\rho\{\gamma_\mu, \gamma_\sigma\}\gamma_\nu\\
&= 2g_{\nu\rho}\gamma_\mu\gamma_\sigma + 2g_{\nu\sigma}\gamma_\rho\gamma_\mu - 2g_{\mu\sigma}\gamma_\rho\gamma_\nu - 2g_{\mu\rho}\gamma_\nu\gamma_\sigma \ .
\end{aligned}$$

From the above result we obtain:

$$[\sigma_{\mu\nu}, \sigma_{\rho\sigma}] = 2\mathrm{i}(g_{\nu\rho}\sigma_{\mu\sigma} + g_{\mu\sigma}\sigma_{\nu\rho} - g_{\mu\rho}\sigma_{\nu\sigma} - g_{\nu\sigma}\sigma_{\mu\rho}) \ .$$

The matrices $\frac{1}{2}\sigma_{\mu\nu}$ are generators of the Lorentz group in the spinor representation.

3.18 Let M be a matrix which commutes with all γ–matrices. Using the Problem 3.11, we can write ($\Gamma^b \neq I$)

$$M = c_b\Gamma^b + \sum_{a\neq b} c_a\Gamma^a \ . \tag{3.4}$$

On the other hand, we know that there is always a matrix Γ^d which anticommute with $\Gamma^b \neq \mathrm{I}$. Multiplying the expression (3.4) by matrix Γ_d from the left, and by Γ^d from the right, we get:

$$\Gamma_d M \Gamma^d = -c_b\Gamma^b + \sum_{a\neq b} \eta c_a\Gamma^a \ . \tag{3.5}$$

The matrix M commutes with γ_μ, and therefore with Γ^d, so we get

$$M = -c_b\Gamma^b + \sum_{a\neq b} \eta c_a\Gamma^a \ . \tag{3.6}$$

If we now multiply equations (3.4) and (3.6) by Γ_b and take trace of the resulting expressions, we get $c_b = 0$. So, each of the coefficients in the expansion (3.4) is equal to zero except the unit matrix coefficient.

3.19 By applying the Baker–Hausdorff formula

$$\mathrm{e}^B A \mathrm{e}^{-B} = A + [B, A] + \frac{1}{2!}[B, [B, A]] + \cdots$$

we get

$$\begin{aligned} U\boldsymbol{\alpha}U^\dagger &= \boldsymbol{\alpha} + 2\beta\boldsymbol{n} - 2(\boldsymbol{n}\cdot\boldsymbol{\alpha})\boldsymbol{n} - \frac{8}{3!}\beta\boldsymbol{n} + \frac{16}{4!}(\boldsymbol{\alpha}\cdot\boldsymbol{n})\boldsymbol{n} + \cdots \\ &= \boldsymbol{\alpha} + \sum_{k=1}^{\infty} \frac{(-1)^k 2^{2k}}{(2k)!}(\boldsymbol{\alpha}\cdot\boldsymbol{n})\boldsymbol{n} + \sum_{k=0}^{\infty} \frac{(-1)^k 2^{2k+1}}{(2k+1)!}\beta\boldsymbol{n} \ , \end{aligned} \tag{3.7}$$

since

$$\begin{aligned} [\beta\boldsymbol{\alpha}\cdot\boldsymbol{n}, \alpha^i] &= n^j(\beta\{\alpha^j, \alpha^i\} - \{\beta, \alpha^i\}\alpha^j) = 2\beta n^i \ , \\ [\beta\boldsymbol{\alpha}\cdot\boldsymbol{n}, [\beta\boldsymbol{\alpha}\cdot\boldsymbol{n}, \alpha^i]] &= -4(\boldsymbol{\alpha}\cdot\boldsymbol{n})n^i \ , \\ [\beta\boldsymbol{\alpha}\cdot\boldsymbol{n}, [\beta\boldsymbol{\alpha}\cdot\boldsymbol{n}, [\beta\boldsymbol{\alpha}\cdot\boldsymbol{n}, \alpha^i]]] &= -8\beta n^i \ , \\ [\beta\boldsymbol{\alpha}\cdot\boldsymbol{n}, [\beta\boldsymbol{\alpha}\cdot\boldsymbol{n}, [\beta\boldsymbol{\alpha}\cdot\boldsymbol{n}, [\beta\boldsymbol{\alpha}\cdot\boldsymbol{n}, \alpha^i]]]] &= 16(\boldsymbol{\alpha}\cdot\boldsymbol{n})n^i \ , \text{etc.} \end{aligned}$$

On the other hand, we have the following identities $(\beta\boldsymbol{\alpha}\cdot\boldsymbol{n})^2 = -1$, $(\beta\boldsymbol{\alpha}\cdot\boldsymbol{n})^3 = -(\beta\boldsymbol{\alpha}\boldsymbol{n})$, $(\beta\boldsymbol{\alpha}\cdot\boldsymbol{n})^4 = 1, \ldots$ so that

$$\begin{aligned} \boldsymbol{\alpha} + (U^2 - I)(\boldsymbol{\alpha}\cdot\boldsymbol{n})\boldsymbol{n} &= \boldsymbol{\alpha} + 2\beta\boldsymbol{n} - 2(\boldsymbol{\alpha}\cdot\boldsymbol{n})\boldsymbol{n} - \frac{8}{3!}\beta\boldsymbol{n} + \cdots \\ &= \boldsymbol{\alpha} + \sum_{k=1}^{\infty} \frac{(-1)^k 2^{2k}}{(2k)!}(\boldsymbol{\alpha}\cdot\boldsymbol{n})\boldsymbol{n} + \sum_{k=0}^{\infty} \frac{(-1)^k 2^{2k+1}}{(2k+1)!}\beta\boldsymbol{n} \ . \end{aligned} \tag{3.8}$$

It is clear that the results (3.7) and (3.8) are equal.

3.20 It is straightforward to show that the γ–matrices satisfy the relation $\{\gamma_\mu, \gamma_\nu\} = 2g_{\mu\nu}$. The connection with Dirac representation $\gamma_\mu^{\mathrm{Dirac}}$ is given by

$$\gamma_\mu S = S\gamma_\mu^{\mathrm{Dirac}} \,. \tag{3.9}$$

This statement is known as the fundamental (Pauli) theorem. If we substitute $S = \begin{pmatrix} a & b \\ c & d \end{pmatrix}$, where a, b, c, d are 2×2 matrices, into (3.9) we find

$$\begin{pmatrix} c & d \\ a & b \end{pmatrix} = \begin{pmatrix} a & -b \\ c & -d \end{pmatrix}, \quad \begin{pmatrix} -\sigma^i c & -\sigma^i d \\ \sigma^i a & \sigma^i b \end{pmatrix} = \begin{pmatrix} b\sigma^i & -a\sigma^i \\ d\sigma^i & -c\sigma^i \end{pmatrix} . \tag{3.10}$$

The solution of (3.10) is $a = -b = c = d = \mathrm{I}$. A particular solution for S is given by

$$S = \frac{1}{\sqrt{2}} \begin{pmatrix} \mathrm{I} & -\mathrm{I} \\ \mathrm{I} & \mathrm{I} \end{pmatrix} .$$

The matrices $\sigma_{\mu\nu}$ are

$$\sigma_{oi} = -\mathrm{i} \begin{pmatrix} -\sigma^i & 0 \\ 0 & \sigma^i \end{pmatrix}, \quad \sigma_{ij} = \epsilon_{ijk} \begin{pmatrix} \sigma^k & 0 \\ 0 & \sigma^k \end{pmatrix} , \tag{3.11}$$

while

$$\gamma^5 = \mathrm{i}\gamma^0\gamma^1\gamma^2\gamma^3 = \begin{pmatrix} -\mathrm{I} & 0 \\ 0 & \mathrm{I} \end{pmatrix} . \tag{3.12}$$

3.21 Matrices

$$\gamma^0 = \sigma^1 = \begin{pmatrix} 0 & 1 \\ 1 & 0 \end{pmatrix}$$

and

$$\gamma^1 = -\mathrm{i}\sigma^2 = \begin{pmatrix} 0 & -1 \\ 1 & 0 \end{pmatrix}$$

have the following properties:

$$(\gamma^0)^2 = 1, \ (\gamma^1)^2 = -1, \ \gamma^0\gamma^1 = -\gamma^1\gamma^0 ,$$

hence, they satisfy the Clifford algebra (3.A). The matrix γ^5 is defined by

$$\gamma^5 = \gamma^0\gamma^1 = \begin{pmatrix} 1 & 0 \\ 0 & -1 \end{pmatrix} .$$

$\mathrm{tr}(\gamma^5\gamma^\mu\gamma^\nu)$ is an antisymmetric tensor and it should be proportional to $\epsilon^{\mu\nu}$:

$$\mathrm{tr}(\gamma^5\gamma^\mu\gamma^\nu) = C\epsilon^{\mu\nu} .$$

By fixing $\mu = 0, \ \nu = 1$ we obtain[1] $C = 2$. One can easily show that

$$\gamma^5\gamma^\mu = \epsilon^{\mu\nu}\gamma_\nu .$$

[1] Our sign convention is $\epsilon^{01} = +1$.

4

The Dirac equation

4.1 In terms of α and β matrices, the Dirac Hamiltonian has the form $H_{\rm D} = \boldsymbol{\alpha}\cdot\boldsymbol{p} + \beta m$, so that:

(a) $[H_{\rm D}, \boldsymbol{p}\,] = 0$,
(b) $[H_{\rm D}, L^i] = \epsilon^{ijk}[\boldsymbol{\alpha}\cdot\boldsymbol{p}+\beta m, x^j p^k] = \epsilon^{ijk}\alpha^l[p^l, x^j]p^k = -\mathrm{i}\epsilon^{ijk}\alpha^j p^k = \mathrm{i}(\boldsymbol{p}\times\boldsymbol{\alpha})^i$,
(c) $[H_{\rm D}, \boldsymbol{L}^2] = -\mathrm{i}\epsilon^{ijk}\alpha^j(L^i p^k + p^k L^i) \neq 0$,
(d) $[H_{\rm D}, S^i] = -\frac{\mathrm{i}}{4}[H_{\rm D}, \epsilon^{ijk}\alpha^j\alpha^k] = \mathrm{i}\epsilon^{ijk}p^k\alpha^j = -\mathrm{i}(\boldsymbol{p}\times\boldsymbol{\alpha})^i$,
(e) By applying (b) and (d) we get that this commutator vanishes.
(f) $[H_{\rm D}, \boldsymbol{J}^2] = 0$,
(g) From (d) we have $[H_{\rm D}, \boldsymbol{\Sigma}\cdot\hat{\boldsymbol{p}}] = -\frac{\mathrm{i}}{2|\boldsymbol{p}|}\epsilon^{ijk}p^j\alpha^k p^i = 0$,
(h) Only if vectors $\boldsymbol{n}$ and $\boldsymbol{p}$ are collinear the commutator vanishes. In the opposite case it is not zero.

4.2 The plane wave

$$\psi = \begin{pmatrix}\varphi\\ \chi\end{pmatrix} \mathrm{e}^{-\mathrm{i}p\cdot x} , \tag{4.1}$$

is a particular solution of the Dirac equation,

$$(\mathrm{i}\gamma^\mu\partial_\mu - m)\psi(x) = 0 . \tag{4.2}$$

By substituting (4.1) in (4.2) (in the Dirac representation of γ–matrices) we obtain

$$\begin{pmatrix} E-m & -\boldsymbol{\sigma}\cdot\boldsymbol{p}\\ \boldsymbol{\sigma}\cdot\boldsymbol{p} & -E-m\end{pmatrix}\begin{pmatrix}\varphi\\ \chi\end{pmatrix} = 0 , \tag{4.3}$$

where E and $\boldsymbol{p}$ are the energy and momentum of the particle, respectively. Nontrivial solutions of the homogeneous system (4.3) exist if and only if its determinant vanishes. This gives the following relation between energy and momentum: $E = \pm\sqrt{\boldsymbol{p}^2+m^2} = \pm E_p$, which tells us that there are solutions of positive and negative energy as we expected.

For the positive energy solution, $E = E_p$, the system (4.3) has the following form:

$$(E_p - m)\varphi - (\boldsymbol{\sigma}\cdot\boldsymbol{p})\chi = 0 \;,$$
$$(\boldsymbol{\sigma}\cdot\boldsymbol{p})\varphi - (E_p + m)\chi = 0 \;. \tag{4.4}$$

These relations imply:

$$\chi = \frac{\boldsymbol{\sigma}\cdot\boldsymbol{p}}{E_p + m}\varphi \;, \tag{4.5}$$

or

$$u(E_p, \boldsymbol{p}) = \begin{pmatrix} \varphi \\ \chi \end{pmatrix} = \begin{pmatrix} \varphi \\ \frac{\boldsymbol{\sigma}\cdot\boldsymbol{p}}{E_p+m}\varphi \end{pmatrix} \;, \tag{4.6}$$

where φ is arbitrary. For the negative energy solution, $E = -E_p$, the system (4.3) is solved by

$$u(-E_p, \boldsymbol{p}) = \begin{pmatrix} \varphi \\ \chi \end{pmatrix} = \begin{pmatrix} -\frac{\boldsymbol{\sigma}\cdot\boldsymbol{p}}{E_p+m}\chi \\ \chi \end{pmatrix} \;. \tag{4.7}$$

If we introduce the notation $v(\boldsymbol{p}) = u(-E_p, -\boldsymbol{p})$ and $u(\boldsymbol{p}) = u(E_p, \boldsymbol{p})$, linearly independent solutions of Equation (4.2), for fixed $\boldsymbol{p}$, are given as

$$u(\boldsymbol{p})\mathrm{e}^{-\mathrm{i}p\cdot x}, \quad v(\boldsymbol{p})\mathrm{e}^{\mathrm{i}p\cdot x},$$

where $p^\mu = (E_p, \boldsymbol{p})$. Note the change of sign in the negative energy solution. The energy and momentum of the solution $u(\boldsymbol{p})\mathrm{e}^{-\mathrm{i}p\cdot x}$ are E_p and $\boldsymbol{p}$, respectively, while for $v(\boldsymbol{p})\mathrm{e}^{\mathrm{i}p\cdot x}$, they are $-E_p$ and $-\boldsymbol{p}$. In order to find the additional degrees of freedom, let us recall that the helicity operator $\frac{1}{2}\boldsymbol{\Sigma}\cdot\hat{\boldsymbol{p}}$, where $\hat{\boldsymbol{p}} = \boldsymbol{p}/|\boldsymbol{p}|$, commutes with the Dirac Hamiltonian [see Problem 4.1 (g)]. From the eigenequation

$$\boldsymbol{\sigma}\cdot\hat{\boldsymbol{p}}\varphi = \pm\varphi \;,$$

(and a similar equation for χ) we obtain

$$\varphi_1 = \frac{1}{\sqrt{2(1+\hat{p}_3)}}\begin{pmatrix} \hat{p}_3 + 1 \\ \hat{p}_1 + i\hat{p}_2 \end{pmatrix} \;, \quad \varphi_2 = \frac{1}{\sqrt{2(1+\hat{p}_3)}}\begin{pmatrix} -\hat{p}_1 + i\hat{p}_2 \\ \hat{p}_3 + 1 \end{pmatrix} \;, \tag{4.8}$$

(and similarly for χ_r, $r = 1, 2$). If we take $\boldsymbol{p} = p\boldsymbol{e}_z$, the basis vectors become

$$\begin{pmatrix} 1 \\ 0 \end{pmatrix} \;, \quad \begin{pmatrix} 0 \\ 1 \end{pmatrix} \;. \tag{4.9}$$

Then, the basis bispinors are

$$u_1(\boldsymbol{p}) = \mathcal{N}_p \begin{pmatrix} \begin{pmatrix} 1 \\ 0 \end{pmatrix} \\ \frac{\boldsymbol{\sigma}\cdot\boldsymbol{p}}{E_p+m}\begin{pmatrix} 1 \\ 0 \end{pmatrix} \end{pmatrix}, \quad u_2(\boldsymbol{p}) = \mathcal{N}_p \begin{pmatrix} \begin{pmatrix} 0 \\ 1 \end{pmatrix} \\ \frac{\boldsymbol{\sigma}\cdot\boldsymbol{p}}{E_p+m}\begin{pmatrix} 0 \\ 1 \end{pmatrix} \end{pmatrix},$$
$$v_1(\boldsymbol{p}) = \mathcal{N}_p \begin{pmatrix} \frac{\boldsymbol{\sigma}\cdot\boldsymbol{p}}{E_p+m}\begin{pmatrix} 0 \\ 1 \end{pmatrix} \\ \begin{pmatrix} 0 \\ 1 \end{pmatrix} \end{pmatrix}, \quad v_2(\boldsymbol{p}) = \mathcal{N}_p \begin{pmatrix} \frac{\boldsymbol{\sigma}\cdot\boldsymbol{p}}{E_p+m}\begin{pmatrix} 1 \\ 0 \end{pmatrix} \\ \begin{pmatrix} 1 \\ 0 \end{pmatrix} \end{pmatrix}, \tag{4.10}$$

where $\mathcal{N}_p = \sqrt{\frac{E_p+m}{2m}}$ is the normalization factor. Do not forget that $\boldsymbol{p} = p\boldsymbol{e}_z$ i.e. $\boldsymbol{p}\cdot\boldsymbol{\sigma} = p\sigma_3$. In this case, the bispinors (4.10) form the helicity basis. For arbitrary momentum $\boldsymbol{p}$ we have to use (4.8) instead of (4.9), if we want to construct the helicity basis. Although, in that case vectors in (4.10) are also a base, but not the helicity one. Spinors u and v are normalized according to (4.D).

General solution of (4.2) is given by

$$\psi = \frac{1}{(2\pi)^{3/2}}\sum_{r=1}^{2}\int \mathrm{d}^3\boldsymbol{p}\sqrt{\frac{m}{E_p}}\left(u_r(\boldsymbol{p})c_r(\boldsymbol{p})\mathrm{e}^{-\mathrm{i}p\cdot x} + v_r(\boldsymbol{p})d_r^\dagger(\boldsymbol{p})\mathrm{e}^{\mathrm{i}p\cdot x}\right) . \tag{4.11}$$

The Dirac spinor (bispinor) ψ contains two SL(2, C) spinors, as is easily seen in the chiral (Weyl) representation. The Dirac spinor is transformed according to the $(1/2,0)\oplus(0,1/2)$ reducible representation of the quantum Lorentz group (i.e. SL(2, C) group, which is universally covering group for the Lorentz group).

4.3 The states $u_s(\boldsymbol{p}), v_s(\boldsymbol{p})$ are eigenstates of the energy operator, $\mathrm{i}\frac{\partial}{\partial t}$ with eigenvalues E_p and $-E_p$, respectively.

4.4 By using the expressions for the Dirac spinors found in Problem 4.2, we obtain

$$\sum_r u_r(\boldsymbol{p})\bar{u}_r(\boldsymbol{p}) =$$
$$\frac{E_p+m}{2m}\begin{pmatrix} \varphi_1\varphi_1^\dagger + \varphi_2\varphi_2^\dagger & -(\varphi_1\varphi_1^\dagger + \varphi_2\varphi_2^\dagger)\frac{\boldsymbol{\sigma}\cdot\boldsymbol{p}}{E_p+m} \\ \frac{\boldsymbol{\sigma}\cdot\boldsymbol{p}}{E_p+m}(\varphi_1\varphi_1^\dagger + \varphi_2\varphi_2^\dagger) & -\frac{\boldsymbol{\sigma}\cdot\boldsymbol{p}}{E_p+m}(\varphi_1\varphi_1^\dagger + \varphi_2\varphi_2^\dagger)\frac{\boldsymbol{\sigma}\cdot\boldsymbol{p}}{E_p+m} \end{pmatrix},$$

where φ_r $(r = \{1,2\})$ are given by (4.8). They satisfy the completeness relation $\varphi_1\varphi_1^\dagger + \varphi_2\varphi_2^\dagger = \mathrm{I}$. Using also $(\boldsymbol{p}\cdot\boldsymbol{\sigma})^2 = \boldsymbol{p}^2 = E_p^2 - m^2$, we get

$$\sum_{r=1}^{2} u_r(\boldsymbol{p})\bar{u}_r(\boldsymbol{p}) = \frac{1}{2m}\begin{pmatrix} E_p + m & -\boldsymbol{\sigma}\cdot\boldsymbol{p} \\ \boldsymbol{\sigma}\cdot\boldsymbol{p} & -E_p + m \end{pmatrix} = \frac{\not{p} + m}{2m} .$$

The second identity can be shown in a similar manner.

4.5 Using the expressions for the projectors given in Problem 4.4, we see that

$$\Lambda_+^2 = \frac{1}{4m^2}(\not{p}^2 + 2m\not{p} + m^2) = \Lambda_+ \ ,$$

where we have used $\not{p}^2 = p^2 = m^2$. Similarly, we obtain $\Lambda_-^2 = \Lambda_-$. Orthogonality of the projectors follows from the identity

$$(\not{p} + m)(\not{p} - m) = p^2 - m^2 = 0 \ .$$

At this stage we apply the Dirac equation in momentum space (4.C). Namely,

$$\Lambda_+ u_r(\boldsymbol{p}) = \frac{1}{2m}(\not{p} + m)u_r(\boldsymbol{p}) = \frac{1}{2m}(m + m)u_r(\boldsymbol{p}) = u_r(\boldsymbol{p}) \ ,$$

$$\Lambda_- u_r(\boldsymbol{p}) = \frac{1}{2m}(\not{p} - m)u_r(\boldsymbol{p}) = \frac{1}{2m}(m - m)u_r(\boldsymbol{p}) = 0 \ .$$

Similarly, one can prove the identities $\Lambda_- v_r(\boldsymbol{p}) = 0$, $\Lambda_+ v_r(\boldsymbol{p}) = v_r(\boldsymbol{p})$.

4.6

(a) We can directly prove this property. For example, the x–component of the vector $\boldsymbol{\Sigma}$ is

$$\Sigma^1 = \frac{\mathrm{i}}{2}(\gamma^2\gamma^3 - \gamma^3\gamma^2) = \mathrm{i}\gamma^2\gamma^3 \ .$$

On the other hand, $\gamma_5\gamma_0\gamma^1 = \mathrm{i}\gamma_1\gamma_2\gamma_3\gamma^1 = \mathrm{i}\gamma^2\gamma^3$. The corresponding identities for the y and z–components can be proven in a similar way.

(b) By applying the definition of $\boldsymbol{\Sigma}$, we have

$$\begin{aligned}[\Sigma^i, \Sigma^j] &= -\frac{1}{4}\epsilon^{ilm}\epsilon^{jpq}[\gamma^l\gamma^m, \gamma^p\gamma^q] \\ &= -\frac{1}{4}\epsilon^{ilm}\epsilon^{jpq}\left([\gamma^l\gamma^m, \gamma^p]\gamma^q + \gamma^p[\gamma^l\gamma^m, \gamma^q]\right) \ . \end{aligned} \tag{4.12}$$

Next step is to expand the commutators in terms of the anticommutators:

$$\begin{aligned}[\Sigma^i, \Sigma^j] = -\frac{1}{4}\epsilon^{ilm}\epsilon^{jpq}\big(&\gamma^l\{\gamma^m, \gamma^p\}\gamma^q - \{\gamma^l, \gamma^p\}\gamma^m\gamma^q \\ &+\gamma^p\gamma^l\{\gamma^m, \gamma^q\} - \gamma^p\{\gamma^l, \gamma^q\}\gamma^m\big) \ . \end{aligned} \tag{4.13}$$

Then, using anticommutation relations (3.A) we get

$$[\Sigma^i, \Sigma^j] = -\frac{1}{2}\epsilon^{ilm}\epsilon^{jpq}\left(g^{mp}\gamma^l\gamma^q - g^{lp}\gamma^m\gamma^q + g^{mq}\gamma^p\gamma^l - g^{lq}\gamma^p\gamma^m\right) \ . \tag{4.14}$$

The first term in (4.14) has the form

$$\epsilon^{ilm}\epsilon^{jpq}g^{mp}\gamma^l\gamma^q = (\delta^{ij}\delta^{lq} - \delta^{iq}\delta^{lj})\gamma^l\gamma^q = -3\delta^{ij} - \gamma^j\gamma^i \ .$$

Other terms in (4.14) can be transformed in the same way. Finally,

$$[\Sigma^i, \Sigma^j] = \gamma^j\gamma^i - \gamma^i\gamma^j \ .$$

On the other hand,

$$2\mathrm{i}\epsilon^{ijk}\Sigma^k = -\epsilon^{ijk}\epsilon^{klm}\gamma^l\gamma^m = \gamma^j\gamma^i - \gamma^i\gamma^j \ ,$$

so that

$$[\Sigma^i, \Sigma^j] = 2\mathrm{i}\epsilon^{ijk}\Sigma^k \ .$$

We conclude that operators $\frac{1}{2}\boldsymbol{\Sigma}$ are the generators of SU(2) subgroup of the Lorentz group[1]

(c) $S^2 = -\frac{1}{4}\boldsymbol{\Sigma}^2 = -\frac{1}{4}(\gamma_5\gamma_0\boldsymbol{\gamma})^2 = \frac{1}{4}\boldsymbol{\gamma}\cdot\boldsymbol{\gamma} = -\frac{3}{4}$.

4.7 Use the expressions $\boldsymbol{\sigma}\cdot\hat{\boldsymbol{p}}\varphi_r = (-1)^{r+1}\varphi_r$ and $\boldsymbol{\sigma}\cdot\hat{\boldsymbol{p}}\chi_r = (-1)^r\chi_r$ from Problem 4.2. For example:

$$\begin{aligned}
\frac{\boldsymbol{\Sigma}\cdot\boldsymbol{p}}{|\boldsymbol{p}|}u_r(\boldsymbol{p}) &= \frac{\boldsymbol{\Sigma}\cdot\boldsymbol{p}}{|\boldsymbol{p}|}\mathcal{N}\begin{pmatrix}\varphi_r \\ \frac{\boldsymbol{\sigma}\cdot\boldsymbol{p}}{E_p+m}\varphi_r\end{pmatrix} \\
&= \mathcal{N}\begin{pmatrix}\boldsymbol{\sigma}\cdot\hat{\boldsymbol{p}} & 0 \\ 0 & \boldsymbol{\sigma}\cdot\hat{\boldsymbol{p}}\end{pmatrix}\begin{pmatrix}\varphi_r \\ \frac{\boldsymbol{\sigma}\cdot\boldsymbol{p}}{E_p+m}\varphi_r\end{pmatrix} \\
&= \mathcal{N}\begin{pmatrix}\boldsymbol{\sigma}\cdot\hat{\boldsymbol{p}}\varphi_r \\ \frac{(\boldsymbol{\sigma}\cdot\boldsymbol{p})(\boldsymbol{\sigma}\cdot\hat{\boldsymbol{p}})}{E_p+m}\varphi_r\end{pmatrix} \\
&= (-1)^{r+1}\mathcal{N}\begin{pmatrix}\varphi_r \\ \frac{\boldsymbol{\sigma}\cdot\boldsymbol{p}}{E_p+m}\varphi_r\end{pmatrix} \\
&= (-1)^{r+1}u_r(\boldsymbol{p}) \ ,
\end{aligned}$$

where $\mathcal{N}$ is the normalization factor. It is easy to see that the spinors $u_r(\boldsymbol{p})$ and $v_r(\boldsymbol{p})$ are not eigenspinors of the operator $\boldsymbol{\Sigma}\cdot\boldsymbol{n}$, unless vectors $\boldsymbol{n}$ and $\boldsymbol{p}$ are parallel.

4.8 The transformation operator from the rest frame to the frame moving along the z–axis with velocity v, is $S(\Lambda(v\boldsymbol{e}_z)) = \mathrm{e}^{-\frac{\mathrm{i}}{2}\omega_{03}\sigma^{03}}$. By using the relation $\omega_{03} = -\varphi = -\operatorname{arctan}(v)$, we obtain

$$\begin{aligned}
S(\Lambda) &= \cosh\left(\frac{\varphi}{2}\right)\mathrm{I} - \sinh\left(\frac{\varphi}{2}\right)\begin{pmatrix}0 & \sigma_3 \\ \sigma_3 & o\end{pmatrix} \\
&= \sqrt{\frac{E_p+m}{2m}}\begin{pmatrix}\mathrm{I} & -\frac{p\sigma_3}{E_p+m} \\ -\frac{p\sigma_3}{E_p+m} & \mathrm{I}\end{pmatrix} .
\end{aligned}$$

For arbitrary boost, $\sigma_3 p$ should be replaced by $\boldsymbol{\sigma}\cdot\boldsymbol{p}$. The operator $S(\Lambda)$ is not unitary one. Since the Lorentz group is noncompact, it does not have finite dimensional irreducible unitary representations.

4.9 In this case we have

$$S = \begin{pmatrix}\cos\left(\frac{\theta}{2}\right) + \mathrm{i}\sin\left(\frac{\theta}{2}\right)\sigma^3 & 0 \\ 0 & \cos\left(\frac{\theta}{2}\right) + \mathrm{i}\sin\left(\frac{\theta}{2}\right)\sigma^3\end{pmatrix} .$$

---

[1] Recall that $\Sigma^k = \frac{1}{2}\epsilon^{kij}\sigma^{ij}$.

This operator is unitary because SO(3) is a compact subgroup of the Lorentz group.

4.10 The Pauli–Lubanski vector is

$$W^{\mu}=\frac{1}{2}\epsilon^{\mu\nu\rho\sigma}(\mathrm{i}x_{\nu}\partial_{\rho}-\mathrm{i}x_{\rho}\partial_{\nu}+\frac{1}{2}\sigma_{\nu\rho})\mathrm{i}\partial_{\sigma}=\frac{\mathrm{i}}{4}\epsilon^{\mu\nu\rho\sigma}\sigma_{\nu\rho}\partial_{\sigma}\ , \tag{4.15}$$

since the contraction of a symmetric and an antisymmetric tensors vanishes. Then

$$\begin{aligned}
W^2\psi(x) &= -\frac{1}{16}\epsilon^{\mu\nu\rho\sigma}\epsilon_{\mu\alpha\beta\gamma}\sigma_{\nu\rho}\sigma^{\alpha\beta}\partial_{\sigma}\partial^{\gamma}\psi(x)\\
&= \frac{1}{16}\Big(\delta^{\nu}_{\alpha}\delta^{\rho}_{\beta}\delta^{\sigma}_{\gamma}-\delta^{\nu}_{\alpha}\delta^{\sigma}_{\beta}\delta^{\rho}_{\gamma}+\delta^{\rho}_{\alpha}\delta^{\sigma}_{\beta}\delta^{\nu}_{\gamma}-\\
&\quad -\delta^{\rho}_{\alpha}\delta^{\nu}_{\beta}\delta^{\sigma}_{\gamma}+\delta^{\sigma}_{\alpha}\delta^{\nu}_{\beta}\delta^{\rho}_{\gamma}-\delta^{\sigma}_{\alpha}\delta^{\rho}_{\beta}\delta^{\nu}_{\gamma}\Big)\sigma_{\nu\rho}\sigma^{\alpha\beta}\partial_{\sigma}\partial^{\gamma}\psi(x)\\
&= \frac{1}{16}\left(2\sigma^{\alpha\beta}\sigma_{\alpha\beta}\Box-4\sigma^{\alpha\gamma}\sigma_{\alpha\rho}\partial^{\rho}\partial_{\gamma}\right)\psi\\
&= \frac{3}{4}\Box\psi\\
&= -\frac{3}{4}m^2\psi\ ,
\end{aligned}$$

where we used identity

$$\sigma_{\mu\sigma}\sigma^{\mu\nu}=2\gamma_{\sigma}\gamma^{\nu}+\delta^{\nu}_{\sigma}$$

and the results of Problems 1.5 and 3.5.

4.11 It is easy to see (Problem 3.16 and the condition $s\cdot p=0$) that

$$\begin{aligned}
\frac{W_{\mu}s^{\mu}}{m} &= \frac{1}{4m}\epsilon_{\mu\nu\rho\sigma}\sigma^{\nu\rho}P^{\sigma}s^{\mu}=\frac{1}{2m}\gamma_5\sigma_{\mu\sigma}s^{\mu}\partial^{\sigma}\\
&= \frac{\mathrm{i}}{2m}\gamma_5(\gamma_{\mu}\gamma_{\sigma}-g_{\mu\sigma})(\mp\mathrm{i}p^{\sigma})s^{\mu}=\pm\frac{1}{2m}\gamma_5\not{s}\not{p}=\frac{1}{2}\gamma_5\not{s}\ .
\end{aligned}$$

The previous equation holds on space of plane wave solutions; upper (lower) sing is related to positive (negative) energy solutions. In the rest frame, the vector s^{μ} becomes $(0,\boldsymbol{n})$, so $\not{s}=-\boldsymbol{n}\cdot\boldsymbol{\gamma}$, and we can use $\frac{\not{p}}{m}=\frac{p^0\gamma^0}{m}=\gamma^0$, so that

$$\frac{W\cdot s}{m}=\pm\frac{1}{2}\gamma_5\gamma_0\boldsymbol{n}\cdot\boldsymbol{\gamma}=\pm\frac{1}{2}\boldsymbol{\Sigma}\cdot\boldsymbol{n}\ .$$

where Problem 4.6 has been used.

4.12 Positive energy solutions satisfy

$$\gamma_5\not{s}u(\boldsymbol{p},\pm s)=\pm u(\boldsymbol{p},\pm s)\ . \tag{4.16}$$

If we choose that polarization vector s^{μ} in the rest frame equals $(0,\boldsymbol{n}=\frac{\boldsymbol{p}}{|\boldsymbol{p}|})$, according to the formulation of this problem, then in the frame in which electron has momentum $\boldsymbol{p}$, the polarization vector is obtained by applying a Lorentz boost:

$$s^\mu = \begin{pmatrix} \frac{E_p}{m} & \frac{p^j}{m} \\ \frac{p^i}{m} & \delta_{ij} + \frac{p^i p^j}{m(E_p+m)} \end{pmatrix} \begin{pmatrix} 0 \\ n^j \end{pmatrix}$$
$$= \begin{pmatrix} \frac{\boldsymbol{p}\cdot\boldsymbol{n}}{m} \\ \boldsymbol{n} + \frac{(\boldsymbol{n}\cdot\boldsymbol{p})\boldsymbol{p}}{m(E_p+m)} \end{pmatrix} .$$

For $\boldsymbol{n} = \boldsymbol{p}/|\boldsymbol{p}|$ we get $s^\mu = (\frac{|\boldsymbol{p}|}{m}, \frac{E_p}{m}\boldsymbol{n})$. Using that, we find

$$\gamma_5 \not{s} u(\boldsymbol{p}, \pm s) = \frac{1}{m}\gamma_5 \not{s} \not{p} u(\boldsymbol{p}, \pm s)$$
$$= \frac{1}{m}\gamma_5 \left(\frac{|\boldsymbol{p}|}{m}\gamma_0 - \frac{E_p}{m}\boldsymbol{\gamma}\cdot\boldsymbol{n}\right)(E_p\gamma_0 - \boldsymbol{p}\cdot\boldsymbol{\gamma})u(\boldsymbol{p}, \pm s) .$$

If we insert $(\boldsymbol{p}\cdot\boldsymbol{\gamma})^2 = -\boldsymbol{p}^2$ in the previous formula we obtain:

$$\gamma_5 \not{s} u(\boldsymbol{p}, \pm s) = \gamma_5\gamma_0\boldsymbol{\gamma}\cdot\frac{\boldsymbol{p}}{|\boldsymbol{p}|}u(\boldsymbol{p}, \pm s) = \frac{\boldsymbol{\Sigma}\cdot\boldsymbol{p}}{|\boldsymbol{p}|}u(\boldsymbol{p}, \pm s) . \tag{4.17}$$

From the expressions (4.16) and (4.17) we get

$$\frac{\boldsymbol{\Sigma}\cdot\boldsymbol{p}}{|\boldsymbol{p}|}u(\boldsymbol{p}, \pm s) = \pm u(\boldsymbol{p}, \pm s) .$$

The similar procedure can be done for negative energy solutions. Starting from

$$\gamma_5 \not{s} v(\boldsymbol{p}, \pm s) = \pm v(\boldsymbol{p}, \pm s) ,$$

one gets

$$\frac{\boldsymbol{\Sigma}\cdot\boldsymbol{p}}{|\boldsymbol{p}|}v(\boldsymbol{p}, \pm s) = \mp v(\boldsymbol{p}, \pm s) .$$

4.13 In the ultrarelativistic limit, $m \ll E_p$, the vector s^μ is given by

$$s^\mu \approx \left(\frac{E_p}{m}, \frac{\boldsymbol{p}}{m}\right) \approx \frac{p^\mu}{m} .$$

Then we have

$$\gamma_5 \not{s} u(\boldsymbol{p}, \pm s) \approx \gamma_5 \frac{\not{p}}{m} u(\boldsymbol{p}, \pm s) = \gamma_5 u(\boldsymbol{p}, \pm s) , \tag{4.18}$$

where we used the Dirac equation $\not{p}u(\boldsymbol{p}, \pm s) = mu(\boldsymbol{p}, \pm s)$. From (4.18) we conclude that the helicity operator $\boldsymbol{\Sigma}\cdot\boldsymbol{p}/|\boldsymbol{p}|$ is equal to the chirality operator γ_5. The eigenequation becomes

$$\gamma_5 u(\boldsymbol{p}, \pm s) = \pm u(\boldsymbol{p}, \pm s) .$$

For v spinors the situation is similar. So, for the particles of high energy (i.e. neglected mass) helicity and chirality are approximatively equal, while for massless particles these two quantities exactly are equal.

4.14 The commutator between $\gamma_5 \not{s}$ and $\not{p}$ is

$$\begin{aligned}[\gamma_5 \not{s}, \not{p}] &= \gamma_5 \not{s}\not{p} - \not{p}\gamma_5\not{s} \\ &= \gamma_5(\not{s}\not{p} + \not{p}\not{s}) \\ &= \gamma_5 s^\mu p^\nu \{\gamma_\mu, \gamma_\nu\} \\ &= 2s\cdot p\gamma_5 = 0 \ .\end{aligned}$$

From $(\gamma_5\not{s})^2 = -s^2 = 1$ it follows that eigenvalues of the operator $\gamma_5\not{s}$ are ± 1. Then the eigen projectors are

$$\Sigma(\pm s) = \frac{1 \pm \gamma_5\not{s}}{2} \ .$$

4.15 The average value of $\boldsymbol{\Sigma}\cdot\boldsymbol{n}$ in state

$$\psi(x) = \sqrt{\frac{E_p + m}{2m}} \begin{pmatrix} \varphi \\ \frac{\boldsymbol{\sigma}\cdot\boldsymbol{p}}{E_p+m}\varphi \end{pmatrix} \mathrm{e}^{-\mathrm{i}p\cdot x} \ , \tag{4.19}$$

is

$$\begin{aligned}\langle\boldsymbol{\Sigma}\cdot\boldsymbol{n}\rangle &= \frac{\int \mathrm{d}^3x \psi^\dagger(x)\boldsymbol{\Sigma}\cdot\boldsymbol{n}\psi(x)}{\int \mathrm{d}^3x\psi^\dagger(x)\psi(x)} \\ &= \frac{E_p+m}{2E_p}\left(\varphi^\dagger\boldsymbol{\sigma}\cdot\boldsymbol{n}\varphi + \frac{\varphi^\dagger(\boldsymbol{\sigma}\cdot\boldsymbol{p})(\boldsymbol{\sigma}\cdot\boldsymbol{n})(\boldsymbol{\sigma}\cdot\boldsymbol{p})\varphi}{(E_p+m)^2}\right) \ .\end{aligned} \tag{4.20}$$

Since

$$(\boldsymbol{\sigma}\cdot\boldsymbol{A})(\boldsymbol{\sigma}\cdot\boldsymbol{B}) = \boldsymbol{A}\cdot\boldsymbol{B} + \mathrm{i}(\boldsymbol{A}\times\boldsymbol{B})\cdot\boldsymbol{\sigma} \tag{4.21}$$

it follows that

$$(\boldsymbol{\sigma}\cdot\boldsymbol{p})(\boldsymbol{\sigma}\cdot\boldsymbol{n})(\boldsymbol{\sigma}\cdot\boldsymbol{p}) = |\boldsymbol{p}|^2(n_3\sigma_3 - n_2\sigma_2 - n_1\sigma_1) \ . \tag{4.22}$$

By substituting (4.22) into (4.20) we get:

$$\begin{aligned}\langle\boldsymbol{\Sigma}\cdot\boldsymbol{n}\rangle &= \frac{1}{|a|^2+|b|^2} \\ &\times \left[\frac{E_p+m}{2E_p}(n_3|a|^2 + (n_1+\mathrm{i}n_2)b^*a + (n_1-\mathrm{i}n_2)a^*b - n_3|b|^2)\right. \\ &+ \left.\frac{E_p-m}{2E_p}(n_3|a|^2 + (-n_1+\mathrm{i}n_2)a^*b - (n_1+\mathrm{i}n_2)b^*a - n_3|b|^2)\right] \ .\end{aligned}$$

In the nonrelativistic limit we obtain

$$\langle\boldsymbol{\Sigma}\cdot\boldsymbol{n}\rangle = \varphi^\dagger\boldsymbol{\sigma}\cdot\boldsymbol{n}\varphi = \frac{n_3|a|^2 + (n_1+\mathrm{i}n_2)b^*a + (n_1-\mathrm{i}n_2)a^*b - n_3|b|^2}{|a|^2+|b|^2} \ .$$

4.16 In the rest frame a spinor takes the following form $\begin{pmatrix} \varphi \\ 0 \end{pmatrix} \mathrm{e}^{-imt}$, where φ satisfies

$$\frac{1}{2}\boldsymbol{\Sigma}\cdot\boldsymbol{n}\begin{pmatrix} \varphi \\ 0 \end{pmatrix} = \frac{1}{2}\begin{pmatrix} \varphi \\ 0 \end{pmatrix} .$$

The last condition becomes

$$\begin{pmatrix} \cos\theta & -\mathrm{i}\sin\theta \\ \mathrm{i}\sin\theta & -\cos\theta \end{pmatrix}\begin{pmatrix} a \\ b \end{pmatrix} = \begin{pmatrix} a \\ b \end{pmatrix} , \tag{4.23}$$

where we put $\varphi = \begin{pmatrix} a \\ b \end{pmatrix}$. From the last expression we obtain

$$\varphi = \begin{pmatrix} \cos\frac{\theta}{2} \\ \mathrm{i}\sin\frac{\theta}{2} \end{pmatrix} . \tag{4.24}$$

In the rest frame the Dirac spinor takes the form

$$\psi_0 = \begin{pmatrix} \cos\frac{\theta}{2} \\ \mathrm{i}\sin\frac{\theta}{2} \\ 0 \\ 0 \end{pmatrix} \mathrm{e}^{-imt} . \tag{4.25}$$

Applying the boost along z–axis, we obtain

$$\psi(x) = S(-p\boldsymbol{e}_z)\psi_0 , \tag{4.26}$$

where S is given in Problem 4.8. Note a minus sign appearing in $S(-p\boldsymbol{e}_z)$! After a simple calculation, we obtain

$$\psi(x) = \sqrt{\frac{E_p + m}{2m}} \begin{pmatrix} \cos\frac{\theta}{2} \\ \mathrm{i}\sin\frac{\theta}{2} \\ \frac{\boldsymbol{p}\cdot\boldsymbol{\sigma}}{E_p+m}\begin{pmatrix} \cos\frac{\theta}{2} \\ \mathrm{i}\sin\frac{\theta}{2} \end{pmatrix} \end{pmatrix} \mathrm{e}^{-\mathrm{i}p\cdot x} . \tag{4.27}$$

The mean value of the operator $\frac{1}{2}\gamma_5 \not{s}$ is

$$\left\langle \frac{1}{2}\gamma_5 \not{s} \right\rangle = \frac{1}{2}\frac{\int \mathrm{d}^3x \psi^\dagger \gamma_5 \not{s} \psi}{\int \mathrm{d}^3x \psi^\dagger \psi} , \tag{4.28}$$

where the vector s^μ is obtained from $(0, \boldsymbol{n})$ by the Lorentz boost along the z–axis. The components of vector s^μ are (see Problem 4.12)

$$s_0 = \frac{\boldsymbol{n}\cdot\boldsymbol{p}}{m}, \quad \boldsymbol{s} = \boldsymbol{n} + \frac{(\boldsymbol{n}\cdot\boldsymbol{p})\boldsymbol{p}}{m(E_p + m)} .$$

In our case we have

$$s^\mu = \left(\frac{p}{m} \cos\theta, 0, \sin\theta, \frac{E_p}{m} \cos\theta \right) .$$

Thus, in the Dirac representation of γ–matrices, $\gamma_5 \not{s}$ is given by

$$\gamma_5 \not{s} = \begin{pmatrix} \boldsymbol{s} \cdot \boldsymbol{\sigma} & -s_0 \mathrm{I} \\ s_0 \mathrm{I} & -\boldsymbol{s} \cdot \boldsymbol{\sigma} \end{pmatrix} , \tag{4.29}$$

and finally

$$\gamma_5 \not{s} = \begin{pmatrix} \frac{E_p}{m} \cos\theta & -\mathrm{i} \sin\theta & -\frac{p}{m} \cos\theta & 0 \\ \mathrm{i} \sin\theta & -\frac{E_p}{m} \cos\theta & 0 & -\frac{p}{m} \cos\theta \\ \frac{p}{m} \cos\theta & 0 & -\frac{E_p}{m} \cos\theta & \mathrm{i} \sin\theta \\ 0 & \frac{p}{m} \cos\theta & -\mathrm{i} \sin\theta & \frac{E_p}{m} \cos\theta \end{pmatrix} . \tag{4.30}$$

By substituting (4.30) and (4.27) in the formula (4.28), we obtain:

$$\left\langle \frac{1}{2} \gamma_5 \not{s} \right\rangle = \frac{1}{2} ,$$

as we expected, because $\psi(x)$ is the eigenstate of the operator $\frac{1}{2}\gamma_5 \not{s}$, with eigenvalue $\frac{1}{2}$.

4.17 The Dirac Hamiltonian can be rewritten in terms of γ–matrices so that

$$[H_\mathrm{D}, \gamma_5] = [\gamma^0 \boldsymbol{\gamma} \cdot \boldsymbol{p} + \gamma^0 m, \gamma_5] = 2m\gamma^0\gamma_5 .$$

Thus, the operator γ_5 is a constant of motion in the case of massless Dirac particle. Its eigenvalues and eigen projectors are ± 1, $\Sigma_\pm = \frac{1}{2}(1 \pm \gamma_5)$, respectively. The operator γ_5 is known as the chirality operator.

4.18 By multiplying the Dirac equation from the left by γ_5, we obtain $(\mathrm{i}\not{\partial} + m)\gamma_5 \psi = 0$. By adding and subtracting the previous equations and the Dirac equation, we get

$$\mathrm{i}\not{\partial}\psi_\mathrm{L} - m\psi_\mathrm{R} = 0,$$

$$\mathrm{i}\not{\partial}\psi_\mathrm{R} - m\psi_\mathrm{L} = 0 .$$

4.19

(a) The system of equations can be rewritten as the Dirac equation. The Dirac spinor takes form

$$\psi = \begin{pmatrix} \psi_\mathrm{L} \\ \psi_\mathrm{R} \end{pmatrix} ,$$

while

$$\gamma^\mu = \begin{pmatrix} 0 & \sigma^\mu \\ \bar{\sigma}^\mu & 0 \end{pmatrix} ,$$

are γ–matrices (see Problem 3.20).

(b) In order to be covariant, these equations have to have the following form

$$i\sigma^\mu \partial'_\mu \psi'_R(x') = m\psi'_L(x') , \tag{4.31}$$

$$i\bar{\sigma}^\mu \partial'_\mu \psi'_L(x') = m\psi'_R(x') , \tag{4.32}$$

in the primed frame ($x' = \Lambda x$). If we assume that the new spinors take the form $\psi'_L(x') = S_L\psi_L(x)$ and $\psi'_R(x') = S_R\psi_R(x)$, where S_L and S_R are nonsingular 2×2 matrices, Equations (4.31) and (4.32) become

$$i\sigma^\mu S_R \Lambda_\mu{}^\nu \partial_\nu \psi_R(x) = mS_L\psi_L(x) , \tag{4.33}$$

$$i\bar{\sigma}^\mu S_L \Lambda_\mu{}^\nu \partial_\nu \psi_L(x) = mS_R\psi_R(x) . \tag{4.34}$$

By multiplying Equation (4.33) by S_L^{-1} from left, and (4.34) by S_R^{-1} also from left we obtain

$$iS_L^{-1}\sigma^\mu S_R \Lambda_\mu{}^\nu \partial_\nu \psi_R(x) = m\psi_L(x) , \tag{4.35}$$

$$iS_R^{-1}\bar{\sigma}^\mu S_L \Lambda_\mu{}^\nu \partial_\nu \psi_L(x) = m\psi_R(x) . \tag{4.36}$$

The system of equations is covariant if the conditions

$$S_R^{-1}\bar{\sigma}^\mu S_L = \Lambda^\mu{}_\nu \bar{\sigma}^\nu ,$$

$$S_L^{-1}\sigma^\mu S_R = \Lambda^\mu{}_\nu \sigma^\nu$$

hold. The solution for matrices S_L and S_R is given as

$$S_L = \exp\left(\frac{1}{2}\varphi_i\sigma^i + \frac{i}{2}\theta_k\sigma^k\right) \approx 1 + \frac{1}{2}\varphi_i\sigma^i + \frac{i}{2}\theta_k\sigma^k , \tag{4.37}$$

$$S_R = \exp\left(-\frac{1}{2}\varphi_i\sigma^i + \frac{i}{2}\theta_k\sigma^k\right) \approx 1 - \frac{1}{2}\varphi_i\sigma^i + \frac{i}{2}\theta_k\sigma^k . \tag{4.38}$$

The parameters θ_i and φ_i were defined in Problem 1.8. Boost along the x–axis is defined by :

$$S_L = \cosh\left(\frac{\varphi_1}{2}\right) + \sigma_1 \sinh\left(\frac{\varphi_1}{2}\right) \tag{4.39}$$

$$S_R = \cosh\left(\frac{\varphi_1}{2}\right) - \sigma_1 \sinh\left(\frac{\varphi_1}{2}\right) . \tag{4.40}$$

Note that ψ_L and ψ_R transform in the same way under rotations, but differently under boosts. The left ψ_L, and right ψ_R spinors transform under $(\frac{1}{2}, 0)$ and $(0, \frac{1}{2})$ irreducible representation of the Lorentz group respectively.

4.20 First note that

$$[H_{\rm D}, K] = [\boldsymbol{\alpha}\cdot\boldsymbol{p}, \beta(\boldsymbol{\Sigma}\cdot\boldsymbol{L})] + [\boldsymbol{\alpha}\cdot\boldsymbol{p}, \beta] + m[\beta, \beta(\boldsymbol{\Sigma}\cdot\boldsymbol{L})] \ . \tag{4.41}$$

The first term in the expression (4.41) is

$$\begin{aligned}[\boldsymbol{\alpha}\cdot\boldsymbol{p}, \beta(\boldsymbol{\Sigma}\cdot\boldsymbol{L})] &= \beta[\boldsymbol{\alpha}\cdot\boldsymbol{p}, \boldsymbol{\Sigma}\cdot\boldsymbol{L}] + [\boldsymbol{\alpha}\cdot\boldsymbol{p}, \beta]\boldsymbol{\Sigma}\cdot\boldsymbol{L} \\ &= -\frac{\mathrm{i}}{2}\epsilon^{mnp}\epsilon^{mjl}\beta\left(p^i\{\alpha^i, \alpha^n\}\alpha^p x^j p^l - p^i\alpha^n\{\alpha^i, \alpha^p\}x^j p^l + \right. \\ &\left. + \ \alpha^n\alpha^p\alpha^i[p^i, x^j]p^l - 2\alpha^i\alpha^n\alpha^p p^i x^j p^l\right) \ .\end{aligned}$$

Using the relations $\{\alpha^i, \alpha^j\} = 2\delta_{ij}$ and $[x^i, p^j] = \mathrm{i}\delta_{ij}$, we obtain

$$\begin{aligned}[\boldsymbol{\alpha}\cdot\boldsymbol{p}, \beta(\boldsymbol{\Sigma}\cdot\boldsymbol{L})] &= -\frac{\mathrm{i}}{2}\beta\left(4\alpha^l p^n x^n p^l - 4\alpha^j p^l x^j p^l - \right. \\ &\left. - \ \mathrm{i}\alpha^j\alpha^l\alpha^j p^l + 3\mathrm{i}\alpha^i p^i - 2\alpha^i\alpha^j\alpha^l p^i x^j p^l + 2\alpha^i\alpha^l\alpha^j p^i x^j p^l\right) \\ &= \mathrm{i}\beta\left(2\alpha^i p^l x^i p^l - 2\mathrm{i}\boldsymbol{\alpha}\cdot\boldsymbol{p} - \alpha^j\alpha^i\alpha^l p^i x^j p^l - \alpha^i\alpha^l\alpha^j p^i x^j p^l\right) \ ,\end{aligned}$$

where we used $\alpha^i\alpha^j\alpha^i = -\alpha^j$. By substituting $p^i x^j = x^j p^i - \mathrm{i}\delta^{ij}$ into the last line of previous formula, we have

$$[\boldsymbol{\alpha}\cdot\boldsymbol{p}, \beta(\boldsymbol{\Sigma}\cdot\boldsymbol{L})] = 2\beta(\boldsymbol{\alpha}\cdot\boldsymbol{p}) \ . \tag{4.42}$$

The second term in (4.41) is $-2\beta(\boldsymbol{\alpha}\cdot\boldsymbol{p})$, while the third term vanishes. Thus,

$$[H_D, K] = 0 \ .$$

4.21 From (3.E) we have

$$\begin{aligned}\mathrm{i}\bar{u}(\boldsymbol{p}_1)\sigma^{\mu\nu}(p_1 - p_2)_\nu u(\boldsymbol{p}_2) &= \frac{1}{2}\bar{u}(\boldsymbol{p}_1)(\gamma^\nu\gamma^\mu - \gamma^\mu\gamma^\nu)(p_1 - p_2)_\nu u(\boldsymbol{p}_2) \\ &= \frac{1}{2}\bar{u}(\boldsymbol{p}_1)[-\gamma^\mu(\not{p}_1 - \not{p}_2) + (\not{p}_1 - \not{p}_2)\gamma^\mu]u(\boldsymbol{p}_2) \\ &= \frac{1}{2}\bar{u}(\boldsymbol{p}_1)[-\gamma^\mu(\not{p}_1 - m) + (m - \not{p}_2)\gamma^\mu]u(\boldsymbol{p}_2) \ .\end{aligned}$$

By using $\gamma^\mu\not{p}_1 = 2p_1^\mu - \not{p}_1\gamma^\mu$ and $\not{p}_2\gamma^\mu = 2p_2^\mu - \gamma^\mu\not{p}_2$ we obtain

$$\mathrm{i}\bar{u}(\boldsymbol{p}_1)\sigma^{\mu\nu}(p_1 - p_2)_\nu u(\boldsymbol{p}_2) = 2m\bar{u}(\boldsymbol{p}_1)\gamma^\mu u(\boldsymbol{p}_2) - (p_1 + p_2)^\mu\bar{u}(\boldsymbol{p}_1)u(p_2) \ ,$$

where we used that $u(\boldsymbol{p})$ and $\bar{u}(\boldsymbol{p})$ satisfy the Dirac equation. The last expression is the requested identity. The second identity can be proven similarly.

4.23 It is easy to see that

$$\gamma_\alpha\gamma_\mu\gamma_\beta = 2g_{\alpha\mu}\gamma_\beta - 2g_{\alpha\beta}\gamma_\mu + 2g_{\mu\beta}\gamma_\alpha - \gamma_\beta\gamma_\mu\gamma_\alpha \ . \tag{4.43}$$

From (4.43) we have

$$\bar{u}(\boldsymbol{p}_2)\not{p}_1\gamma_\mu\not{p}_2 u(\boldsymbol{p}_1) = \bar{u}(\boldsymbol{p}_2)[2m(p_1+p_2)_\mu - (2p_1\cdot p_2 + m^2)\gamma_\mu]u(\boldsymbol{p}_1) \ , \quad (4.44)$$

where we used the Dirac equation (4.C). The first term in (4.44) can be transformed by using the Gordon identity (Problem 4.21)

$$\begin{aligned}\bar{u}(\boldsymbol{p}_2)\not{p}_1\gamma_\mu\not{p}_2 u(\boldsymbol{p}_1) &= \bar{u}(\boldsymbol{p}_2)[-2p_1\cdot p_2 + 3m^2]\gamma_\mu u(\boldsymbol{p}_1) - 2m\mathrm{i}\bar{u}(\boldsymbol{p}_2)\sigma_{\mu\nu}q^\nu u(\boldsymbol{p}_1)\\ &= \bar{u}(\boldsymbol{p}_2)\left\{(q^2+m^2)\gamma_\mu - 2\mathrm{i}m\sigma_{\mu\nu}q^\nu\right\}u(\boldsymbol{p}_1). \quad (4.45)\end{aligned}$$

From the last expression we can make the following identifications: $F_1 = q^2 + m^2$ and $F_2 = -2\mathrm{i}m$.

4.24 By using $u(\boldsymbol{p}) = \not{p}u(\boldsymbol{p})/m$ and

$$\{\gamma_\mu, \gamma_5\} = 0 \ ,$$

we have

$$\bar{u}(\boldsymbol{p})\gamma_5 u(\boldsymbol{p}) = \frac{1}{m}\bar{u}(\boldsymbol{p})\gamma_5\not{p}u(\boldsymbol{p}) = -\frac{1}{m}\bar{u}(\boldsymbol{p})\not{p}\gamma_5 u(\boldsymbol{p}) \ .$$

By applying the Dirac equation (3.C) we obtain

$$\bar{u}(\boldsymbol{p})\gamma_5 u(\boldsymbol{p}) = -\bar{u}(\boldsymbol{p})\gamma_5 u(\boldsymbol{p}) \ .$$

Thus $\bar{u}(\boldsymbol{p})\gamma_5 u(\boldsymbol{p}) = 0$. By using the Gordon identity (for $\mu = 0$) it finally follows that

$$\frac{1}{2}\bar{u}(\boldsymbol{p})(1-\gamma_5)u(\boldsymbol{p}) = \frac{m}{2E_p}N \ .$$

4.25 $F_1 = -\mathrm{i}q^2$, $F_2 = -2\mathrm{i}m$, $F_3 = -2m$.

4.26 By applying the operator $(\mathrm{i}\not{\partial} + m)$ to the Dirac equation we obtain

$$(\mathrm{i}\not{\partial} + m)(\mathrm{i}\not{\partial} - m)\psi = -(\Box + m^2)\psi = 0 \ .$$

4.27 The probability density is $\rho(x) = \psi^\dagger(x)\psi(x)$. By using the expression for the wave function from Problem 4.2, we easily get $\rho = \frac{E_p}{m}$. The current density is $\boldsymbol{j} = \bar{\psi}\boldsymbol{\gamma}\psi = \frac{\boldsymbol{p}}{m}\bar{\psi}\psi$, where the Gordon identity (for $\mu = i$) has been applied. Finally $\boldsymbol{j} = \frac{\boldsymbol{p}}{m}$.

4.28 The position operator in the Heisenberg picture satisfies the following equation

$$\frac{\mathrm{d}\boldsymbol{r}_\mathrm{H}}{\mathrm{d}t} = -\mathrm{i}[\boldsymbol{r}_\mathrm{H}, H] = \boldsymbol{\alpha}_\mathrm{H}.$$

In order to integrate the last equation we have to find the Dirac matrices in the Heisenberg picture

$$\boldsymbol{\alpha}_\mathrm{H} = \mathrm{e}^{\mathrm{i}Ht}\boldsymbol{\alpha}\mathrm{e}^{-\mathrm{i}Ht} = \sum_{n=0}^{\infty}\frac{(\mathrm{i}t)^n}{n!}[H,[H,\ldots[H,\boldsymbol{\alpha}]\ldots]] \ .$$

Since

$$[H,\boldsymbol{\alpha}] = 2(\boldsymbol{p} - \boldsymbol{\alpha}H) \,, \tag{4.46}$$
$$[H,[H,\boldsymbol{\alpha}]] = -2^2(\boldsymbol{p} - \boldsymbol{\alpha}H)H \,, \tag{4.47}$$
$$[H,[H,[H,\boldsymbol{\alpha}]]] = 2^3(\boldsymbol{p} - \boldsymbol{\alpha}H)H^2 \,, \text{etc.} \tag{4.48}$$

we get

$$\begin{aligned}\boldsymbol{\alpha}_{\mathrm{H}} &= \boldsymbol{\alpha} + (\boldsymbol{\alpha}H - \boldsymbol{p})\left(-2\mathrm{i}t + \frac{(2\mathrm{i}t)^2}{2!}H - \frac{(2\mathrm{i}t)^3}{3!}H^2 + \ldots\right) \\ &= \frac{\boldsymbol{p}}{H} + \left(\boldsymbol{\alpha} - \frac{\boldsymbol{p}}{H}\right)\mathrm{e}^{-2\mathrm{i}tH} \,. \end{aligned} \tag{4.49}$$

Then, equation

$$\frac{\mathrm{d}\boldsymbol{r}_{\mathrm{H}}}{\mathrm{d}t} = \frac{\boldsymbol{p}}{H} + \left(\boldsymbol{\alpha} - \frac{\boldsymbol{p}}{H}\right)\mathrm{e}^{-2\mathrm{i}tH} \tag{4.50}$$

implies

$$\boldsymbol{r}_{\mathrm{H}} = \boldsymbol{r} + \frac{\boldsymbol{p}}{H}t - \mathrm{i}\left(\boldsymbol{\alpha} - \frac{\boldsymbol{p}}{H}\right)\frac{1}{2H} + \mathrm{i}\left(\boldsymbol{\alpha} - \frac{\boldsymbol{p}}{H}\right)\frac{1}{2H}\mathrm{e}^{-2\mathrm{i}Ht}.$$

The integration constant is determined using the condition $\boldsymbol{r}_{\mathrm{H}}(t=0) = \boldsymbol{r}$. As we see "the motion of particle" is a superposition of classical uniform and rapid oscillatory motions.

4.29 We should calculate the coefficients $c_r(\boldsymbol{p})$ and $d_r^*(\boldsymbol{p})$ in the expansion

$$\psi(0,\boldsymbol{x}) = \frac{1}{(2\pi)^{3/2}}\sum_r \int \mathrm{d}^3\boldsymbol{p}\sqrt{\frac{m}{E_p}}(c_r(\boldsymbol{p})u_r(\boldsymbol{p})\mathrm{e}^{\mathrm{i}\boldsymbol{p}\cdot\boldsymbol{x}} + d_r^*(\boldsymbol{p})v_r(\boldsymbol{p})\mathrm{e}^{-\mathrm{i}\boldsymbol{p}\cdot\boldsymbol{x}}) \,. \tag{4.51}$$

If we multiply this expression by $u_s^\dagger(\boldsymbol{q})\mathrm{e}^{-\mathrm{i}\boldsymbol{q}\cdot\boldsymbol{x}}$ from left and integrate over $\boldsymbol{x}$, we get

$$c_s(\boldsymbol{q}) = \frac{1}{(2\pi)^{3/2}}\sqrt{\frac{m}{E_q}}\int \mathrm{d}^3\boldsymbol{x}\, u_s^\dagger(\boldsymbol{q})\psi(0,\boldsymbol{x})\mathrm{e}^{-\mathrm{i}\boldsymbol{q}\cdot\boldsymbol{x}} \,,$$

where we applied the relations

$$u_r^\dagger(\boldsymbol{p})u_s(\boldsymbol{p}) = v_r^\dagger(\boldsymbol{p})v_s(\boldsymbol{p}) = \frac{E_p}{m}\delta_{rs},\; v_r^\dagger(-\boldsymbol{p})u_s(\boldsymbol{p}) = u_r^\dagger(-\boldsymbol{p})v_s(\boldsymbol{p}) = 0 \,. \tag{4.52}$$

These relations can be obtained from (4.D) by using the Gordon identity. Similarly for d coefficients we get

$$d_s^*(\boldsymbol{q}) = \frac{1}{(2\pi)^{3/2}}\sqrt{\frac{m}{E_q}}\int \mathrm{d}^3\boldsymbol{x}\, v_s^\dagger(\boldsymbol{q})\psi(0,\boldsymbol{x})\mathrm{e}^{\mathrm{i}\boldsymbol{q}\cdot\boldsymbol{x}} \,.$$

Carrying out the integrations, we find

$$
\begin{aligned}
c_1(\boldsymbol{p}) &= \frac{1}{(2\pi)^{3/2}}\sqrt{\frac{E_p+m}{2E_p}}\,,\\
c_2(\boldsymbol{p}) &= 0,\\
d_1^*(\boldsymbol{p}) &= \frac{1}{(2\pi)^{3/2}}\frac{1}{\sqrt{2E_p(E_p+m)}}(p_x+\mathrm{i}p_y)\,,\\
d_2^*(\boldsymbol{p}) &= \frac{1}{(2\pi)^{3/2}}\frac{1}{\sqrt{2E_p(E_p+m)}}p_z\,.
\end{aligned}
\tag{4.53}
$$

The wave function at time $t > 0$ is

$$
\psi(x) = \frac{1}{(2\pi)^{3/2}}\sum_r\int \mathrm{d}^3\boldsymbol{p}\sqrt{\frac{m}{E_p}}(c_r(\boldsymbol{p})u_r(\boldsymbol{p})_r\mathrm{e}^{-\mathrm{i}p\cdot x}+d_r^*(\boldsymbol{p})v_r(\boldsymbol{p})\mathrm{e}^{\mathrm{i}p\cdot x})\,, \tag{4.54}
$$

where the coefficients $c_r(\boldsymbol{p})$ and $d_r^*(\boldsymbol{p})$ are given in (4.53).

4.30 In this case the coefficients $c_r(\boldsymbol{p})$ and $d_r^*(\boldsymbol{p})$ in expansion (4.51) are:

$$
\begin{aligned}
c_1(\boldsymbol{p}) &= \left(\frac{d^2}{\pi}\right)^{3/4}\sqrt{\frac{E_p+m}{2E_p}}\mathrm{e}^{-d^2\boldsymbol{p}^2/2}\,,\\
c_2(\boldsymbol{p}) &= 0\,,\\
d_1^*(\boldsymbol{p}) &= \left(\frac{d^2}{\pi}\right)^{3/4}\frac{1}{\sqrt{2E_p(E_p+m)}}\mathrm{e}^{-d^2\boldsymbol{p}^2/2}(p_x+\mathrm{i}p_y)\,,\\
d_2^*(\boldsymbol{p}) &= \left(\frac{d^2}{\pi}\right)^{3/4}\frac{1}{\sqrt{2E_p(E_p+m)}}p_z\mathrm{e}^{-d^2\boldsymbol{p}^2/2}\,.
\end{aligned}
$$

4.31 The equation for spin 1/2 particle in the electromagnetic field has the following form

$$
[\mathrm{i}\gamma^\mu(\partial_\mu-\mathrm{i}eA_\mu)-m]\psi = 0\,. \tag{4.55}
$$

If we assume that a wave function for $z > 0$ has the form

$$
\psi = \begin{pmatrix}\varphi\\ \chi\end{pmatrix}\mathrm{e}^{-\mathrm{i}Et+\mathrm{i}qz}\,, \tag{4.56}
$$

then (4.55) becomes

$$
\begin{pmatrix} E-m-V & -\sigma_3 q\\ \sigma_3 q & -E-m+V\end{pmatrix}\begin{pmatrix}\varphi\\ \chi\end{pmatrix} = 0\,. \tag{4.57}
$$

The system of equations (4.57) has a nontrivial solution if and only if

$$
E = V \pm \sqrt{q^2+m^2}\,. \tag{4.58}
$$

The wave function[2] is

[2] From the boundary conditions it follows that there is no spin flip.

$$\psi_I = a \begin{pmatrix} 1 \\ 0 \\ \frac{p\sigma_3}{(E+m)} \begin{pmatrix} 1 \\ 0 \end{pmatrix} \end{pmatrix} \mathrm{e}^{-\mathrm{i}Et+\mathrm{i}pz}$$

$$+ b \begin{pmatrix} 1 \\ 0 \\ \frac{-p\sigma_3}{(E+m)} \begin{pmatrix} 1 \\ 0 \end{pmatrix} \end{pmatrix} \mathrm{e}^{-\mathrm{i}Et-\mathrm{i}pz} , \quad z < 0 , \tag{4.59}$$

$$\psi_{II} = d \begin{pmatrix} 1 \\ 0 \\ \frac{q\sigma_3}{(E+m-V)} \begin{pmatrix} 1 \\ 0 \end{pmatrix} \end{pmatrix} \mathrm{e}^{-\mathrm{i}Et+\mathrm{i}qz}, \quad z > 0 ,$$

where $p = \sqrt{E^2 - m^2}$. The terms proportional to the coefficient a, b and d in (4.59) are the initial ψ_{in}, reflected ψ_{r} and transmitted wave ψ_{t}. Since the Dirac equation is the first order equation, the continuity condition is satisfied for the wave function only. The condition $\psi_I(0) = \psi_{II}(0)$ gives

$$a + b = d , \tag{4.60}$$

$$a - b = rd , \tag{4.61}$$

where $r = \frac{E+m}{E+m-V}\frac{q}{p}$. Now, we will consider three cases:
1. If $|E - V| \leq m$, the momentum q is imaginary, $q = \mathrm{i}\kappa$ so that the wave function exponentially decreases in the region $z > 0$, as is the case in nonrelativistic quantum mechanics. The transmitted, reflected and incident currents are:

$$\boldsymbol{j}_{\mathrm{r}} = \bar{\psi}_{\mathrm{tr}}\gamma^3\psi_{\mathrm{tr}}\boldsymbol{e}_z = 0 , \tag{4.62}$$

$$\boldsymbol{j}_{\mathrm{r}} = \bar{\psi}_{\mathrm{r}}\gamma^3\psi_{\mathrm{r}}\boldsymbol{e}_z = -\frac{2p}{E+m}|b|^2\boldsymbol{e}_z , \tag{4.63}$$

$$\boldsymbol{j}_{\mathrm{in}} = \bar{\psi}_{\mathrm{in}}\gamma^3\psi_{\mathrm{in}}\boldsymbol{e}_z = \frac{2p}{E+m}|a|^2\boldsymbol{e}_z . \tag{4.64}$$

Since $\boldsymbol{j}_{\mathrm{tr}} = 0$ the transmission coefficient is zero. The reflection coefficient is

$$R = \frac{-j_r}{j_{\mathrm{in}}} = \left|\frac{p(E+m-V) - \mathrm{i}\kappa(E+m)}{p(E+m-V) + \mathrm{i}\kappa(E+m)}\right|^2 = 1 . \tag{4.65}$$

2. If $V < E - m$, the momentum q is real. The currents are:

$$\boldsymbol{j}_{\mathrm{tr}} = \frac{2q}{E+m-V}|d|^2\boldsymbol{e}_z , \tag{4.66}$$

$$\boldsymbol{j}_{\mathrm{r}} = -\frac{2p}{E+m}|b|^2\boldsymbol{e}_z , \tag{4.67}$$

$$\boldsymbol{j}_{\mathrm{in}} = \frac{2p}{E+m}|a|^2\boldsymbol{e}_z . \tag{4.68}$$

The transmission coefficient is

$$T = \frac{j_{\rm tr}}{j_{\rm in}} = r\left|\frac{d}{a}\right|^2 = \frac{4r}{(1+r)^2}\,, \tag{4.69}$$

while the reflection coefficient is

$$R = \frac{-j_{\rm r}}{j_{\rm in}} = \left(\frac{1-r}{1+r}\right)^2 . \tag{4.70}$$

3. If $E+m < V$, the momentum q is real, which implies that the wave function in region $z > 0$ becomes oscillating. This is caused by the fact that there are two parts of electron spectrum separated by a gap, whose width is equal to 2m. The expressions for the coefficients of reflection and transmission are the same as in the second case. But in this case, the coefficient of reflection is greater then 1, while $T < 0$. The described effect is known as *the Klein paradox*. The explanation of this paradox is given in Problem 2.9.

4.32 The solution of the Dirac equation is

$$\psi_I = \begin{pmatrix} 1 \\ 0 \\ \frac{p\sigma_3}{(E+m)}\begin{pmatrix} 1 \\ 0 \end{pmatrix} \end{pmatrix} \mathrm{e}^{\mathrm{i}pz}$$

$$+ B \begin{pmatrix} 1 \\ 0 \\ \frac{-p\sigma_3}{(E+m)}\begin{pmatrix} 1 \\ 0 \end{pmatrix} \end{pmatrix} \mathrm{e}^{-\mathrm{i}pz}\,, \quad z < 0\,,$$

$$\psi_{II} = C \begin{pmatrix} 1 \\ 0 \\ \frac{q\sigma_3}{(E+m-V)}\begin{pmatrix} 1 \\ 0 \end{pmatrix} \end{pmatrix} \mathrm{e}^{\mathrm{i}qz}$$

$$+ D \begin{pmatrix} 1 \\ 0 \\ \frac{-q\sigma_3}{(E+m-V)}\begin{pmatrix} 1 \\ 0 \end{pmatrix} \end{pmatrix} \mathrm{e}^{-\mathrm{i}qz}\,, \quad 0 < z < a\,,$$

$$\psi_{III} = F \begin{pmatrix} 1 \\ 0 \\ \frac{p\sigma_3}{(E+m)}\begin{pmatrix} 1 \\ 0 \end{pmatrix} \end{pmatrix} \mathrm{e}^{\mathrm{i}pz}\,, \quad z > a\,,$$

where $p = \sqrt{E^2 - m^2}$ and $q = \sqrt{(E-V)^2 - m^2}$. From the boundary conditions $\psi_I(0) = \psi_{II}(0)$ and $\psi_{II}(a) = \psi_{III}(a)$, we obtain the transmission coefficient

$$T = |F|^2 = 16\frac{|r|^2}{|(1+r)^2\mathrm{e}^{-\mathrm{i}qa} - (1-r)^2\mathrm{e}^{\mathrm{i}qa}|^2}\,,$$

where $r = \frac{q}{p}\frac{E+m}{E+m-V}$.

4.33

(a) The wave function is

$$\begin{aligned}
\psi_I &= \begin{pmatrix} B \\ B' \\ \frac{-i\kappa\sigma_3}{(E+m)}\begin{pmatrix} B \\ B' \end{pmatrix} \end{pmatrix} e^{\kappa z}, \quad z < -a, \\
\psi_{II} &= \begin{pmatrix} C \\ C' \\ \frac{q\sigma_3}{(E+m+V)}\begin{pmatrix} C \\ C' \end{pmatrix} \end{pmatrix} e^{iqz} \\
&+ \begin{pmatrix} D \\ D' \\ \frac{-q\sigma_3}{(E+m+V)}\begin{pmatrix} D \\ D' \end{pmatrix} \end{pmatrix} e^{-iqz}, \quad -a < z < a, \\
\psi_{III} &= \begin{pmatrix} F \\ F' \\ \frac{i\kappa\sigma_3}{(E+m)}\begin{pmatrix} F \\ F' \end{pmatrix} \end{pmatrix} e^{-\kappa z}, \quad z > a,
\end{aligned} \tag{4.71}$$

where $\kappa = \sqrt{m^2 - E^2}$ and $q = \sqrt{(E+V)^2 - m^2}$. Since there is no spin flip, we can take $B' = C' = D' = F' = 0$. From the boundary conditions $\psi_I(-a) = \psi_{II}(-a)$ and $\psi_{II}(a) = \psi_{III}(a)$, it follows that

$$\begin{aligned}
e^{-\kappa a} B &= e^{-iqa} C + e^{iqa} D \\
e^{-\kappa a} F &= e^{iqa} C + e^{-iqa} D \\
-ir e^{-\kappa a} B &= e^{-iqa} C - e^{iqa} D \\
ir e^{-\kappa a} F &= e^{iqa} C - e^{-iqa} D \, ,
\end{aligned}$$

where $r = \frac{\kappa}{q}\frac{E+m+V}{E+m}$. By combining previous equations we obtain

$$\begin{aligned}
e^{-\kappa a}(B - F) &= 2i\sin(qa)(D - C) \\
ir e^{-\kappa a}(B - F) &= 2\cos(qa)(D - C) \\
e^{-\kappa a}(B + F) &= 2\cos(qa)(D + C) \\
r e^{-\kappa a}(B + F) &= 2\sin(qa)(D + C) \, .
\end{aligned}$$

Further, we will distinguish two classes of solutions: odd and even. If $B = F$ and $C = D$, the third and the fourth equations give the following dispersion relation:

$$\tan(qa) = \frac{\kappa}{q}\frac{E + m + V}{E + m} \, .$$

These solutions satisfy the following property: $\psi'(z) = \gamma_0\psi(-z) = \psi(z)$; thus they are even. On the other hand, if $B = -F$ and $C = -D$, the dispersion relation is

$$\cot(qa) = -\frac{\kappa}{q}\frac{E+m+V}{E+m} .$$

This class of solutions satisfy $\psi'(z) = \gamma_0\psi(-z) = -\psi(z)$, and therefore they are odd.

(b) The dispersion relations are transcendental equations and they cannot be solved analytically. We can analyze them graphically.
For even solutions, the dispersion relation has the form

$$q\tan(qa) = f(q) , \tag{4.72}$$

where

$$f(q) = \sqrt{2V\sqrt{q^2+m^2} - q^2 - V^2}\,\frac{m+\sqrt{q^2+m^2}}{m+\sqrt{q^2+m^2}-V} ,$$

and its graphical solution is given in Fig. 4.1.

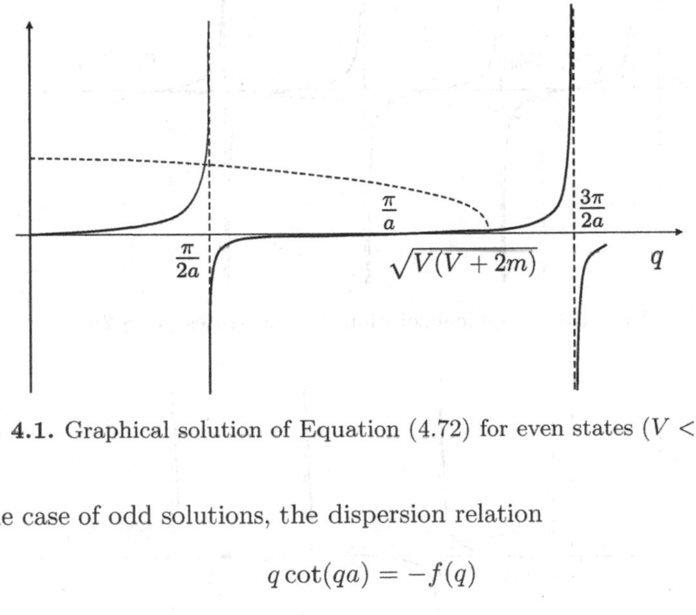

Fig. 4.1. Graphical solution of Equation (4.72) for even states ($V < 2m$)

In the case of odd solutions, the dispersion relation

$$q\cot(qa) = -f(q) \tag{4.73}$$

is shown in Fig. 4.2. From these figures we see that the spectrum of electron bound states will contain N states if the condition

$$\frac{(N-1)\pi}{2a} \leq \sqrt{V(V+2m)} < \frac{N\pi}{2a}$$

is satisfied. It is easy to see that if $N=1$ then this solution is even.

(c) Graphical solutions for odd and even part of spectrum are given in Fig. 4.3 and Fig. 4.4.

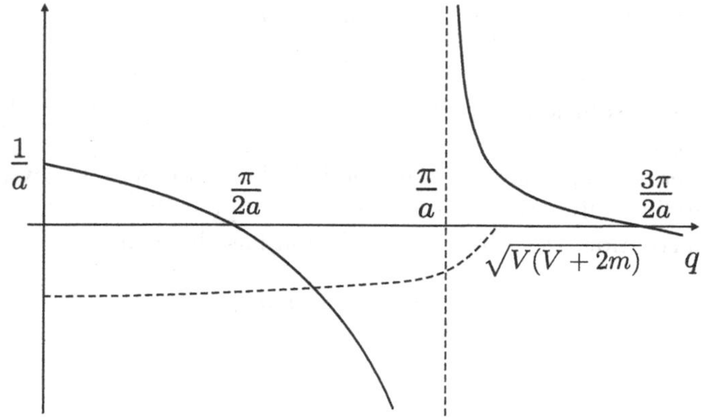

Fig. 4.2. Graphical solution of Equation (4.73) for odd states ($V < 2m$)

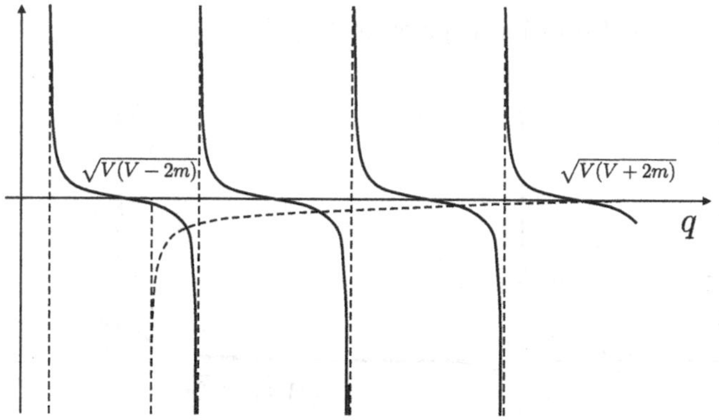

Fig. 4.3. Graphical solution for odd states ($V > 2m$)

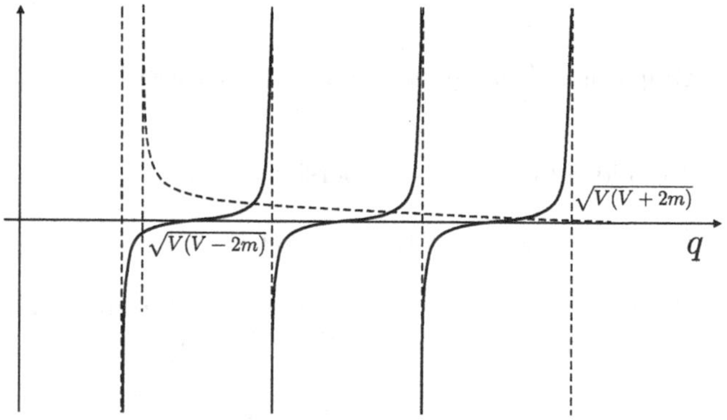

Fig. 4.4. Graphical solution for even states ($V > 2m$)

4.34 The Dirac equation in this case has following form

$$\left[\mathrm{i}\gamma^0\frac{\partial}{\partial t}+\mathrm{i}\gamma^1\left(\frac{\partial}{\partial x}-\mathrm{i}eBy\right)+\mathrm{i}\gamma^2\frac{\partial}{\partial y}+\mathrm{i}\gamma^3\frac{\partial}{\partial z}-m\right]\psi=0\,. \tag{4.74}$$

A particular solution of (4.74) is

$$\psi=\mathrm{e}^{-\mathrm{i}Et+\mathrm{i}p_x x+\mathrm{i}p_z z}\begin{pmatrix}\varphi(y)\\ \chi(y)\end{pmatrix}\,. \tag{4.75}$$

By substituting (4.75) in (4.74) we obtain

$$\begin{pmatrix} E-m & (eBy-p_x)\sigma_1-p_z\sigma_3+\mathrm{i}\sigma_2\frac{\mathrm{d}}{\mathrm{d}y}\\ (p_x-eBy)\sigma_1+p_z\sigma_3-\mathrm{i}\sigma_2\frac{\mathrm{d}}{\mathrm{d}y} & -E-m\end{pmatrix}\begin{pmatrix}\varphi\\ \chi\end{pmatrix}=0\,. \tag{4.76}$$

From the second equation in (4.76), follows

$$\chi(y)=\frac{1}{E+m}\left(p_x\sigma_1+p_z\sigma_3-eBy\sigma_1-\mathrm{i}\sigma_2\frac{\mathrm{d}}{\mathrm{d}y}\right)\varphi(y)\,, \tag{4.77}$$

and plugging it into the first equation of (4.76), we get

$$\left(\frac{\mathrm{d}^2}{\mathrm{d}y^2}-(p_x-eBy)^2+E^2-m^2-p_z^2-eB\sigma_3\right)\varphi=0\,, \tag{4.78}$$

where we used the following identity

$$\sigma_i\sigma_j=\delta_{ij}+\mathrm{i}\epsilon_{ijk}\sigma_k\,.$$

By introducing new variable $\xi=p_x-eBy$, Equation (4.78) becomes the Schrödinger equation for a linear oscillator (parameters M, ω and ϵ), where

$$M^2\omega^2=\frac{1}{(eB)^2},\quad 2M\epsilon=\frac{E^2-m^2-p_z^2\mp eB}{(eB)^2}\,.$$

We assumed that the spinor φ is an eigenstate of $\sigma_3/2$, i.e.

$$\frac{1}{2}\sigma_3\varphi=\pm\frac{1}{2}\varphi\,.$$

The energy eigenvalues are

$$E_{n,p_z}=\sqrt{m^2+p_z^2\pm eB+(2n+1)eB}\,, \tag{4.79}$$

where $n=0,1,2,\ldots$

4.35 Acting by $(\mathrm{i}\not{\partial}+e\not{A}+m)$ on $(\mathrm{i}\not{\partial}+e\not{A}-m)\psi(x)=0$, we get

$$[\Box-\mathrm{i}e\gamma^\mu\gamma^\nu\partial_\mu A_\nu-2\mathrm{i}eA^\mu\partial_\mu-e^2A^2+m^2]\psi=0\,.$$

On the other hand, one can show that

$$-\frac{e}{2}\sigma_{\mu\nu}F^{\mu\nu} = \mathrm{i}e(\partial_\mu A^\mu - \gamma^\mu\gamma^\nu\partial_\mu A_\nu) \; .$$

The requested result can be obtained by combining these expressions.

4.36 By substituting

$$\psi = \begin{pmatrix} \varphi \\ \chi \end{pmatrix} \mathrm{e}^{-imt}$$

in the Dirac equation

$$(\mathrm{i}\partial\!\!\!/ + eA\!\!\!/ - m)\psi(x) = 0,$$

we obtain the following equations:

$$\left(\mathrm{i}\frac{\partial}{\partial t} + eA_0\right)\varphi = c\boldsymbol{\sigma}\cdot(\boldsymbol{p} + e\boldsymbol{A})\chi \; ,$$

$$\left(\mathrm{i}\frac{\partial}{\partial t} + 2mc^2 + eA_0\right)\chi = c\boldsymbol{\sigma}\cdot(\boldsymbol{p} + e\boldsymbol{A})\varphi \; .$$

In the case $\boldsymbol{A} = 0$, the second equation yields:

$$\chi = \frac{1}{2mc}\left(\boldsymbol{\sigma}\cdot\boldsymbol{p}\varphi - \frac{\mathrm{i}}{2mc^2}\boldsymbol{\sigma}\cdot\boldsymbol{p}\frac{\partial\varphi}{\partial t} - \frac{eA_0}{2mc^2}\boldsymbol{\sigma}\cdot\boldsymbol{p}\varphi\right) \; .$$

Combining this relation with the first equation, we obtain

$$\mathrm{i}\frac{\partial\varphi}{\partial t} = H'\varphi \; ,$$

where

$$H' = \left[\frac{\boldsymbol{p}^2}{2m} - eA_0 - \frac{\boldsymbol{p}^4}{8m^3c^2} + \frac{e}{4m^2c^2}(2\mathrm{i}\boldsymbol{E}\cdot\boldsymbol{p} - \Delta A_0) - \frac{e}{4m^2c^2}(\mathrm{i}\boldsymbol{E}\cdot\boldsymbol{p} + \boldsymbol{\sigma}\cdot(\boldsymbol{E}\times\boldsymbol{p}))\right] \; .$$

The operator H' is not the Hamiltonian, since it is not hermitian. This is related to the fact that $\varphi^\dagger\varphi$ is not the probability density. Actually, the probability density should be taken in the following form:

$$\rho = \bar{\psi}\psi = \varphi^\dagger\varphi - \chi^\dagger\chi = \varphi^\dagger(1 + \frac{\boldsymbol{p}^2}{4m^2c^2})\varphi + o\left(\frac{v^2}{c^2}\right) \; .$$

We introduce the new wave function

$$\varphi_s = \left(1 + \frac{\boldsymbol{p}^2}{8m^2c^2}\right)\varphi \; .$$

Then, the new Hamiltonian is given by

$$H = \left(1 + \frac{\boldsymbol{p}^2}{8m^2c^2}\right) H' \left(1 - \frac{\boldsymbol{p}^2}{8m^2c^2}\right) .$$

After that, we obtain

$$H = \frac{\boldsymbol{p}^2}{2m} - eA_0 - \frac{\boldsymbol{p}^4}{8m^3c^2} - \frac{e}{8m^2c^2}\Delta A_0 + \frac{e}{4m^2c^2}\boldsymbol{\sigma}\cdot(\boldsymbol{E}\times\boldsymbol{p}) .$$

In the case $\boldsymbol{A} \neq 0$, the Hamiltonian is

$$\begin{aligned} H = & \frac{(\boldsymbol{p}+e\boldsymbol{A})^2}{2m} - eA_0 + \frac{e}{2mc}\boldsymbol{\sigma}\cdot\boldsymbol{B} - \frac{\boldsymbol{p}^4}{8m^3c^2} \\ & - \frac{e}{8m^2c^2}\Delta A_0 + \frac{e}{4m^2c^2}\boldsymbol{\sigma}\cdot(\boldsymbol{E}\times(\boldsymbol{p}+e\boldsymbol{A})) . \end{aligned}$$

4.37 First, we are going to show that $V_\mu(x)$ is a real quantity:

$$\begin{aligned} V_\mu^* = V_\mu^\dagger &= (\bar{\psi}\gamma_\mu\psi)^\dagger \\ &= \psi^\dagger\gamma_\mu^\dagger(\psi^\dagger\gamma^0)^\dagger \\ &= \psi^\dagger\gamma^0\gamma_\mu\gamma^0\gamma^0\psi \\ &= \bar{\psi}\gamma_\mu\psi \\ &= V_\mu . \end{aligned} \tag{4.80}$$

Under proper orthochronous Lorentz transformations, V_μ is transformed in the following way:

$$V'_\mu(x') = \bar{\psi}'(x')\gamma_\mu\psi'(x') = \psi^\dagger(x)\gamma_0 S^{-1}\gamma_\mu S\psi(x) ,$$

where we used the fact that $\gamma_0 S^{-1} = S^\dagger\gamma_0$. Using $S^{-1}\gamma_\mu S = \Lambda_\mu{}^\nu\gamma_\nu$, we obtain $V'_\mu(x') = \Lambda_\mu{}^\nu V_\nu(x)$. So, the quantity V_μ is a Lorentz four-vector.
Under parity we have

$$V_\mu(t,\boldsymbol{x}) \to V'_\mu(t,-\boldsymbol{x}) = \bar{\psi}(t,\boldsymbol{x})\gamma_0\gamma_\mu\gamma_0\psi(t,\boldsymbol{x}) .$$

This implies

$$V'_0(t,\boldsymbol{x}) = V_0(t,-\boldsymbol{x}),\ \ V'_i(t,\boldsymbol{x}) = -V_i(t,-\boldsymbol{x}) .$$

As we know, under charge conjugation the spinors transform according to:

$$\begin{aligned} \psi(x) &\to \psi_c(x) = C\bar{\psi}^T , \\ \bar{\psi} = \psi^\dagger\gamma_0 &\to (C\bar{\psi}^T)^\dagger\gamma_0 \\ &= (C(\gamma^0)^T\psi^*)^\dagger\gamma_0 \\ &= \psi^T((\gamma^0)^T C(\gamma^0)^\dagger)^T \\ &= -\psi^T(C\gamma^0\gamma^0)^T \\ &= \psi^T C . \end{aligned} \tag{4.81}$$

Then, we can find the transformation law for V_μ:

$$V_\mu \to -\psi^T C\gamma_\mu C^{-1}\bar{\psi}^T = (\bar{\psi}\gamma_\mu\psi)^T = V_\mu \ .$$

The following formulae $C\gamma_\mu C^{-1} = -\gamma_\mu^T$, $C = -C^{-1}$ have been used (Prove the last one).
For time reversal we have $\psi(x) \to \psi'(-t, \boldsymbol{x}) = T\psi^*(t, \boldsymbol{x})$, where matrix T satisfies $T\gamma_\mu T^{-1} = \gamma^{\mu *} = \gamma_\mu^T$ and $T^\dagger = T^{-1} = T = -T^*$. It is easy to see that

$$\bar{\psi}(x) \to \bar{\psi}'(-t, \boldsymbol{x}) = \psi^T(t, \boldsymbol{x})T\gamma_0 \ .$$

Then

$$\begin{aligned} V^\mu(t, \boldsymbol{x}) &\to \psi^T T\gamma^0\gamma^\mu T\psi^* \\ &= \psi^T T\gamma^0 T^{-1} T\gamma^\mu T^{-1}\psi^* \\ &= \psi^T(\gamma^0)^T(\gamma^\mu)^T\psi^* \\ &= (\psi^\dagger\gamma^\mu\gamma^0\psi)^T \\ &= \psi^\dagger\gamma^\mu\gamma^0\psi \ . \end{aligned} \tag{4.82}$$

Therefore,

$$V_0'(-t, \boldsymbol{x}) = V_0(t, \boldsymbol{x}),\ V_i'(-t, \boldsymbol{x}) = -V_i(t, \boldsymbol{x}) \ .$$

4.38 The quantity A^μ transforms under Lorentz transformations in the following way:

$$\begin{aligned} A'^\mu(x') &= \Lambda^\mu{}_\nu\bar{\psi}(x)\gamma^\nu S^{-1}\gamma_5 S\psi(x) \\ &= \det\Lambda\ \Lambda^\mu{}_\nu\bar{\psi}(x)\gamma^\nu\gamma_5\psi(x) = \det\Lambda\ \Lambda^\mu{}_\nu A^\nu(x) \ , \end{aligned}$$

where we used

$$\begin{aligned} S^{-1}\gamma_5 S &= -\frac{\mathrm{i}}{4!}\epsilon_{\mu\nu\rho\sigma}S^{-1}\gamma^\mu S S^{-1}\gamma^\nu S S^{-1}\gamma^\rho S S^{-1}\gamma^\sigma S \\ &= -\frac{\mathrm{i}}{4!}\epsilon_{\mu\nu\rho\sigma}\Lambda^\mu{}_\alpha\Lambda^\nu{}_\beta\Lambda^\rho{}_\gamma\Lambda^\sigma{}_\delta\gamma^\alpha\gamma^\beta\gamma^\gamma\gamma^\delta \\ &= -\frac{\mathrm{i}}{4!}\epsilon_{\alpha\beta\gamma\delta}\det\Lambda\ \gamma^\alpha\gamma^\beta\gamma^\gamma\gamma^\delta \\ &= \det\Lambda\ \gamma_5 \ . \end{aligned}$$

The charge conjugation changes the sign of A^μ. The parity changes the sign of the time component, but does not change the sign of spatial components. The effect of time reversal is exactly opposite.

4.39 The quantity $\bar{\psi}\gamma^\mu\partial_\mu\psi$ transforms as a scalar under Lorentz transformations. The parity does not change it. The action of the charge conjugation yields $(\partial_\mu\bar{\psi})\gamma^\mu\psi$, while the time reversal produces $-(\partial_\mu\bar{\psi})\gamma^\mu\psi$.

4.40 By transposing the Dirac equation,

$$\bar{u}(p,s)(\not{p}-m)=0\,,$$

and using $C^{-1}\gamma^\mu C=-(\gamma^\mu)^T$, one gets the requested result.

4.41 Let us assume that there are two different matrices C' and C'', which both satisfy the relation $C\gamma^\mu C^{-1}=-(\gamma^\mu)^T$. Then from $C''\gamma_\mu C''^{-1}=C'\gamma_\mu C'^{-1}$ follows that $[C'^{-1}C'',\gamma_\mu]=0$, whereupon (see Problem 3.18) the requested relation follows.

4.42 We directly obtain:

(a)

$$\psi_c(x)=N_p\begin{pmatrix}-\frac{p}{E+m}\begin{pmatrix}0\\1\end{pmatrix}\\ \begin{pmatrix}0\\1\end{pmatrix}\end{pmatrix}\mathrm{e}^{-\mathrm{i}Et-\mathrm{i}pz}\,.$$

(b)

$$\psi'(x')=\begin{pmatrix}1\\0\\0\\0\end{pmatrix}\mathrm{e}^{-imt'}\,.$$

(c)

$$\psi_p(t,\boldsymbol{x})=N_p\begin{pmatrix}\begin{pmatrix}1\\0\end{pmatrix}\\ -\frac{p}{E_p+m}\begin{pmatrix}1\\0\end{pmatrix}\end{pmatrix}\mathrm{e}^{-\mathrm{i}(Et+pz)}\,.$$

Momentum is inverted under parity. Time reversal transforms the wave function into

$$\psi_t(t,\boldsymbol{x})=-\mathrm{i}N_p\begin{pmatrix}\begin{pmatrix}0\\1\end{pmatrix}\\ \frac{p}{E_p+m}\begin{pmatrix}0\\1\end{pmatrix}\end{pmatrix}\mathrm{e}^{\mathrm{i}(-Et-pz)}\,,$$

and we see that spin and the direction of the momentum are inverted.

(d) The wave function for S' observer is

$$\psi'(x')=N_p\begin{pmatrix}\varphi\\ \frac{p}{E_p+m}\varphi\end{pmatrix}\mathrm{e}^{\mathrm{i}(Et-p'z')}$$

where

$$\varphi=\begin{pmatrix}\cos\left(\frac{\theta}{2}\right)\\ \mathrm{i}\sin\left(\frac{\theta}{2}\right)\end{pmatrix}\,.$$

4.43 $P=\gamma_0=\begin{pmatrix}0 & I\\ I & 0\end{pmatrix}$, $C=\mathrm{i}\gamma^2\gamma^0=\mathrm{i}\begin{pmatrix}\sigma^2 & 0\\ 0 & -\sigma^2\end{pmatrix}$.

4.44 Multiplying the equation

$$\frac{\boldsymbol{\Sigma}\cdot\boldsymbol{p}}{|p|}u_r(p) = (-1)^{r+1}u_r(p)\ , \tag{4.83}$$

by γ_0 from left, we obtain

$$\frac{\boldsymbol{\Sigma}\cdot(-\boldsymbol{p})}{|p|}u_r(-p) = (-1)^r u_r(-p)\ , \tag{4.84}$$

since $\gamma_0 u_r(\boldsymbol{p}) = u_r(-\boldsymbol{p})$. From (4.84) we see that the helicity is inverted. Under the time reversal, the wave function of the Dirac particle (4.6) becomes

$$\begin{aligned}\psi_t(t,\boldsymbol{x}) &= \mathrm{i}\gamma^1\gamma^3\psi_r^*(-t,\boldsymbol{x})\\ &= -\mathcal{N}\begin{pmatrix}\sigma^2\varphi_r^*\\ \frac{\sigma^2(\boldsymbol{\sigma}^*\cdot\boldsymbol{p})}{E_p+m}\varphi_r^*\end{pmatrix}\mathrm{e}^{\mathrm{i}(-E_pt-\boldsymbol{p}\cdot\boldsymbol{x})}\\ &= -\mathcal{N}\begin{pmatrix}\sigma^2\varphi_r^*\\ -\frac{(\boldsymbol{\sigma}\cdot\boldsymbol{p})\sigma^2}{E_p+m}\varphi_r^*\end{pmatrix}\mathrm{e}^{\mathrm{i}(-E_pt-\boldsymbol{p}\cdot\boldsymbol{x})}\ ,\end{aligned} \tag{4.85}$$

where we used $\sigma^2\boldsymbol{\sigma}^* = -\boldsymbol{\sigma}\sigma^2$ in the second step. From the last expression, we conclude that the momentum changes its direction, i.e. $\boldsymbol{p}\to-\boldsymbol{p}$. Prove that $\sigma^2\varphi_1^* = \mathrm{i}\varphi_2$ and $\sigma^2\varphi_2^* = -\mathrm{i}\varphi_1$. Now, we consider the case $r=1$ (the other case $r=2$ is similar). From (4.85) it follows that

$$\psi_t(t,\boldsymbol{x}) = -\mathrm{i}\mathcal{N}\begin{pmatrix}\varphi_2\\ -\frac{\boldsymbol{p}\cdot\boldsymbol{\sigma}}{E_p+m}\varphi_2\end{pmatrix}\mathrm{e}^{\mathrm{i}(-E_pt-\boldsymbol{p}\cdot\boldsymbol{x})}\ . \tag{4.86}$$

By applying $\frac{\boldsymbol{\Sigma}\cdot(-\boldsymbol{p})}{|\boldsymbol{p}|}$ on (4.86), we see that the helicity is unchanged. The same result can be obtained by complex conjugation and multiplication of Equation (4.83) from left by $\mathrm{i}\gamma^1\gamma^3$. You can prove the same for v spinors.

4.45 The transformed Hamiltonian is

$$H' = \boldsymbol{\alpha}\cdot\boldsymbol{p}\left(\cos(2p\theta) - \frac{m}{p}\sin(2p\theta)\right) + m\beta\left(\cos(2p\theta) + \frac{p}{m}\sin(2p\theta)\right)\ ,$$

where $p = |\boldsymbol{p}|$. In order to have even form of the Hamiltonian, the coefficient multiplying $\boldsymbol{\alpha}\cdot\boldsymbol{p}$ has to be zero. This is satisfied if $\tan(2p\theta) = p/m$.

4.47 First prove that:

$$U = \cos(p\theta) + \frac{\beta\boldsymbol{\alpha}\cdot\boldsymbol{p}}{p}\sin(p\theta) = \sqrt{\frac{E_p+m}{2E_p}} + \frac{\beta\boldsymbol{\alpha}\cdot\boldsymbol{p}}{\sqrt{2E_p(E_p+m)}}\ ,$$

hence

$$\boldsymbol{x}_{\mathrm{FW}} = \left(\sqrt{\frac{E_p+m}{2E_p}} + \frac{\beta\boldsymbol{\alpha}\cdot\boldsymbol{p}}{\sqrt{2E_p(E_p+m)}}\right)\boldsymbol{x}\left(\sqrt{\frac{E_p+m}{2E_p}} - \frac{\beta\boldsymbol{\alpha}\cdot\boldsymbol{p}}{\sqrt{2E_p(E_p+m)}}\right).$$

From the well known identity $[\boldsymbol{x}, f(\boldsymbol{p})] = \mathrm{i}\nabla f(\boldsymbol{p})$ we get two auxiliary results:

$$\boldsymbol{x}\sqrt{\frac{E_p+m}{2E_p}} = -\frac{\mathrm{i}}{2}\sqrt{\frac{E_p}{2(E_p+m)}}\frac{m}{E_p^3}\boldsymbol{p} + \sqrt{\frac{E_p+m}{2E_p}}\boldsymbol{x},$$

$$\boldsymbol{x}\frac{\beta\boldsymbol{\alpha}\cdot\boldsymbol{p}}{\sqrt{2E_p(E_p+m)}} = \frac{\mathrm{i}\beta\boldsymbol{\alpha}}{\sqrt{2E_p(E_p+m)}} - \frac{\mathrm{i}\beta(\boldsymbol{\alpha}\cdot\boldsymbol{p})(2E_p+m)}{2\sqrt{2}(E_p(E_p+m))^{3/2}}\frac{\boldsymbol{p}}{E_p} + \frac{\beta\boldsymbol{\alpha}\cdot\boldsymbol{p}}{\sqrt{2E_p(E_p+m)}}\boldsymbol{x} \,.$$

Using these formulae we get

$$\boldsymbol{x}_{\mathrm{FW}} = \boldsymbol{x} - \mathrm{i}\frac{\boldsymbol{p}}{2E_p(E_p+m)} + \mathrm{i}\frac{\boldsymbol{p}(\beta\boldsymbol{\alpha}\cdot\boldsymbol{p})}{2E_p^2(E_p+m)} - \mathrm{i}\frac{\beta\boldsymbol{\alpha}}{2E_p} + \mathrm{i}\frac{\boldsymbol{\alpha}(\boldsymbol{\alpha}\cdot\boldsymbol{p})}{2E_p(E_p+m)} \,.$$

The last expression can be rewritten in the form

$$\boldsymbol{x}_{\mathrm{FW}} = \boldsymbol{x} + \mathrm{i}\frac{\boldsymbol{p}(\beta\boldsymbol{\alpha}\cdot\boldsymbol{p})}{2E_p^2(E_p+m)} - \mathrm{i}\frac{\beta\boldsymbol{\alpha}}{2E_p} - \frac{\boldsymbol{\Sigma}\times\boldsymbol{p}}{2E_p(E_p+m)} \,.$$

The Foldy–Wouthuysen transformation does not change the momentum, so that

$$[x^k_{\mathrm{FW}}, p^l_{\mathrm{FW}}] = \mathrm{i}\delta^{kl} \,.$$

5

Classical fields and symmetries

5.1 We apply the definition of functional derivative (5.A).

(a) From

$$\delta F_\mu = \partial_\mu \delta\phi = \int d^4y (\partial_\mu \delta\phi)_y \delta^{(4)}(y-x) = -\int d^4y \partial_\mu^y \delta^{(4)}(y-x)\delta\phi(y) ,$$

we have

$$\frac{\delta F_\mu[\phi(x)]}{\delta\phi(y)} = -\partial_\mu^y \delta^{(4)}(y-x) ,$$

(b) The first functional derivative of the action with respect to ϕ is

$$\frac{\delta S}{\delta\phi(x)} = -\Box\phi - \frac{\partial V}{\partial\phi} .$$

Then

$$\begin{aligned}\delta\left(\frac{\delta S}{\delta\phi(x)}\right) &= -\Box\delta\phi(x) - \frac{\partial^2 V}{\partial\phi^2(x)}\delta\phi(x) \\ &= \int d^4y \Big[-\Box_y \delta^{(4)}(x-y) - \\ &\quad - \frac{\partial^2 V}{\partial\phi(x)\partial\phi(y)}\delta^{(4)}(x-y)\Big]\delta\phi(y) .\end{aligned}$$

Hence,

$$\frac{\delta^2 S}{\delta\phi(x)\delta\phi(y)} = -\Box_y \delta^{(4)}(y-x) - \frac{\partial^2 V}{\partial\phi(x)\partial\phi(y)}\delta^{(4)}(x-y) .$$

5.2 In this problem we use the Euler–Lagrange equations of motion (5.B).

(a) First note that $\frac{\partial\mathcal{L}}{\partial A_\rho} = m^2 A^\rho$ and $\frac{\partial\mathcal{L}}{\partial(\partial_\sigma A_\rho)} = -2\partial^\rho A^\sigma + \lambda g^{\rho\sigma}(\partial_\mu A^\mu)$ so that the equations of motion are given by

$$(\lambda - 2)\partial_\sigma\partial^\rho A^\sigma - m^2 A^\rho = 0 .$$

(b) The derivative of the Lagrangian density with respect to $\partial_\sigma A_\rho$ is

$$\frac{\partial \mathcal{L}}{\partial(\partial_\sigma A_\rho)} = -\frac{1}{2}F^{\mu\nu}\frac{\partial F_{\mu\nu}}{\partial(\partial_\sigma A_\rho)} = -\frac{1}{2}F^{\mu\nu}(\delta^\sigma_\mu \delta^\rho_\nu - \delta^\sigma_\nu \delta^\rho_\mu) = -F^{\sigma\rho} .$$

In the last step we used the fact that $F_{\rho\sigma}$ is an antisymmetric tensor, i.e. $F_{\rho\sigma} = -F_{\sigma\rho}$. The Euler–Lagrange equations of motion are

$$\partial_\sigma F^{\sigma\rho} + m^2 A^\rho = 0 .$$

By using the definition of field strength $F^{\rho\sigma}$, the Euler–Lagrange equations become

$$\left(\delta^\rho_\sigma \Box - \partial_\sigma \partial^\rho + m^2 \delta^\rho_\sigma\right) A^\sigma = 0 .$$

(c) $(\Box + m^2)\phi = -\lambda\phi^3$.

(d) The equations of motion are:

$$-\Box A^\rho + \partial_\sigma \partial^\rho A^\sigma = -\mathrm{i}e[\phi(\partial^\rho \phi^* + \mathrm{i}eA^\rho \phi^*) - \phi^*(\partial^\rho \phi - \mathrm{i}eA^\rho \phi)] ,$$

$$\Box\phi^* + 2\mathrm{i}eA^\rho \partial_\rho \phi^* + \mathrm{i}e\phi^* \partial_\rho A^\rho - e^2 A^2 \phi^* + m^2 \phi^* = 0 ,$$

$$\Box\phi - 2\mathrm{i}eA^\rho \partial_\rho \phi - \mathrm{i}e\phi \partial_\rho A^\rho - e^2 A^2 \phi + m^2 \phi = 0 .$$

(e) The equations are:

$$(\mathrm{i}\gamma^\mu \partial_\mu - m)\psi = \mathrm{i}g\gamma_5 \psi\phi , \quad \bar\psi(\mathrm{i}\gamma^\mu \overleftarrow{\partial_\mu} + m) = -\mathrm{i}g\bar\psi\gamma_5\phi ,$$

$$\Box\phi + m^2\phi = \lambda\phi^3 - \mathrm{i}g\bar\psi\gamma_5\psi .$$

5.3 The variation of the action is

$$\begin{aligned}
\delta S &= \int_{-\infty}^{\infty} \mathrm{d}t \int_0^L \mathrm{d}x \left(\partial_\mu \phi \partial^\mu (\delta\phi) - m^2 \phi \delta\phi\right) \\
&= \int_{-\infty}^{\infty} \mathrm{d}t \int_0^L \mathrm{d}x [\partial_\mu(\partial^\mu \phi \delta\phi) - (\Box + m^2)\phi\delta\phi] \\
&= \int_0^L \mathrm{d}x \partial_0 \phi \delta\phi \Big|_{t=-\infty}^{t=\infty} - \int_{-\infty}^{\infty} \mathrm{d}t \frac{\partial\phi}{\partial x}\delta\phi \Big|_{x=0}^{x=L} \\
&- \int_{-\infty}^{\infty} \mathrm{d}t \int_0^L \mathrm{d}x (\Box\phi + m^2\phi)\delta\phi ,
\end{aligned}$$

where we integrated by parts. As the first term vanishes, from Hamiltonian principe one obtains the equation of motion

$$(\Box + m^2)\phi = 0 ,$$

and the boundary conditions:

$$\delta\phi(t, x = 0) = \delta\phi(t, x = L) = 0 , \quad \text{(Dirichlet boundary conditions)}$$

or

$$\phi'(t, x=0) = \phi'(t, x=L) = 0 \ , \quad \text{(Neumann boundary conditions)},$$

where prime denote the partial derivative with respect to x. Here, we see that beside the equation of motion we get the boundary conditions in order to eliminate the surface term. Let us mention that the mixed boundary conditions can be imposed.

5.4 In order to show that the change $\mathcal{L} \to \mathcal{L} + \partial_\mu F^\mu(\phi_r)$ does not change the equations of motion, we have to prove that

$$\delta \int_\Omega \mathrm{d}^4x \partial_\mu F^\mu(\phi_r) = 0 \ .$$

Applying the Gauss theorem we get

$$\delta \int_\Omega \mathrm{d}^4x \partial_\mu F^\mu(\phi_r) = \oint_{\partial\Omega} d\Sigma^\mu \delta F_\mu = \oint_{\partial\Omega} d\Sigma^\mu \frac{\partial F_\mu}{\partial \phi_r} \delta\phi_r = 0 \ ,$$

since the variation of fields on the boundary is equal to zero.

5.5 Add to the Lagrangian density the term $-\frac{1}{2}\partial_\mu(\phi\partial^\mu\phi)$. Note that it does not have the form as in Problem 5.4, because here the function F^μ depends on the field derivatives. However,

$$\delta \int_\Omega \mathrm{d}^4x \partial_\mu(\phi\partial^\mu\phi) = \oint_{\partial\Omega} \mathrm{d}\Sigma^\mu \delta(\phi\partial_\mu\phi) = \oint_{\partial\Omega} \mathrm{d}\Sigma^\mu (\delta\phi\partial_\mu\phi + \phi\delta\partial_\mu\phi) \ .$$

The first term is zero since $\delta\phi|_{\partial\Omega} = 0$. If we take that the boundary is at infinity ($r \to \infty$), the second term is also zero because the fields tend to zero at infinity.

5.6 Use the similar reasoning as in the previous problem.

5.7 The equation of motion for the vector field was derived in Problem 5.2 (b). Acting by ∂_ρ on this equation we obtain $m^2\partial_\rho A^\rho = 0$. Since $m \neq 0$, we conclude that $\partial_\rho A^\rho = 0$.

5.8 The field strength tensor, $F_{\mu\nu}$ is invariant under the gauge transformations. From this, it follows that the Lagrangian is also invariant. The condition $\partial_\mu A^\mu = 0$ does not follow from the equations of motion, but by using gauge symmetry we can transform the potential so that it satisfies this condition. This condition is called the Lorentz gauge.

5.9 Firstly, show that

$$\frac{\partial \mathcal{L}}{\partial(\partial_\alpha h_{\rho\sigma})} = \partial^\alpha h^{\rho\sigma} - \partial^\sigma h^{\rho\alpha} - \partial^\rho h^{\sigma\alpha} + \frac{1}{2} g^{\rho\alpha}\partial^\sigma h$$
$$+ \frac{1}{2} g^{\sigma\alpha}\partial^\rho h + g^{\rho\sigma}\partial_\mu h^{\mu\alpha} - g^{\rho\sigma}\partial^\alpha h \ .$$

The equations of motion are

$$\Box h_{\rho\sigma} - \partial^\alpha \partial_\sigma h_{\rho\alpha} - \partial^\alpha \partial_\rho h_{\sigma\alpha} + \partial_\rho \partial_\sigma h \\ + g_{\rho\sigma}\partial_\mu \partial_\nu h^{\mu\nu} - g_{\rho\sigma}\Box h = 0.$$

In order to prove gauge invariance of the action show that the Lagrangian density is changed up to four–divergence term.

5.11 This transformation is an internal one, so it is enough to prove the invariance of the Lagrangian density. The transformation law for the kinetic term is

$$\begin{aligned}\frac{1}{2}[(\partial\phi_1)^2 + (\partial\phi_2)^2] \to \frac{1}{2}[(\partial\phi_1')^2 + (\partial\phi_2')^2] \\ = \frac{1}{2}[(\partial\phi_1 \cos\theta - \partial\phi_2 \sin\theta)^2 + (\partial\phi_1 \sin\theta + \partial\phi_2 \cos\theta)^2] \\ = \frac{1}{2}[(\partial\phi_1)^2 + (\partial\phi_2)^2] \,.\end{aligned}$$

Similarly, we can prove that the other two terms are invariant. The infinitesimal variations of the fields ϕ_i are $\delta\phi_1 = -\theta\phi_2$ and $\delta\phi_2 = \theta\phi_1$, so that

$$j_\mu = \frac{\partial\mathcal{L}}{\partial(\partial^\mu\phi_i)}\delta\phi_i = \theta(\phi_1\partial_\mu\phi_2 - \phi_2\partial_\mu\phi_1) \,.$$

The parameter θ can be dropped out since it is a constant. The charge corresponding to the SO(2) symmetry is $Q = \int \mathrm{d}^3x(\phi_1\dot{\phi}_2 - \phi_2\dot{\phi}_1)$.

5.12 Under the SU(2) transformations, the fields are transformed according to $\phi' = \mathrm{e}^{\frac{\mathrm{i}}{2}\tau^a\theta^a}\phi$, where τ^a $(a = 1, 2, 3)$ are the Pauli matrices. For an infinitesimal transformation we obtain

$$\delta\phi_i = \frac{\mathrm{i}}{2}\tau^a_{ij}\theta^a\phi_j, \quad \delta\phi^*_i = -\frac{\mathrm{i}}{2}\phi^*_j\tau^a_{ji}\theta^a \,.$$

The Noether current is determined by

$$\begin{aligned}j_\mu &= \frac{\partial\mathcal{L}}{\partial(\partial^\mu\phi_i)}\delta\phi_i + \delta\phi^*_i\frac{\partial\mathcal{L}}{\partial(\partial^\mu\phi^*_i)} \\ &= \frac{\mathrm{i}}{2}\theta^a\left(\partial_\mu\phi^*_i\tau^a_{ij}\phi_j - \phi^*_i\tau^a_{ij}\partial_\mu\phi_j\right) \,.\end{aligned}$$

From the previous relation (θ^a are constant independent parameters) it follows that the conserved currents are:

$$j^a_\mu = -\frac{\mathrm{i}}{2}\left(\partial_\mu\phi^*_i\tau^a_{ij}\phi_j - \phi^*_i\tau^a_{ij}\partial_\mu\phi_j\right) \,.$$

The charges are

$$Q^a = -\frac{\mathrm{i}}{2}\int \mathrm{d}^3\boldsymbol{x}(\partial_0\phi^*_i\tau^a_{ij}\phi_j - \phi^*_i\tau^a_{ij}\partial_0\phi_j) \,.$$

5.13 The currents and charges are

$$j^a_\mu = \frac{1}{2}\bar{\psi}_i\gamma_\mu\tau^a_{ij}\psi_j \;, \quad Q^a = \frac{1}{2}\int \mathrm{d}^3x\psi^\dagger_i\tau^a_{ij}\psi_j \;.$$

The equations of motion are $(\mathrm{i}\gamma^\mu\partial_\mu - m)\psi_i = 0$ and $\bar{\psi}_i(\mathrm{i}\gamma^\mu\overleftarrow{\partial_\mu} + m) = 0$. The current conversation law, $\partial_\mu j^{\mu a} = 0$ can be proved easily:

$$2\partial_\mu j^{\mu a} = (\partial_\mu\bar{\psi}_i)\gamma^\mu\tau^a_{ij}\psi_j + \bar{\psi}_i\gamma^\mu\tau^a_{ij}\partial_\mu\psi_j = \mathrm{i}m\bar{\psi}_i\tau^a_{ij}\psi_j + \bar{\psi}_i\tau^a_{ij}(-\mathrm{i}m\psi_j) = 0 \;,$$

where we used the equations of motion. The Noether theorem is valid on–shell.

5.14

(a) The phase invariance is the $U(1)$ symmetry, where $\psi \to \psi' = \mathrm{e}^{\mathrm{i}\theta}\psi$ and $\bar{\psi} \to \bar{\psi}' = \mathrm{e}^{-\mathrm{i}\theta}\bar{\psi}$. The Noether current is $j_\mu = \bar{\psi}\gamma_\mu\psi$, while the charge is given by $Q = -e\int \mathrm{d}^3\boldsymbol{x}\psi^\dagger\psi$. Note that the current does not have additional indices since U(1) is a one–parameter group.
(b) $j_\mu = \mathrm{i}(\phi^*\partial_\mu\phi - \phi\partial_\mu\phi^*)$, $\quad Q = \mathrm{i}q\int \mathrm{d}^3\boldsymbol{x}(\phi^*\partial_0\phi - \phi\partial_0\phi^*)$.

5.15 The equations of motion are $(\Box + m^2)\phi_i = 0$. The expression $\phi^T\phi$ is invariant under SO(3) transformations, hence the Lagrangian density has the same symmetry. The generators of SO(3) group are

$$J^1 = \begin{pmatrix} 0 & 0 & 0 \\ 0 & 0 & -\mathrm{i} \\ 0 & \mathrm{i} & 0 \end{pmatrix}, \quad J^2 = \begin{pmatrix} 0 & 0 & \mathrm{i} \\ 0 & 0 & 0 \\ -\mathrm{i} & 0 & 0 \end{pmatrix}, \quad J^3 = \begin{pmatrix} 0 & -\mathrm{i} & 0 \\ \mathrm{i} & 0 & 0 \\ 0 & 0 & 0 \end{pmatrix} . \tag{5.1}$$

Note that we can write

$$(J^k)_{ij} = -\mathrm{i}\epsilon_{kij} \;.$$

Under SO(3) transformations, the infinitesimal variations of the fields are $\delta\phi_i = \mathrm{i}(J^k)_{ij}\theta^k\phi_j = \epsilon_{kij}\theta_k\phi_j$ and the Noether current is

$$\begin{aligned} j_\mu &= \frac{\partial\mathcal{L}}{\partial(\partial^\mu\phi_i)}\delta\phi_i \\ &= \epsilon_{kij}\phi_j\partial_\mu\phi_i\theta_k \\ &= -\boldsymbol{\theta}\cdot(\boldsymbol{\phi}\times\partial_\mu\boldsymbol{\phi}) \;. \end{aligned}$$

The parameters of rotations θ_k, are arbitrary and therefore the currents

$$j^\mu_k = -\epsilon_{kij}\phi_j\partial^\mu\phi_i$$

are also conserved.

5.16 First, derive the following formula $\mathrm{e}^{\mathrm{i}\alpha\gamma_5} = \cos\alpha + \mathrm{i}\gamma_5\sin\alpha$. The transformation law for the Dirac Lagrangian density under the chiral transformation is given by

$$\begin{aligned}\mathcal{L} &\to \psi^\dagger \mathrm{e}^{-\mathrm{i}\alpha\gamma_5}\gamma_0(\mathrm{i}\gamma_\mu\partial^\mu - m)\mathrm{e}^{\mathrm{i}\alpha\gamma_5}\psi \\ &= (\cos^2\alpha + \sin^2\alpha)\bar{\psi}\mathrm{i}\gamma_\mu\partial^\mu\psi \;\; - m\bar{\psi}(\cos\alpha + \mathrm{i}\gamma_5\sin\alpha)^2\psi \\ &= \bar{\psi}\mathrm{i}\gamma_\mu\partial^\mu\psi \;\; - m\bar{\psi}(\cos 2\alpha + \mathrm{i}\gamma_5\sin 2\alpha)\psi \; .\end{aligned}$$

From the previous expression we can conclude that the Lagrangian density is invariant only for massless fermions. The Noether current is $j_\mu = \bar{\psi}\gamma_\mu\gamma_5\psi$. Prove that $\partial_\mu j^\mu$ is proportional to the mass m of the field.

5.17 The current is given by

$$\begin{aligned}j_\mu &= \frac{\partial\mathcal{L}}{\partial(\partial^\mu\sigma)}\delta\sigma + \frac{\partial\mathcal{L}}{\partial(\partial^\mu\pi^a)}\delta\pi^a + \frac{\partial\mathcal{L}}{\partial(\partial^\mu\Psi_i)}\delta\Psi_i + \delta\bar{\Psi}_i\frac{\partial\mathcal{L}}{\partial(\partial^\mu\bar{\Psi}_i)} \\ &= -\epsilon^{abc}\alpha^b\partial_\mu\pi^a\pi^c - \frac{1}{2}\bar{\Psi}_i\gamma_\mu\alpha^a\tau^a_{ij}\Psi_j \; .\end{aligned}$$

The final result has the form

$$\boldsymbol{j}_\mu = \boldsymbol{\pi}\times\partial_\mu\boldsymbol{\pi} + \frac{1}{2}\bar{\Psi}\gamma_\mu\boldsymbol{\tau}\Psi \; .$$

5.18

(a) For translations, we have $\delta x^\mu = \epsilon^\mu$, while the total variations of the fields equal zero. The Noether current is

$$T^{\mu\nu} = \frac{\partial\mathcal{L}}{\partial(\partial_\mu\phi_r)}\frac{\partial\phi_r}{\partial x_\nu} - \mathcal{L}g^{\mu\nu} \; . \tag{5.2}$$

The index ν in (5.2) comes from the group of translations. For a real scalar field, from (5.2) we obtain

$$T_{\mu\nu} = \partial_\mu\phi\partial_\nu\phi - \frac{1}{2}\left[(\partial\phi)^2 - m^2\phi^2\right]g_{\mu\nu} \; . \tag{5.3}$$

The conserved charges are the Hamiltonian (for $\nu = 0$),

$$H = \int \mathrm{d}^3\boldsymbol{x}T^{00} = \frac{1}{2}\int \mathrm{d}^3\boldsymbol{x}\left[(\partial_0\phi)^2 + (\nabla\phi)^2 + m^2\phi^2\right] , \tag{5.4}$$

and the momentum (for $\nu = i$)

$$P^i = \int \mathrm{d}^3\boldsymbol{x}T^{0i} = \int \mathrm{d}^3\boldsymbol{x}\partial_0\phi\partial^i\phi \; . \tag{5.5}$$

For the Dirac field the energy–momentum tensor is given by

$$T^{\mu\nu} = \mathrm{i}\bar{\psi}\gamma^\mu\partial^\nu\psi - \mathcal{L}g^{\mu\nu} \; .$$

The Hamiltonian and momentum are given by

$$H = \int d^3\boldsymbol{x}\bar{\psi}[-i\gamma\nabla + m]\psi \ , \tag{5.6}$$

$$\boldsymbol{P} = -i\int d^3\boldsymbol{x}\psi^\dagger\nabla\psi \ . \tag{5.7}$$

For electromagnetic field the energy–momentum tensor is

$$T^{\mu\nu} = \frac{\partial\mathcal{L}}{\partial(\partial_\mu A_\rho)}\frac{\partial A_\rho}{\partial x_\nu} - \mathcal{L}g^{\mu\nu}$$

from which we obtain

$$T^{\mu\nu} = -F^{\mu\rho}\partial^\nu A_\rho + \frac{1}{4}F^2 g^{\mu\nu} \ . \tag{5.8}$$

For the Lorentz transformations $\delta x^\nu = \omega^{\nu\rho}x_\rho$ and

$$\delta\phi = 0 \ , \delta\psi = -\frac{i}{4}\sigma_{\nu\rho}\omega^{\nu\rho}\psi, \ \delta A_\mu = \omega_\mu{}^\nu A_\nu \ ,$$

The Noether currents for scalar, spinor and electromagnetic field are

$$\begin{aligned} j_\mu &= [x_\nu T_{\mu\rho} - x_\rho T_{\mu\nu}]\omega^{\nu\rho} \ , \\ j_\mu &= [\frac{1}{2}\bar{\psi}\gamma_\mu\sigma_{\nu\rho}\psi + \ x_\nu T_{\mu\rho} - x_\rho T_{\mu\nu}]\omega^{\nu\rho} \ , \\ j_\mu &= [F_{\mu\rho}A_\nu - F_{\mu\nu}A_\rho + (x_\nu T_{\mu\rho} - x_\rho T_{\mu\nu})]\omega^{\nu\rho} \ . \end{aligned} \tag{5.9}$$

Dropping the parameters of the Lorentz transformations $\omega^{\nu\rho}$, the conserved currents have the form $M_{\mu\nu\rho}$, and they are given by the expression in square brackets in (5.9). The angular-momentum is $M_{\nu\rho} = \int d^3\boldsymbol{x}M_{0\nu\rho}$.

(b) As we see, the energy–momentum tensors for Dirac and electromagnetic fields are not symmetric. To find the symmetrized energy–momentum tensors we employ the procedure given in the problem. For the Dirac field we have

$$\begin{aligned} \chi_{\rho\mu\nu} &= \frac{1}{4}(-\bar{\psi}\gamma_\mu\sigma_{\rho\nu}\psi + \bar{\psi}\gamma_\rho\sigma_{\mu\nu}\psi + \bar{\psi}\gamma_\nu\sigma_{\mu\rho}\psi) \\ &= \frac{i}{8}\bar{\psi}(4g_{\mu\nu}\gamma_\rho - 4g_{\rho\nu}\gamma_\mu + \gamma_\mu\gamma_\nu\gamma_\rho - \gamma_\rho\gamma_\nu\gamma_\mu)\psi \ . \end{aligned}$$

Using (4.43) we find

$$\begin{aligned} \partial^\rho\chi_{\rho\mu\nu} = &-\frac{i}{4}\partial_\nu\bar{\psi}\gamma_\mu\psi - \frac{i}{4}\partial_\mu\bar{\psi}\gamma_\nu\psi - \frac{3i}{4}\bar{\psi}\gamma_\mu\partial_\nu\psi \\ &+ \frac{i}{4}\bar{\psi}\gamma_\nu\partial_\mu\psi + g_{\mu\nu}\frac{i}{2}(\partial_\nu\bar{\psi}\gamma^\nu\psi + \bar{\psi}\partial\!\!\!/\psi) \ . \end{aligned}$$

The symmetrized energy–momentum tensor for Dirac field is

$$\tilde{T}_{\mu\nu} = \frac{\mathrm{i}}{4}(\bar{\psi}\gamma_\nu\partial_\mu\psi + \bar{\psi}\gamma_\mu\partial_\nu\psi - \partial_\mu\bar{\psi}\gamma_\nu\psi - \partial_\nu\bar{\psi}\gamma_\mu\psi)$$
$$- g_{\mu\nu}\left(-\frac{\mathrm{i}}{2}\partial_\nu\bar{\psi}\gamma^\nu\psi + \frac{\mathrm{i}}{2}\bar{\psi}\partial\!\!\!/\psi - m\bar{\psi}\psi\right) .$$

Similarly we determine the symmetrized energy–momentum tensor for the electromagnetic field. From transformation rule of the electromagnetic potential with respect to Lorentz transformations

$$\delta A_\alpha = \omega_{\alpha\beta}A^\beta \equiv \frac{1}{2}\omega^{\mu\nu}(I_{\mu\nu})_{\alpha\beta}A^\beta ,$$

follows that

$$(I_{\mu\nu})_{\alpha\beta} = g_{\mu\alpha}g_{\nu\beta} - g_{\mu\beta}g_{\nu\alpha} .$$

Then $\chi^{\rho\mu\nu} = F^{\mu\rho}A^\nu$ and the new energy–momentum tensor is

$$\tilde{T}^{\mu\nu} = -F^{\mu\rho}F^\nu{}_\rho + \frac{1}{4}F^2g^{\mu\nu} . \tag{5.10}$$

If we introduce the electric and magnetic fields: $F^{0i} = -E^i$, $F_{ij} = -\epsilon_{ijk}B^k$, then the components of energy–momentum tensor are:

$$\begin{aligned}
\tilde{T}^{00} &= -F^{0i}F^0{}_i + \frac{1}{4}(2F_{0i}F^{0i} + F_{ij}F^{ij}) \\
&= \boldsymbol{E}^2 + \frac{1}{4}(-2\boldsymbol{E}^2 + 2\boldsymbol{B}^2) \\
&= \frac{1}{2}(\boldsymbol{E}^2 + \boldsymbol{B}^2) , \\
\tilde{T}^{0i} &= -F^{0j}F^i{}_j \\
&= \epsilon^{ijk}E^jB^k \\
&= (\boldsymbol{E}\times\boldsymbol{B})^i , \\
\tilde{T}^{ij} &= -E^iE^j + \epsilon^{ikl}\epsilon^{jkn}B^lB^n + \frac{1}{2}(\boldsymbol{E}^2 - \boldsymbol{B}^2)\delta_{ij} \\
&= -\left(E^iE^j + B^iB^j - \delta_{ij}\tilde{T}_{00}\right) .
\end{aligned} \tag{5.11}$$

From the expression (5.11) we conclude that $\tilde{T}_{00}$ $\tilde{T}^{0i}$, $-\tilde{T}_{ij}$ are the energy density of electromagnetic field, the Poynting vector, and the components of the Maxwell stress tensor.

5.19 The variation of form is defined by $\delta_0\phi(x) = \phi'(x) - \phi(x)$. From

$$\delta_0\phi = \delta\phi - \partial_\mu\phi\delta x^\mu,$$

where $\delta\phi = \phi'(x') - \phi(x)$ is the total variation of a field, it follows that the infinitesimal form variation of ϕ is

$$\delta_0\phi = \rho(\phi(x) + x^\mu\partial_\mu\phi) . \tag{5.12}$$

The induced change of the action is

$$S' - S = \frac{1}{2}\int \mathrm{d}^4x' \left[(\partial'\phi')^2 - m^2\phi'^2(x')\right] - \frac{1}{2}\int \mathrm{d}^4x \left[(\partial\phi)^2 - m^2\phi^2(x)\right] . \tag{5.13}$$

The transformed volume of integration is given by

$$\mathrm{d}^4x' = |J|\mathrm{d}^4x = \det(\mathrm{e}^{-\rho}I)\mathrm{d}^4x = \mathrm{e}^{-4\rho}\mathrm{d}^4x . \tag{5.14}$$

The field derivative is changed according to the following rule:

$$\partial_\mu\phi(x) \to \frac{\partial\phi'}{\partial x'^\mu} = \frac{\partial x^\nu}{\partial x'^\mu}\frac{\partial}{\partial x^\nu}(\mathrm{e}^{\rho}\phi) = \mathrm{e}^{2\rho}\partial_\mu\phi . \tag{5.15}$$

Thus, the change of the action is

$$\begin{aligned} S' - S &= \frac{1}{2}\int \mathrm{d}^4x \mathrm{e}^{-4\rho}\left[\mathrm{e}^{4\rho}(\partial\phi)^2 - m^2\mathrm{e}^{2\rho}\phi^2(x)\right] \\ &\quad - \frac{1}{2}\int \mathrm{d}^4x \left[(\partial\phi)^2 - m^2\phi^2(x)\right] \\ &= \frac{1}{2}m^2(1-\mathrm{e}^{-2\rho})\int \mathrm{d}^4x\phi^2(x) . \end{aligned}$$

For an infinitesimal dilatation ($\rho \ll 1$), the variation of the action is

$$\delta S = m^2\rho\int \mathrm{d}^4x\phi^2(x) . \tag{5.16}$$

From (5.16) it is clear that the theory of massless scalar field is invariant under dilatations.
The conserved current is

$$j^\mu = -\phi\partial^\mu\phi - x^\nu\partial^\mu\phi\partial_\nu\phi + \mathcal{L}x^\mu . \tag{5.17}$$

By calculating $\partial_\mu j^\mu$ one obtains that $\partial_\mu j^\mu$ is proportional to the mass m.

5.20 From

$$\mathrm{d}^4x' = \mathrm{e}^{-4\rho}\mathrm{d}^4x \approx (1-4\rho)\mathrm{d}^4x , \tag{5.18}$$

and

$$\bar\psi'(x')\gamma^\mu\partial'_\mu\psi'(x') = \mathrm{e}^{4\rho}\bar\psi\gamma^\mu\partial_\mu\psi \approx (1+4\rho)\bar\psi\gamma^\mu\partial_\mu\psi , \tag{5.19}$$

it follows that this transformation leaves the action unchanged. The Noether current is $j^\mu = -\frac{3}{2}\mathrm{i}\bar\psi\gamma^\mu\psi - \mathrm{i}x^\nu\bar\psi\gamma^\mu\partial_\nu\psi + x^\mu\mathcal{L}$.

6

Green functions

6.1 The Green function of the Klein–Gordon equation satisfies the equation

$$(\Box_x + m^2)\Delta(x-y) = -\delta^{(4)}(x-y) \; . \tag{6.1}$$

Fourier transformations of the Green function and the δ-function in (6.1) gives

$$(\Box_x + m^2)\frac{1}{(2\pi)^4}\int d^4k \tilde{\Delta}(k) e^{-ik\cdot(x-y)} = -\frac{1}{(2\pi)^4}\int d^4k e^{-ik\cdot(x-y)} \; . \tag{6.2}$$

From (6.2) follows

$$\tilde{\Delta}(k) = \frac{1}{k^2 - m^2} = \frac{1}{k_0^2 - \mathbf{k}^2 - m^2} \; .$$

Then, the Green function is defined by

$$\Delta(x-y) = \int \frac{d^4k}{(2\pi)^4}\frac{1}{k_0^2 - \mathbf{k}^2 - m^2} e^{-ik\cdot(x-y)} \; . \tag{6.3}$$

The integral (6.3) is divergent, since the integrand has the poles in $k_0 = \pm\omega_k$. We shall modify the contour of integration to make the integral (6.3) convergent. It is clear that we have to give the physical reasons for this modification of integral. The poles can be evaded in four different ways. The first one is from the upper side (Fig. 6.1). The exponential term in (6.3) for large energy k_0 behaves as $e^{(x_0 - y_0)\mathrm{Im}k_0}$, therefore the contour for $x_0 > y_0$ has to be closed from the lower side ($\mathrm{Im}k_0 < 0$), while in the case $x_0 < y_0$ we will close the integration contour on the upper side. By applying the Cauchy theorem we get

$$\Delta(x-y) = -\frac{1}{(2\pi)^4}\int d^3\mathbf{k} e^{i\mathbf{k}\cdot(\mathbf{x}-\mathbf{y})} 2\pi i(\mathrm{Res}_{\omega_k} + \mathrm{Res}_{-\omega_k})\theta(x^0 - y^0) \; . \tag{6.4}$$

From (6.4) follows

$$\Delta_{\mathrm{R}} = -\frac{\mathrm{i}}{(2\pi)^3} \int \frac{\mathrm{d}^3 \boldsymbol{k}}{2\omega_k} \mathrm{e}^{\mathrm{i}\boldsymbol{k}\cdot(\boldsymbol{x}-\boldsymbol{y})} (\mathrm{e}^{-\mathrm{i}\omega_k(x^0-y^0)} - \mathrm{e}^{\mathrm{i}\omega_k(x^0-y^0)})\theta(x^0 - y^0) \ . \quad (6.5)$$

$\Delta_{\mathrm{R}}(x-y)$ is *the retarded Green function.* The solution of the inhomogeneous equation $(\Box + m^2)\phi = J$ is

$$\phi(x) = -\int \mathrm{d}^4 y \Delta(x-y) J(y) + \phi_0 \ , \quad (6.6)$$

where ϕ_0 is a solution of homogeneous equation. From the expressions (6.5) and (6.6) (because of θ–function), we conclude that we integrate over y^0 from $-\infty$ to x^0. The value of the field ϕ at time x_0 is determined by the source J at earlier times. For this reason this function is called the retarded Green function.

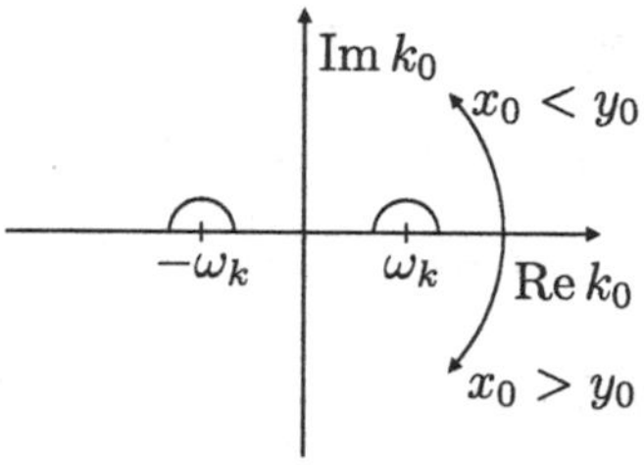

Fig. 6.1. The integration contour for the retarded boundary conditions

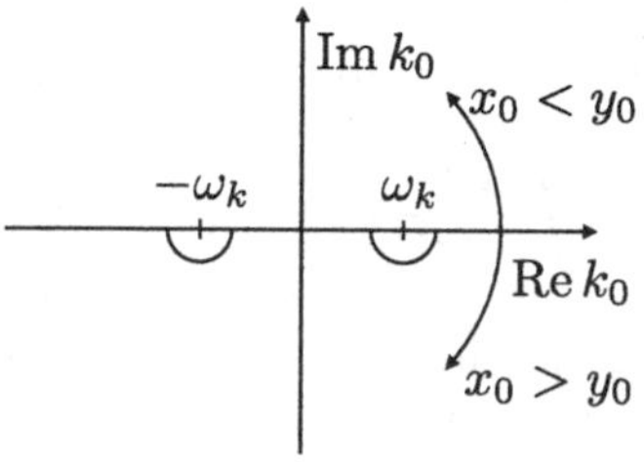

Fig. 6.2. The integration contour for the advanced boundary conditions

By evading poles as in Fig. 6.2 we get the so-called *advanced Green function*

$$\Delta_{\mathrm{A}} = \frac{\mathrm{i}}{(2\pi)^3} \int \frac{\mathrm{d}^3 \boldsymbol{k}}{2\omega_k} \mathrm{e}^{\mathrm{i}\boldsymbol{k}\cdot(\boldsymbol{x}-\boldsymbol{y})} (\mathrm{e}^{-\mathrm{i}\omega_k(x^0-y^0)} - \mathrm{e}^{\mathrm{i}\omega_k(x^0-y^0)})\theta(y^0 - x^0) \ . \quad (6.7)$$

The advanced Green function contributes nontrivially to the field $\phi(x)$ for $y_0 > x_0$. If we evade poles as in Fig. 6.3, we get *the Feynman propagator*:

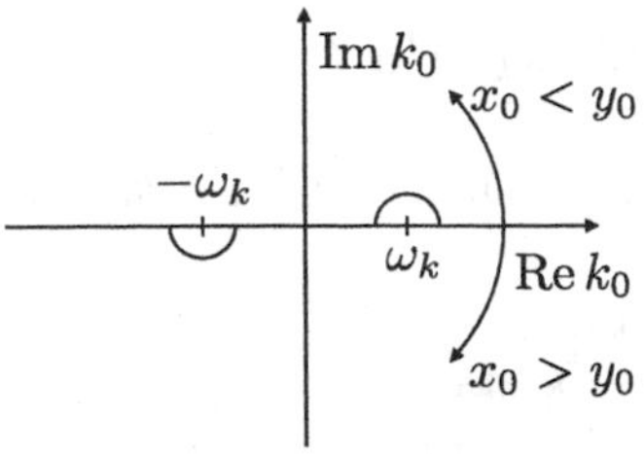

Fig. 6.3. The integration contour which defined the Feynman propagator

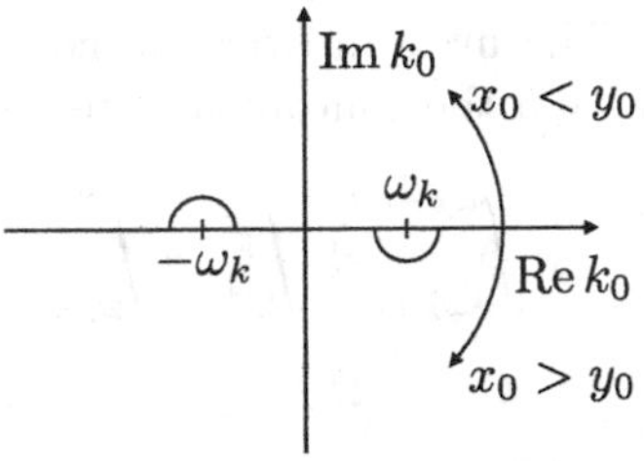

Fig. 6.4. The integration contour for the Dyson Green function

$$
\begin{aligned}
\Delta_{\mathrm{F}} &= \frac{\mathrm{i}}{(2\pi)^3}\int \mathrm{d}^3\boldsymbol{k}\mathrm{e}^{\mathrm{i}\boldsymbol{k}\cdot(\boldsymbol{x}-\boldsymbol{y})}\left[\mathrm{Res}_{-\omega_k}\theta(y^0-x^0)-\mathrm{Res}_{\omega_k}\theta(x^0-y^0)\right] \\
&= -\frac{\mathrm{i}}{(2\pi)^3}\int \frac{\mathrm{d}^3\boldsymbol{k}}{2\omega_k}\mathrm{e}^{\mathrm{i}\boldsymbol{k}\cdot(\boldsymbol{x}-\boldsymbol{y})}\Big[\mathrm{e}^{-\mathrm{i}\omega_k(x^0-y^0)}\theta(x^0-y^0) \qquad (6.8)\\
&\quad +\mathrm{e}^{\mathrm{i}\omega_k(x^0-y^0)}\theta(y^0-x^0)\Big] .
\end{aligned}
$$

We can conclude that positive (negative) energy solutions propagate forward (backward) in spacetime. This is what we need in the relativistic quantum physics in contrast to the classical theory (for example in classical electrodynamics), where all physically relevant information is contained in the retarded Green function. *Dyson Green function* is obtained by evading poles as in Fig. 6.4. This Green function can be evaluated in a way similar to the previous three cases. It is recommended to do this calculation as an exercise.

6.2 From (6.5) and (6.8) it follows that (we take $y = 0$)

$$
\Delta_{\mathrm{F}}(x) - \Delta_{\mathrm{R}}(x) = -\frac{\mathrm{i}}{(2\pi)^3}\int \frac{\mathrm{d}^3\boldsymbol{k}}{2\omega_k}\mathrm{e}^{\mathrm{i}(\omega_k t+\boldsymbol{k}\cdot\boldsymbol{x})} , \qquad (6.9)
$$

since $\theta(t) + \theta(-t) = 1$. By applying $(\Box + m^2)$ on (6.9) we get

$$
(\Box + m^2)[\Delta_{\mathrm{F}}(x) - \Delta_{\mathrm{R}}(x)] = 0.
$$

6.3

$$
I = \int \mathrm{d}^4k\delta(k^2-m^2)\theta(k_0)f(k)
$$

$$
\begin{aligned}
&= \int \mathrm{d}^4 k \delta(k_0^2 - \omega_k^2)\theta(k_0) f(k) \\
&= \int \mathrm{d}^3\boldsymbol{k} \mathrm{d}k_0 \frac{1}{2\omega_k} \left[\delta(k_0 - \omega_k) + \delta(k_0 + \omega_k)\right] \theta(k_0) f(k) \\
&= \int \frac{\mathrm{d}^3\boldsymbol{k}}{2\omega_k} f(k)\bigg|_{k_0=\omega_k} .
\end{aligned}
$$

From this calculation it is clear that the expression $\mathrm{d}^3\boldsymbol{k}/(2\omega_k)$ is a Lorentz invariant measure.

6.5 Let us take $x_0 < 0$. The integral over the contour in Fig. 6.5 vanishes since there are no poles within the contour of integration. So, we get

$$
\int_{-R}^{-\omega_k-\rho} + \int_{C_\rho^-} + \int_{-\omega_k+\rho}^{\omega_k-\rho} + \int_{C_\rho^+} + \int_{\omega_k+\rho}^{R} + \int_{C_R} = 0 . \tag{6.10}
$$

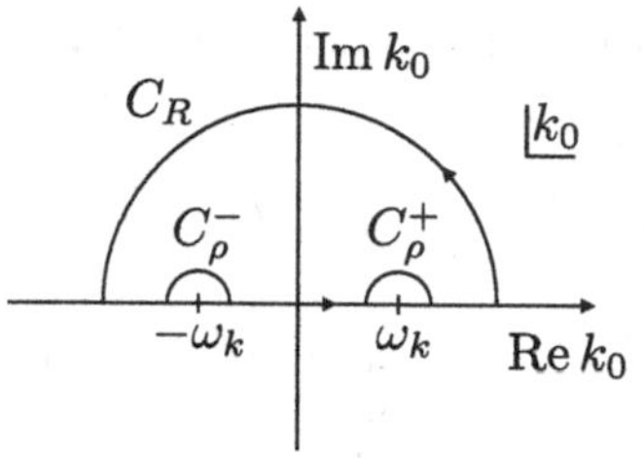

Fig. 6.5. The integration contour that defined the principal-part propagator

The integral along the half–circle, C_R tends to zero for large R, which can be seen if we take that limit in the integrand. If in the integral $\int_{C_\rho^+}$ we take $k_0 = \omega_k + \rho\mathrm{e}^{\mathrm{i}\varphi}$, it becomes

$$
\int_{C_\rho^+} = \int_{\pi}^{0} \mathrm{i}\mathrm{e}^{-\mathrm{i}x_0(\omega_k+\rho\mathrm{e}^{\mathrm{i}\varphi})} \frac{1}{\rho\mathrm{e}^{\mathrm{i}\varphi} + 2\omega_k} \mathrm{d}\varphi . \tag{6.11}
$$

By taking $\rho \to 0$ in (6.11) we get

$$
\int_{C_\rho^+} = -\frac{\mathrm{i}\pi}{2\omega_k} \mathrm{e}^{-\mathrm{i}\omega_k x_0} . \tag{6.12}
$$

In the same way we can show that

$$
\int_{C_\rho^-} = \frac{\mathrm{i}\pi}{2\omega_k} \mathrm{e}^{\mathrm{i}\omega_k x_0} . \tag{6.13}
$$

From (6.10), (6.12) and (6.13) we get (for $x_0 < 0$)

$$\bar{\Delta}(x) = \frac{\mathrm{i}\pi}{(2\pi)^4}\int \frac{\mathrm{d}^3\boldsymbol{k}}{2\omega_k}\mathrm{e}^{\mathrm{i}\boldsymbol{k}\cdot\boldsymbol{x}}\left[\mathrm{e}^{-\mathrm{i}\omega_k x_0} - \mathrm{e}^{\mathrm{i}\omega_k x_0}\right]\theta(-x_0). \tag{6.14}$$

The case $x_0 > 0$ is analogous to the previous one. The result is

$$\bar{\Delta}(x) = -\frac{\mathrm{i}\pi}{(2\pi)^4}\int \frac{\mathrm{d}^3\boldsymbol{k}}{2\omega_k}\mathrm{e}^{\mathrm{i}\boldsymbol{k}\cdot\boldsymbol{x}}\left[\mathrm{e}^{-\mathrm{i}\omega_k x_0} - \mathrm{e}^{\mathrm{i}\omega_k x_0}\right]\theta(x_0)\,. \tag{6.15}$$

By comparing equations (6.14) and (6.15) with the expressions for Δ_R and Δ_A we obtain

$$\bar{\Delta}(x) = \frac{1}{2}(\Delta_\mathrm{R}(x) + \Delta_\mathrm{A}(x))\,.$$

6.6

$$\Delta(x) = -\frac{\mathrm{i}}{(2\pi)^3}\int \frac{\mathrm{d}^3\boldsymbol{k}}{2\omega_k}\mathrm{e}^{\mathrm{i}\boldsymbol{k}\cdot\boldsymbol{x}}(\mathrm{e}^{-\mathrm{i}\omega_k t} - \mathrm{e}^{\mathrm{i}\omega_k t})\,, \tag{6.16}$$

$$\Delta_\pm(x) = \mp\frac{\mathrm{i}}{(2\pi)^3}\int \frac{\mathrm{d}^3\boldsymbol{k}}{2\omega_k}\mathrm{e}^{\mathrm{i}(\boldsymbol{k}\cdot\boldsymbol{x}\mp\omega_k t)}\,. \tag{6.17}$$

6.7 By using the expression for Δ obtained in Problem 6.6 we get

$$\partial_i\Delta(x) = -\frac{\mathrm{i}}{(2\pi)^3}\int \frac{\mathrm{d}^3\boldsymbol{k}}{2\omega_k}\mathrm{i}k_i\mathrm{e}^{\mathrm{i}\boldsymbol{k}\cdot\boldsymbol{x}}(\mathrm{e}^{-\mathrm{i}\omega_k t} - \mathrm{e}^{\mathrm{i}\omega_k t}) = 0\,, \tag{6.18}$$

since the integrand is an odd function of $\boldsymbol{k}$. The second identity can be proven easily.

6.8 By applying the operator $(\Box + m^2)$ to the expression (6.16) we get

$$(\Box+m^2)\Delta(x) = -\frac{\mathrm{i}}{(2\pi)^3}\int \frac{\mathrm{d}^3\boldsymbol{k}}{2\omega_k}(-\omega_k^2+\boldsymbol{k}^2+m^2)\left[\mathrm{e}^{\mathrm{i}(-\omega_k t+\boldsymbol{k}\cdot\boldsymbol{x})} - \mathrm{e}^{\mathrm{i}(\omega_k t+\boldsymbol{k}\cdot\boldsymbol{x})}\right]\,,$$

from which follows that $(\Box + m^2)\Delta(x) = 0$, as $k^2 = m^2$.

6.9 For $m = 0$ from (6.8) it follows that

$$\begin{aligned}\Delta_\mathrm{F}|_{m=0} &= -\frac{\mathrm{i}}{(2\pi)^3}\int \frac{\mathrm{d}^3\boldsymbol{k}}{2k}\mathrm{e}^{\mathrm{i}\boldsymbol{k}\cdot\boldsymbol{x}}\left[\mathrm{e}^{-\mathrm{i}kx^0}\theta(x^0) + \mathrm{e}^{\mathrm{i}kx_0}\theta(-x^0)\right]\\ &= -\frac{\mathrm{i}}{2(2\pi)^2}\int_0^\infty\int_0^\pi k\sin\theta\mathrm{d}k\mathrm{d}\theta\\ &\times\left[\mathrm{e}^{\mathrm{i}k(-t+r\cos\theta)}\theta(t) + \mathrm{e}^{\mathrm{i}k(t+r\cos\theta)}\theta(-t)\right]\,,\end{aligned} \tag{6.19}$$

where in the second line we integrated over the polar angle φ. Integration over θ gives

$$\begin{aligned}\Delta_\mathrm{F}(x)|_{m=0} &= -\frac{1}{2(2\pi)^2 r}\int_0^\infty \mathrm{d}k\left[(\mathrm{e}^{-\mathrm{i}k(t-r)} - \mathrm{e}^{-\mathrm{i}k(t+r)})\theta(t)\right.\\ &\left.+(\mathrm{e}^{\mathrm{i}k(t+r)} - \mathrm{e}^{\mathrm{i}k(t-r)})\theta(-t)\right]\,.\end{aligned} \tag{6.20}$$

Now, we shall consider separately two cases: $t > 0$ and $t < 0$. In the first one, $t > 0$ the second term in the integrand of (6.20) is zero. The first part of the integrand has bad behavior for large k. We regularize it by making substitution $t \to t - \mathrm{i}\epsilon$, where $\epsilon \to 0^+$. In this way we ensure convergence of this integral. Then from (6.20) it follows that

$$\Delta_{\mathrm{F}}|_{m=0} = \frac{\mathrm{i}}{2(2\pi)^2 r}\left(\frac{1}{t-r-\mathrm{i}\epsilon} - \frac{1}{t+r-\mathrm{i}\epsilon}\right) \tag{6.21}$$

$$= \frac{\mathrm{i}}{(2\pi)^2}\frac{1}{t^2-r^2-\mathrm{i}\epsilon} = \frac{\mathrm{i}}{(2\pi)^2}\frac{1}{x^2-\mathrm{i}\epsilon} \,. \tag{6.22}$$

By applying the formula

$$\frac{1}{z \pm \mathrm{i}\epsilon} = \mathrm{P}\frac{1}{z} \mp \mathrm{i}\pi\delta(z) \,, \tag{6.23}$$

in expression (6.22) we get

$$\Delta_{\mathrm{F}}(x)\,|_{m=0} = -\frac{1}{4\pi}\delta(x^2) + \frac{\mathrm{i}}{4\pi^2}\mathrm{P}\frac{1}{x^2} \,. \tag{6.24}$$

For the case $t < 0$ one also obtains the expression (6.24); this is left as an exercise.

6.10 We shall start from (6.5) and use spherical coordinates. Integration over angles θ and φ leads to

$$\Delta_{\mathrm{R}}(x) = -\frac{1}{2(2\pi)^2 r}\int_0^\infty \mathrm{d}k\left[\mathrm{e}^{-\mathrm{i}k(t-r)} - \mathrm{e}^{\mathrm{i}k(t+r)} - \mathrm{e}^{-\mathrm{i}k(t+r)} + \mathrm{e}^{\mathrm{i}k(t-r)}\right]\theta(t) \,. \tag{6.25}$$

The change of variable $k' = -k$ in the third and the fourth integral in expression (6.25) gives

$$\Delta_{\mathrm{R}}(x) = -\frac{1}{2(2\pi)^2 r}\int_{-\infty}^\infty \mathrm{d}k(\mathrm{e}^{-\mathrm{i}k(t-r)} - \mathrm{e}^{\mathrm{i}k(t+r)})\theta(t) \,. \tag{6.26}$$

Note the change of the lower integration limit in the expression (6.26). From (6.26) follows

$$\Delta_{\mathrm{R}}|_{m=0}(x) = -\frac{1}{4\pi r}\left[\delta(t-r) - \delta(t+r)\right]\theta(t) \,. \tag{6.27}$$

The second term in (6.27) has a "wrong" sign but it is irrelevant as this term vanishes ($t > 0$ and $r > 0$). By changing this minus into a plus in (6.27) we finally obtain:

$$\Delta_{\mathrm{R}}|_{m=0}(x) = -\frac{1}{2\pi}\delta(t^2 - r^2)\theta(t) = -\frac{1}{2\pi}\delta(x^2)\theta(t) \,. \tag{6.28}$$

The case of advanced Green function is left for an exercise.

6.11 In the Problem 6.1, we modified the the contour of integration according to the boundary conditions, while the poles were not moved. Sometimes it is useful to do the opposite, i.e. to move the poles and to integrate over the real k_0–axis. For the retarded Green function this can be done by changing

$$k^2 - m^2 \to k^2 - m^2 + \mathrm{i}\eta k_0$$

in the propagator denominator, where η is a small positive number. Therefore,

$$\Delta_\mathrm{R}(x-y) = \int \frac{\mathrm{d}^4k}{(2\pi)^4} \frac{\mathrm{e}^{-\mathrm{i}k\cdot(x-y)}}{k^2 - m^2 + \mathrm{i}\eta k_0} \,. \tag{6.29}$$

Now the poles of the integrand in (6.29) are $k_0 = \pm\omega_k - \mathrm{i}\eta/2$. From (6.6) and (6.29) we have

$$\phi_\mathrm{R}(x) = -\frac{g}{(2\pi)^4} \int \mathrm{d}^4k \frac{\mathrm{e}^{-\mathrm{i}k\cdot x}}{k^2 - m^2 + \mathrm{i}\eta k_0} \int \mathrm{d}y_0 \mathrm{e}^{\mathrm{i}k_0 y_0} \int \mathrm{d}^3\boldsymbol{y}\delta^{(3)}(\boldsymbol{y})\mathrm{e}^{-\mathrm{i}\boldsymbol{k}\cdot\boldsymbol{y}} \,. \tag{6.30}$$

First in (6.30) we shall integrate over y_0, then over $\boldsymbol{y}$ and finally over k_0; this gives

$$\phi_\mathrm{R}(x) = \frac{g}{(2\pi)^3} \int \mathrm{d}^3\boldsymbol{k} \frac{\mathrm{e}^{\mathrm{i}\boldsymbol{k}\cdot\boldsymbol{x}}}{\boldsymbol{k}^2 + m^2} \,. \tag{6.31}$$

In order to compute this three-dimensional momentum integral we introduce spherical coordinates; also we take $\boldsymbol{x} = r\boldsymbol{e}_z$. The angular integrations give (in one integral use the change $k' = -k$)

$$\phi_\mathrm{R}(x) = -\frac{g}{(2\pi)^2 \mathrm{i}r} \int_{-\infty}^{\infty} \frac{k\mathrm{d}k}{k^2 + m^2} \mathrm{e}^{-\mathrm{i}kr} \,. \tag{6.32}$$

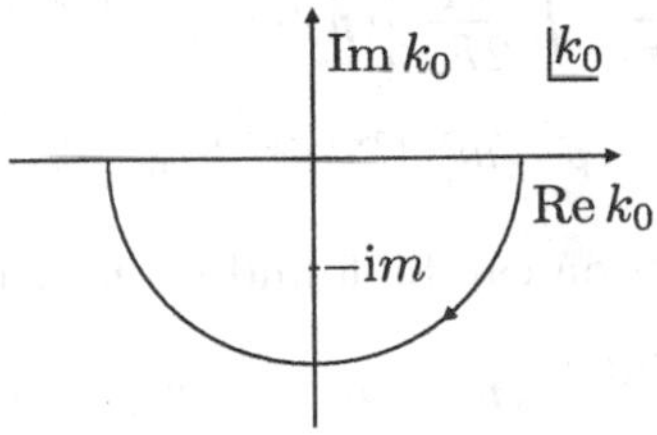

Fig. 6.6.

The integral in (6.32) has the poles at $k_0 = \pm\mathrm{i}m$. The integration contour is given in Fig. 6.6. By applying the Cauchy theorem in (6.32) we obtain:

$$\phi_\mathrm{R}(x) = \frac{g}{4\pi r} \mathrm{e}^{-mr} \,, \tag{6.33}$$

which is the requested result.

6.12 Apply $i\partial\!\!\!/ - m$ on $S(x)$.

6.13 The Fourier transformation of the equation $(i\partial\!\!\!/-m)S(x-y) = \delta^{(4)}(x-y)$ leads to

$$(i\partial\!\!\!/ - m)\frac{1}{(2\pi)^4}\int d^4p\tilde{S}(p)e^{-ip\cdot(x-y)} = \frac{1}{(2\pi)^4}\int d^4pe^{-ip\cdot(x-y)} . \tag{6.34}$$

From (6.34) follows

$$\tilde{S}(p) = \frac{p\!\!\!/ + m}{p^2 - m^2} .$$

Therefore, the Green function is given by

$$S(x-y) = \int \frac{d^4p}{(2\pi)^4}\frac{p\!\!\!/ + m}{p_0^2 - \boldsymbol{p}^2 - m^2}e^{-ip\cdot(x-y)} . \tag{6.35}$$

The poles of the integrand in (6.35) are $p_0 = \pm E_p = \pm\sqrt{\boldsymbol{p}^2 + m^2}$. The propagator is

$$S_F(x-y) = \frac{1}{(2\pi)^4}\int d^3\boldsymbol{p}e^{i\boldsymbol{p}\cdot(\boldsymbol{x}-\boldsymbol{y})}\int_{C_F} dp_0\frac{p_0\gamma^0 + p^i\gamma_i + m}{p_0^2 - E_p^2}e^{-ip_0(x_0-y_0)} , \tag{6.36}$$

where the integration contour C_F is defined in Problem 6.1. Applying the Cauchy theorem we get

$$\begin{aligned}
S_F(x-y) &= -\frac{i}{(2\pi)^3}\int\frac{d^3\boldsymbol{p}}{2E_p}e^{i\boldsymbol{p}\cdot(\boldsymbol{x}-\boldsymbol{y})} \\
&\quad\Big[(E_p\gamma^0 + p_i\gamma^i + m)e^{-iE_p(x_0-y_0)}\theta(x_0-y_0)+ \\
&\quad +(-E_p\gamma^0 + p_i\gamma^i + m)e^{iE_p(x_0-y_0)}\theta(y_0-x_0)\Big] \\
&= -\frac{i}{(2\pi)^3}\int\frac{d^3\boldsymbol{p}}{2E_p}\Big[(p\!\!\!/ + m)e^{-ip\cdot(x-y)}\theta(x_0-y_0)- \\
&\quad -(p\!\!\!/ - m)e^{ip\cdot(x-y)}\theta(y_0-x_0)\Big] .
\end{aligned} \tag{6.37}$$

The advanced Green function can be found in the same way. The result is

$$\begin{aligned}
S_A(x-y) &= \frac{i}{(2\pi)^3}\int\frac{d^3\boldsymbol{p}}{2E_p}e^{i\boldsymbol{p}\cdot(\boldsymbol{x}-\boldsymbol{y})}\Big[(E_p\gamma^0 + p_i\gamma^i + m)e^{-iE_p(x_0-y_0)}- \\
&\quad -(-E_p\gamma^0 + p_i\gamma^i + m)e^{iE_p(x_0-y_0)}\Big]\theta(y_0-x_0) .
\end{aligned} \tag{6.38}$$

For simplicity we take $y = 0$ in (6.37) and (6.38). We have

$$\begin{aligned}
S_F - S_A &= -\frac{i}{(2\pi)^3}\int\frac{d^3\boldsymbol{p}}{2E_p}e^{i(\boldsymbol{p}\cdot\boldsymbol{x}-E_px_0)}(E_p\gamma^0 + p_i\gamma^i + m)(\theta(x_0) + \theta(-x_0)) \\
&= -\frac{i}{(2\pi)^3}\int\frac{d^3\boldsymbol{p}}{2E_p}e^{i(\boldsymbol{p}\cdot\boldsymbol{x}-E_px_0)}(E_p\gamma^0 + p_i\gamma^i + m) .
\end{aligned} \tag{6.39}$$

Thus,

$$S_{\mathrm{F}} - S_{\mathrm{A}} = -\frac{\mathrm{i}}{(2\pi)^3}\int \frac{\mathrm{d}^3\boldsymbol{p}}{2E_p}(E_p\gamma^0 + p^i\gamma_i + m)\mathrm{e}^{-\mathrm{i}p\cdot x}\,. \tag{6.40}$$

By applying $\mathrm{i}\partial\!\!\!/ - m$ on (6.40) we get $(\mathrm{i}\partial\!\!\!/ - m)(S_{\mathrm{F}} - S_{\mathrm{A}}) = 0$, since

$$(p\!\!\!/ + m)(p\!\!\!/ - m) = p^2 - m^2 = 0.$$

6.14 The integration along the curve C_{F} is equivalent to the integration along the real p_0–axis if we make the replacement $p^2 - m^2 \to p^2 - m^2 + \mathrm{i}\epsilon$, where ϵ is a small positive number in the propagator denominator. The simple poles are $p_0 = \pm E_p \mp \mathrm{i}\epsilon$. So we get

$$\psi(x) = \frac{g}{(2\pi)^4}\int \mathrm{d}^4y \int \mathrm{d}p_0 \int \mathrm{d}^3\boldsymbol{p}\frac{p\!\!\!/ + m}{p^2 - m^2 + \mathrm{i}\epsilon}\mathrm{e}^{-\mathrm{i}p\cdot(x-y)}\delta(y_0)\mathrm{e}^{\mathrm{i}\boldsymbol{q}\cdot\boldsymbol{y}}\begin{pmatrix}1\\0\\0\\0\end{pmatrix}.$$

After the integration over the variables y_0 and $\boldsymbol{y}$ we get

$$\psi(x) = \frac{g}{2\pi}\int \mathrm{d}p_0\mathrm{d}^3\boldsymbol{p}\frac{p\!\!\!/ + m}{p^2 - m^2 + \mathrm{i}\epsilon}\mathrm{e}^{-\mathrm{i}(p_0x_0 - \boldsymbol{p}\cdot\boldsymbol{x})}\delta^{(3)}(\boldsymbol{p} - \boldsymbol{q})\begin{pmatrix}1\\0\\0\\0\end{pmatrix}. \tag{6.41}$$

Integration over the momentum $\boldsymbol{p}$ is simple and it gives

$$\psi(x) = \frac{g}{2\pi}\mathrm{e}^{\mathrm{i}\boldsymbol{q}\cdot\boldsymbol{x}}\int_{-\infty}^{\infty}\mathrm{d}p_0\frac{p_0\gamma_0 - \boldsymbol{q}\cdot\boldsymbol{\gamma} + m}{p_0^2 - \boldsymbol{q}^2 - m^2 + \mathrm{i}\epsilon}\mathrm{e}^{-\mathrm{i}p_0x_0}\begin{pmatrix}1\\0\\0\\0\end{pmatrix}. \tag{6.42}$$

Employing the Cauchy theorem we find that

$$\psi(x) = -\frac{\mathrm{i}g}{2E_q}\mathrm{e}^{\mathrm{i}\boldsymbol{q}\cdot\boldsymbol{x}}\left[(-E_q\gamma_0 - \boldsymbol{q}\cdot\boldsymbol{\gamma} + m)\mathrm{e}^{\mathrm{i}E_qx_0}\theta(-x_0)\right.$$
$$\left. + (E_q\gamma_0 - \boldsymbol{q}\cdot\boldsymbol{\gamma} + m)\mathrm{e}^{-\mathrm{i}E_qx_0}\theta(x_0)\right]\begin{pmatrix}1\\0\\0\\0\end{pmatrix}, \tag{6.43}$$

which finally gives:

$$\psi(x) = -\frac{\mathrm{i}g}{2E_q}\mathrm{e}^{\mathrm{i}\boldsymbol{q}\cdot\boldsymbol{x}}$$
$$\times\left[\mathrm{e}^{\mathrm{i}E_qx_0}\begin{pmatrix}-E_q + m\\0\\q_3\\q_1 + \mathrm{i}q_2\end{pmatrix}\theta(-x_0) + \mathrm{e}^{-\mathrm{i}E_qx_0}\begin{pmatrix}E_q + m\\0\\q_3\\q_1 + \mathrm{i}q_2\end{pmatrix}\theta(x_0)\right]. \tag{6.44}$$

6.15 The equation for the free massive vector field A_μ is given by

$$(g^{\rho\sigma}\Box - \partial^\rho\partial^\sigma + m^2 g^{\rho\sigma})A_\sigma = 0 \, . \tag{6.45}$$

The Green function (it is in fact the inverse kinetic operator) is defined by

$$(g^{\rho\sigma}\Box - \partial^\rho\partial^\sigma + m^2 g^{\rho\sigma})_x G_{\sigma\nu}(x-y) = \delta^{(4)}(x-y)\delta^\rho_\nu \, . \tag{6.46}$$

If we introduce

$$G_{\sigma\nu} = \frac{1}{(2\pi)^4}\int \mathrm{d}^4 k \mathrm{e}^{-\mathrm{i}k\cdot(x-y)}\tilde{G}_{\sigma\nu}(k) \, ,$$

in (6.46), we get

$$(-k^2 g^{\rho\sigma} + k^\rho k^\sigma + m^2 g^{\rho\sigma})\tilde{G}_{\sigma\nu} = \delta^\rho_\nu \, . \tag{6.47}$$

We shall assume that the solution of (6.47) has the form $\tilde{G}_{\rho\sigma} = Ak^2 g_{\rho\sigma} + Bk_\rho k_\sigma$, where A and B are scalars, i.e. they depend on k^2 and m^2. Inserting the solution into (6.47), after comparing of the appropriate coefficients, we get

$$A = \frac{1}{-k^4 + k^2 m^2} \, , \qquad B = -\frac{1}{m^2(m^2 - k^2)} \, .$$

The final result takes the following form

$$\tilde{G}_{\mu\nu} = \frac{1}{k^2 - m^2}\left(-g_{\mu\nu} + \frac{k_\mu k_\nu}{m^2}\right) \, . \tag{6.48}$$

6.16 Use the same procedure as in the previous problem. The result is

$$\tilde{G}_{\mu\nu} = -\frac{g_{\mu\nu}}{k^2} + \frac{1+\lambda}{\lambda k^4}k_\mu k_\nu \, .$$

7

Canonical quantization of the scalar field

7.1 Starting from the expressions for scalar field ϕ and its canonical momentum $\pi = \dot{\phi}$,

$$\phi = \int \frac{d^3\boldsymbol{k}}{\sqrt{2(2\pi)^3\omega_k}} \left[a(\boldsymbol{k})e^{-ik\cdot x} + a^\dagger(\boldsymbol{k})e^{ik\cdot x}\right] ,$$

$$\dot{\phi} = i \int \frac{d^3\boldsymbol{k}}{\sqrt{(2\pi)^3 2\omega_k}} \omega_k \left[-a(\boldsymbol{k})e^{-ik\cdot x} + a^\dagger(\boldsymbol{k})e^{ik\cdot x}\right] ,$$

we have

$$\int d^3\boldsymbol{x}\phi(x)e^{-i\boldsymbol{k}'\cdot\boldsymbol{x}} = \frac{(2\pi)^{3/2}}{\sqrt{2\omega_{k'}}} \left[a(\boldsymbol{k}')e^{-i\omega_{k'}t} + a^\dagger(-\boldsymbol{k}')e^{i\omega_{k'}t}\right] , \tag{7.1}$$

$$\int d^3\boldsymbol{x}\dot{\phi}(x)e^{-i\boldsymbol{k}'\cdot\boldsymbol{x}} = i(2\pi)^{3/2}\sqrt{\frac{\omega_{k'}}{2}} \left[a^\dagger(-\boldsymbol{k}')e^{i\omega_{k'}t} - a(\boldsymbol{k}')e^{-i\omega_{k'}t}\right] . \tag{7.2}$$

From (7.1) and (7.2) it follows that

$$a(\boldsymbol{k}) = \frac{1}{(2\pi)^{3/2}} \frac{1}{\sqrt{2\omega_k}} \int d^3\boldsymbol{x}e^{ik\cdot x} \left[\omega_k\phi(x) + i\dot{\phi}(x)\right] , \tag{7.3}$$

$$a^\dagger(\boldsymbol{k}) = \frac{1}{(2\pi)^{3/2}} \frac{1}{\sqrt{2\omega_k}} \int d^3\boldsymbol{x}e^{-ik\cdot x} \left[\omega_k\phi(x) - i\dot{\phi}(x)\right] . \tag{7.4}$$

By using the expressions (7.3) and (7.4), we find:

$$\begin{aligned}
[a(\boldsymbol{k}), a^\dagger(\boldsymbol{k}')] &= \frac{i}{2(2\pi)^3} \frac{1}{\sqrt{\omega_k\omega_{k'}}} \int d^3\boldsymbol{x}d^3\boldsymbol{y}e^{i(k\cdot x - k'\cdot y)} \Big(-\omega_k[\phi(x), \dot{\phi}(y)] + \\
&\qquad + \omega_{k'}[\dot{\phi}(x), \phi(y)]\Big) \\
&= \frac{1}{2(2\pi)^3} \frac{1}{\sqrt{\omega_k\omega_{k'}}} \int d^3\boldsymbol{x}e^{i(\omega_k - \omega_{k'})t + i(\boldsymbol{k}' - \boldsymbol{k})\cdot\boldsymbol{x}} (\omega_k + \omega_{k'}) \\
&= \delta^{(3)}(\boldsymbol{k} - \boldsymbol{k}') .
\end{aligned} \tag{7.5}$$

In the previous formula, we used the equal–time commutation relations for real scalar field (7.C) i.e. we took[1] $x^0 = y^0$. We can do this because the creation and annihilation operators are time independent. This can be proven directly:

$$\frac{\mathrm{d}a(\boldsymbol{k})}{\mathrm{d}t} = \frac{1}{(2\pi)^{3/2}}\frac{1}{\sqrt{2\omega_k}}\int \mathrm{d}^3\boldsymbol{x}\mathrm{e}^{\mathrm{i}k\cdot x}\left[\mathrm{i}\omega_k^2\phi + \mathrm{i}\nabla^2\phi - \mathrm{i}m^2\phi\right] .$$

After two partial integrations in the second term we get

$$\frac{\mathrm{d}a(\boldsymbol{k})}{\mathrm{d}t} = \frac{\mathrm{i}}{(2\pi)^{3/2}}\frac{1}{\sqrt{2\omega_k}}\int \mathrm{d}^3\boldsymbol{x}\mathrm{e}^{\mathrm{i}k\cdot x}\left[\omega_k^2 - \boldsymbol{k}^2 - m^2\right]\phi .$$

The dispersion relation, $\omega_k^2 = m^2 + \boldsymbol{k}^2$ gives $\mathrm{d}a(\boldsymbol{k})/\mathrm{d}t = 0$. It is clear that $a^\dagger(\boldsymbol{k})$ is also time independent.
Similarly, we can prove that:

$$[a(\boldsymbol{k}), a(\boldsymbol{k}')] = [a^\dagger(\boldsymbol{k}), a^\dagger(\boldsymbol{k}')] = 0 .$$

7.2 In this problem, $\phi(x)$ is a classical field, so that $a(\boldsymbol{k})$ and $a^\dagger(\boldsymbol{k})$ are the coefficients rather then operators. We can calculate them from expressions (7.3) and (7.4) inserting $\phi(t=0,\boldsymbol{x}) = 0$ and $\dot{\phi}(t=0,\boldsymbol{x}) = c$:

$$\begin{aligned} a(\boldsymbol{k}) &= \frac{1}{(2\pi)^{3/2}}\frac{1}{\sqrt{2\omega_k}}\int \mathrm{d}^3\boldsymbol{x}\mathrm{e}^{-\mathrm{i}\boldsymbol{k}\cdot\boldsymbol{x}}\mathrm{i}c \\ &= \frac{\mathrm{i}c}{\sqrt{2m}}(2\pi)^{3/2}\delta^{(3)}(\boldsymbol{k}) . \end{aligned}$$

Then, the scalar field is

$$\phi(t,\boldsymbol{x}) = \frac{c}{m}\sin(mt) .$$

Generally, if we know a field and its normal derivative on some space–like surface σ, then the field at an arbitrary point is given by

$$\phi(y) = \int_\sigma [\phi(x)\partial_\mu^x \Delta(x-y) - \Delta(y-x)\partial_\mu\phi(x)]\mathrm{d}\Sigma^\mu .$$

Solve this problem using the previous theorem.

7.3 The results are:

$$:H: = \int \mathrm{d}^3\boldsymbol{k}\omega_k\left[a^\dagger(\boldsymbol{k})a(\boldsymbol{k}) + b^\dagger(\boldsymbol{k})b(\boldsymbol{k})\right] , \tag{7.6}$$

$$:Q: = q\int \mathrm{d}^3\boldsymbol{k}\left[a^\dagger(\boldsymbol{k})a(\boldsymbol{k}) - b^\dagger(\boldsymbol{k})b(\boldsymbol{k})\right] , \tag{7.7}$$

$$:\boldsymbol{P}: = \int \mathrm{d}^3\boldsymbol{k}\boldsymbol{k}\left[a^\dagger(\boldsymbol{k})a(\boldsymbol{k}) + b^\dagger(\boldsymbol{k})b(\boldsymbol{k})\right] . \tag{7.8}$$

[1] This will be done in the forthcoming problems, too.

7.4 $(u_p, u_k) = \delta^{(3)}(\boldsymbol{k}-\boldsymbol{p})$, $(u_p, u_k^*) = 0$.

7.5 From (2.9), we have

$$
\begin{aligned}
\langle 0|\, H\, |0\rangle &= \frac{1}{2}\int \mathrm{d}^3\boldsymbol{k}\omega_k \langle 0|\, (a^\dagger(\boldsymbol{k})a(\boldsymbol{k}) + a(\boldsymbol{k})a^\dagger(\boldsymbol{k}))\, |0\rangle \\
&= \frac{1}{2}\int \mathrm{d}^3\boldsymbol{k}\omega_k \langle 0|\, a(\boldsymbol{k})a^\dagger(\boldsymbol{k})\, |0\rangle \\
&= \frac{1}{2}\int \mathrm{d}^3\boldsymbol{k}\omega_k (\delta^{(3)}(0) - \langle 0|\, a^\dagger(\boldsymbol{k})a(\boldsymbol{k})\, |0\rangle) \\
&= \frac{1}{2}\delta^{(3)}(0)\int \mathrm{d}^3\boldsymbol{k}\sqrt{\boldsymbol{k}^2+m^2} \\
&= 2\pi\delta^{(3)}(0)\int_0^\infty \mathrm{d}k k^2\sqrt{k^2+m^2}\,.
\end{aligned}
$$

By change of variable $k = m\sqrt{t}$, the last integral becomes Euler's beta function

$$
\langle 0|\, H\, |0\rangle = \pi m^4 \delta^{(3)}(0) B(\frac{3}{2}, -2) = -\frac{\pi m^4}{4}\delta^{(3)}(0)\Gamma(-2)\,.
$$

7.6 Use the formulae from Problem 7.3 and the commutation relations (7.D).

(a) Direct calculation yields

$$
\begin{aligned}
[P^\mu, \phi] &= \frac{1}{(2\pi)^{3/2}}\int \frac{\mathrm{d}^3\boldsymbol{k}\mathrm{d}^3\boldsymbol{k}'}{\sqrt{2\omega_{k'}}} k^\mu \left[a^\dagger(\boldsymbol{k})a(\boldsymbol{k}), a(\boldsymbol{k}')\mathrm{e}^{-\mathrm{i}k'\cdot x} + a^\dagger(\boldsymbol{k}')\mathrm{e}^{\mathrm{i}k'\cdot x}\right] \\
&= \frac{1}{(2\pi)^{3/2}}\int \frac{\mathrm{d}^3\boldsymbol{k}}{\sqrt{2\omega_k}} k^\mu \left(-a(\boldsymbol{k})\mathrm{e}^{-\mathrm{i}k\cdot x} + a^\dagger(\boldsymbol{k})\mathrm{e}^{\mathrm{i}k\cdot x}\right) \\
&= -\mathrm{i}\partial^\mu\phi\,.
\end{aligned}
\tag{7.9}
$$

The same result can be obtained if we start from the transformation law of the field ϕ under translations (see Problem 7.20):

$$
\phi(x+\epsilon) = \mathrm{e}^{\mathrm{i}\epsilon\cdot P}\phi(x)\mathrm{e}^{-\mathrm{i}\epsilon\cdot P} = \phi(x) + \mathrm{i}\epsilon^\mu[P_\mu, \phi(x)] + o(\epsilon^2)\,. \tag{7.10}
$$

On the other hand, we have

$$
\phi(x+\epsilon) = \phi(x) + \epsilon^\mu\partial_\mu\phi + o(\epsilon^2)\,. \tag{7.11}
$$

From (7.11) and (7.10) the result (7.9) comes.

(b) First, we calculate the commutator $[P^\mu, \phi^n(x)]$:

$$
\begin{aligned}
[P^\mu, \phi^n(x)] &= \sum_{k=1}^{n} \phi^{k-1}[P^\mu, \phi]\phi^{n-k} \\
&= \sum_{k=1}^{n} \phi^{k-1}(-\mathrm{i}\partial^\mu\phi)\phi^{n-k} \\
&= -\mathrm{i}\partial^\mu\phi^n\,.
\end{aligned}
$$

In the same way one can prove that

$$[P^\mu, \pi^n(x)] = -\mathrm{i}\partial^\mu \pi^n \ .$$

As a consequence,

$$[P_\mu, \phi^n(x)\pi^m(x)] = -\mathrm{i}\partial_\mu(\phi^n(x)\pi^m(x)) \ .$$

An arbitrary analytical function $F(\phi, \pi)$ can be expanded in series as

$$F(\phi, \pi) = \sum_{nm} C_{nm}\phi^n \pi^m \ .$$

Then

$$[P_\mu, F(\phi, \pi)] = -\mathrm{i}\partial_\mu F \ .$$

(c) $[H, a^\dagger(\boldsymbol{k})a(\boldsymbol{q})] = (\omega_k - \omega_q)a^\dagger(\boldsymbol{k})a(\boldsymbol{q})$.
(d) $[Q, P^\mu] = 0$.
(e) $[H, N] = 0$.
(f) $\int \mathrm{d}^3\boldsymbol{x}[H, \phi(x)]\mathrm{e}^{-\mathrm{i}\boldsymbol{p}\cdot\boldsymbol{x}} = (2\pi)^{3/2}\sqrt{\frac{\omega_p}{2}}\left(-a(\boldsymbol{p})\mathrm{e}^{-\mathrm{i}\omega_p t} + a^\dagger(-\boldsymbol{p})\mathrm{e}^{\mathrm{i}\omega_p t}\right)$

7.7 From the Baker–Hausdorff relation follows

$$\mathrm{e}^{\mathrm{i}Q}\phi\mathrm{e}^{-\mathrm{i}Q} = \phi + \mathrm{i}[Q, \phi] + \frac{\mathrm{i}^2}{2!}[Q, [Q, \phi]] + \dots \ . \tag{7.12}$$

The first commutator in the previous expansion is given by

$$\begin{aligned}[Q, \phi] &= \mathrm{i}q \int \mathrm{d}^3\boldsymbol{y}[\phi^\dagger(y)\pi^\dagger(y) - \phi(y)\pi(y), \phi(x)] \\ &= -q \int \mathrm{d}^3\boldsymbol{y}\delta^{(3)}(\boldsymbol{x} - \boldsymbol{y})\phi(y) = -q\phi(x) \ .\end{aligned}$$

Then

$$[Q, [Q, \phi]] = (-q)^2\phi \ , \quad [Q, [Q, [Q, \phi]]] = (-q)^3\phi \ , \dots \tag{7.13}$$

Finally,

$$\mathrm{e}^{\mathrm{i}Q}\phi\mathrm{e}^{-\mathrm{i}Q} = \left(1 - \mathrm{i}q + \frac{(-\mathrm{i}q)^2}{2} + \dots\right)\phi = \mathrm{e}^{-\mathrm{i}q}\phi \ . \tag{7.14}$$

7.8 The angular momentum of a scalar field has the form

$$M^{\mu\nu} = \int \mathrm{d}^3\boldsymbol{x}(x^\mu T^{0\nu} - x^\nu T^{0\mu}) \ .$$

(a) By inserting the previous formula in the commutator, we have

$$[M_{\mu\nu}, \phi(x)] = \int \mathrm{d}^3\boldsymbol{y}[y_\mu(\dot\phi\partial_\nu\phi - g_{0\nu}\mathcal{L}) - y_\nu(\dot\phi\partial_\mu\phi - g_{0\mu}\mathcal{L}), \phi(x)] \ . \tag{7.15}$$

The following equal–time commutators can be easily evaluated:

$$[\mathcal{L}(y), \phi(x)] = -\mathrm{i}\delta^{(3)}(\boldsymbol{x}-\boldsymbol{y})\pi(\boldsymbol{y}) ,$$
$$[\pi(y)\partial_\mu\phi(y), \phi(x)] = -\mathrm{i}\partial_\mu\phi\delta^{(3)}(\boldsymbol{x}-\boldsymbol{y}) - \mathrm{i}\delta_{\mu 0}\pi(y)\delta^{(3)}(\boldsymbol{x}-\boldsymbol{y}) .$$

By substituting these expressions in (7.15) and performing integration, we get

$$[M_{\mu\nu}, \phi(x)] = \mathrm{i}(x_\nu\partial_\mu - x_\mu\partial_\nu)\phi(x) . \tag{7.16}$$

The same result can be obtained if we start from the transformation law for the field $\phi(x)$ under Lorentz transformations,

$$\mathrm{e}^{\frac{\mathrm{i}}{2}\omega_{\mu\nu}M^{\mu\nu}}\phi(x)\mathrm{e}^{-\frac{\mathrm{i}}{2}\omega_{\mu\nu}M^{\mu\nu}} = \phi(\Lambda^{-1}(\omega)x) .$$

(b) We first calculate the commutator $[M_{\mu\nu}, P_0]$:

$$\begin{aligned}
[M_{\mu\nu}, P_0] &= \int \mathrm{d}^3\boldsymbol{x}[x_\mu T_{0\nu} - x_\nu T_{0\mu}, P_0] \\
&= \int \mathrm{d}^3\boldsymbol{x}\,(x_\mu[T_{0\nu}, P_0] - x_\nu[T_{0\mu}, P_0]) \\
&= \mathrm{i}\int \mathrm{d}^3\boldsymbol{x}\,(x_\mu\partial_0 T_{0\nu} - x_\nu\partial_0 T_{0\mu}) \\
&= \mathrm{i}\int \mathrm{d}^3\boldsymbol{x}\,(-x_\mu\partial_i T^i{}_\nu + x_\nu\partial_i T^i{}_\mu) \\
&= \mathrm{i}\int \mathrm{d}^3\boldsymbol{x}\,(g_{\mu i}T^i{}_\nu - g_{i\nu}T^i{}_\mu) \\
&= \mathrm{i}\int \mathrm{d}^3\boldsymbol{x}\,(T_{\mu\nu} - g_{\mu 0}T^0{}_\nu - T_{\nu\mu} + g_{0\nu}T^0{}_\mu) \\
&= -\mathrm{i}(g_{\mu 0}P_\nu - g_{\nu 0}P_\mu) .
\end{aligned} \tag{7.17}$$

In (7.17), we used the results of Problem 7.6 (b), the continuity equation $\partial_\mu T^{\mu\nu} = 0$ and integrated by parts. In the case $\lambda = i$ we can use of a partial integration. The result is $[M_{\mu\nu}, P_i] = -\mathrm{i}(g_{i\mu}P_\nu - g_{i\nu}P_\mu)$. Thus,

$$[M_{\mu\nu}, P_\lambda] = \mathrm{i}(g_{\lambda\nu}P_\mu - g_{\lambda\mu}P_\nu) . \tag{7.18}$$

(c) Let us calculate firstly the commutator $[M_{ij}, M_{kl}]$.

$$\begin{aligned}
[M_{ij}, M_{kl}] &= \int \mathrm{d}^3\boldsymbol{x}\mathrm{d}^3\boldsymbol{y} \Big[x_i\dot{\phi}(x)\partial_j\phi(x) - x_j\dot{\phi}(x)\partial_i\phi(x), \\
&\qquad y_k\dot{\phi}(y)\partial_l\phi(y) - y_l\dot{\phi}(y)\partial_k\phi(y)\Big] \\
&= \int \mathrm{d}^3\boldsymbol{x}\mathrm{d}^3\boldsymbol{y} \Big(x_iy_k[\dot{\phi}(x)\partial_j\phi(x), \dot{\phi}(y)\partial_l\phi(y)] - \\
&\qquad -x_iy_l[\dot{\phi}(x)\partial_j\phi(x), \dot{\phi}(y)\partial_k\phi(y)] \\
&\qquad -x_jy_k[\dot{\phi}(x)\partial_i\phi(x), \dot{\phi}(y)\partial_l\phi(y)] \\
&\qquad +x_jy_l[\dot{\phi}(x)\partial_i\phi(x), \dot{\phi}(y)\partial_k\phi(y)]\Big) \,. \qquad (7.19)
\end{aligned}$$

Applying the equal–time commutation relations, we obtain[2]

$$\begin{aligned}
[M_{ij}, M_{kl}] &= \mathrm{i}\int \mathrm{d}^3\boldsymbol{x}\mathrm{d}^3\boldsymbol{y} \Big[x_iy_k\Big(\dot{\phi}(x)\partial_l\phi(y)\partial_j^x \\
&\quad - \dot{\phi}(y)\partial_j\phi(x)\partial_l^y\Big)\delta^{(3)}(\boldsymbol{x}-\boldsymbol{y}) \\
&\quad - x_iy_l\Big(\dot{\phi}(x)\partial_k\phi(y)\partial_j^x - \dot{\phi}(y)\partial_j\phi(x)\partial_k^y\Big)\delta^{(3)}(\boldsymbol{x}-\boldsymbol{y}) \\
&\quad - x_jy_k\Big(\dot{\phi}(x)\partial_l\phi(y)\partial_i^x - \dot{\phi}(y)\partial_i\phi(x)\partial_l^y\Big)\delta^{(3)}(\boldsymbol{x}-\boldsymbol{y}) \\
&\quad + x_jy_l\Big(\dot{\phi}(x)\partial_k\phi(y)\partial_i^x - \dot{\phi}(y)\partial_i\phi(x)\partial_k^y\Big)\delta^{(3)}(\boldsymbol{x}-\boldsymbol{y})\Big] \,.
\end{aligned}$$

If we use the relation

$$\partial_m^x\delta^{(3)}(\boldsymbol{x}-\boldsymbol{y}) = -\partial_m^y\delta^{(3)}(\boldsymbol{x}-\boldsymbol{y})$$

we obtain

$$\begin{aligned}
&[M_{ij}, M_{kl}] = -\mathrm{i}\int \mathrm{d}^3\boldsymbol{x}\mathrm{d}^3\boldsymbol{y} \\
&\Big[x_iy_k\Big(\dot{\phi}(x)\partial_l\phi(y)\partial_j^y\delta^{(3)}(\boldsymbol{x}-\boldsymbol{y}) - \dot{\phi}(y)\partial_j\phi(x)\partial_l^x\delta^{(3)}(\boldsymbol{x}-\boldsymbol{y})\Big) \\
&-x_iy_l\Big(\dot{\phi}(x)\partial_k\phi(y)\partial_j^y\delta^{(3)}(\boldsymbol{x}-\boldsymbol{y}) - \dot{\phi}(y)\partial_j\phi(x)\partial_k^x\delta^{(3)}(\boldsymbol{x}-\boldsymbol{y})\Big) \\
&-x_jy_k\Big(\dot{\phi}(x)\partial_l\phi(y)\partial_i^y\delta^{(3)}(\boldsymbol{x}-\boldsymbol{y}) - \dot{\phi}(y)\partial_i\phi(x)\partial_l^x\delta^{(3)}(\boldsymbol{x}-\boldsymbol{y})\Big) \\
&+x_jy_l\Big(\dot{\phi}(x)\partial_k\phi(y)\partial_i^y\delta^{(3)}(\boldsymbol{x}-\boldsymbol{y}) - \dot{\phi}(y)\partial_i\phi(x)\partial_k^x\delta^{(3)}(\boldsymbol{x}-\boldsymbol{y})\Big)\Big] \,.
\end{aligned}$$

By performing partial integrations in the last expression, we obtain

[2] We have used the following notation:

$$\partial_m^x = \frac{\partial}{\partial x^m};\ \partial_x^m = \frac{\partial}{\partial x_m} \,.$$

$$\begin{aligned}
[M_{ij}, M_{kl}] &= -\mathrm{i}\int \mathrm{d}^3\boldsymbol{x}\Big[g_{jk}(x_l\dot{\phi}(x)\partial_i\phi(x) - x_i\dot{\phi}(x)\partial_l\phi(x)) \\
&\quad + g_{il}(x_k\dot{\phi}(x)\partial_j\phi(x) - x_j\dot{\phi}(x)\partial_k\phi(x)) \\
&\quad + g_{ik}(x_j\dot{\phi}(x)\partial_l\phi(x) - x_l\dot{\phi}(x)\partial_j\phi(x)) \\
&\quad + g_{jl}(x_i\dot{\phi}(x)\partial_k\phi(x) - x_k\dot{\phi}(x)\partial_i\phi(x))\Big] \\
&= \mathrm{i}(g_{jk}M_{il} + g_{li}M_{jk} - g_{ik}M_{jl} - g_{jl}M_{ik}) \,. \qquad (7.20)
\end{aligned}$$

The next two commutators $[M_{ij}, M_{0k}]$, $[M_{0j}, M_{0k}]$ can be evaluated in the same way. Do this explicitly, please.

7.10

(a) The commutator is given by

$$\begin{aligned}
[Q^a, Q^b] &= -\frac{1}{4}\int \mathrm{d}^3\boldsymbol{x}\mathrm{d}^3\boldsymbol{y}\tau^a_{ij}\tau^b_{mn} \\
&\quad \times \left[\dot{\phi}^\dagger_i(x)\phi_j(x) - \phi^\dagger_i(x)\dot{\phi}_j(x), \dot{\phi}^\dagger_m(y)\phi_n(y) - \phi^\dagger_m(y)\dot{\phi}_n(y)\right] \,.
\end{aligned}$$

Recall that as the charges are time–independent we can work with the equal–time commutators and we have

$$[Q^a, Q^b] = -\frac{\mathrm{i}}{4}\int \mathrm{d}^3\boldsymbol{x}\left(\dot{\phi}^\dagger[\tau^a,\tau^b]\phi - \phi^\dagger[\tau^a,\tau^b]\dot{\phi}\right) \,.$$

By using $[\tau^a, \tau^b] = 2\mathrm{i}\epsilon^{abc}\tau^c$, we get

$$[Q^a, Q^b] = \mathrm{i}\epsilon^{abc}Q^c \,.$$

The second case is similar to the previous one:

$$\begin{aligned}
[Q_i, Q_j] &= \epsilon_{imn}\epsilon_{jpq}\int \mathrm{d}^3\boldsymbol{x}\int \mathrm{d}^3\boldsymbol{y}[\phi_m(x)\dot{\phi}_n(x), \phi_p(y)\dot{\phi}_q(y)] \\
&= \mathrm{i}\int \mathrm{d}^3\boldsymbol{x}(-\epsilon_{imn}\epsilon_{jnq}\phi_m\dot{\phi}_q + \epsilon_{imn}\epsilon_{jpm}\phi_p\dot{\phi}_n) \\
&= \mathrm{i}\int \mathrm{d}^3\boldsymbol{x}(\delta_{ij}\phi_m\dot{\phi}_m - \phi_j\dot{\phi}_i - \delta_{ij}\phi_m\dot{\phi}_m + \phi_i\dot{\phi}_j) \\
&= \mathrm{i}\int \mathrm{d}^3\boldsymbol{x}(\phi_i\dot{\phi}_j - \phi_j\dot{\phi}_i) \\
&= \mathrm{i}\epsilon_{ijk}\epsilon_{kmn}\int \mathrm{d}^3\boldsymbol{x}\phi_m\dot{\phi}_n \\
&= \mathrm{i}\epsilon_{ijk}Q_k \,.
\end{aligned}$$

As in the first part of this problem, we used the equal–time commutation relations and the formula for appropriate product of two three–dimensional ϵ symbols.

(b) The commutator between the charges Q^a and the field ϕ_m can be found similarly:

$$\begin{aligned}
[Q^a, \phi_m(x)] &= -\frac{\mathrm{i}}{2}\int \mathrm{d}^3\boldsymbol{y}\tau^a_{ij}[\dot{\phi}^\dagger_i(y)\phi_j(y) - \phi^\dagger_i(y)\dot{\phi}_j(y), \phi_m(x)] \\
&= -\frac{\mathrm{i}}{2}\tau^a_{ij}\int \mathrm{d}^3\boldsymbol{y}[\dot{\phi}^\dagger_i(y), \phi_m(x)]\phi_j(y) \\
&= -\frac{1}{2}\tau^a_{ij}\int \mathrm{d}^3\boldsymbol{y}\delta^{(3)}(\boldsymbol{x}-\boldsymbol{y})\delta_{im}\phi_j(y) \\
&= -\frac{1}{2}\tau^a_{mj}\phi_j(x)\ .
\end{aligned}$$

In the same way, we find:

$$[Q^a, \phi^\dagger_m(x)] = \frac{1}{2}\tau^a_{im}\phi^\dagger_i\ .$$

The previous two results can be rewritten in the form

$$[\theta^a Q^a, \phi_m(x)] = \mathrm{i}\delta_0\phi_m(x)\ ,$$

$$[\theta^a Q^a, \phi^\dagger_m(x)] = \mathrm{i}\delta_0\phi^\dagger_m(x)\ .$$

In the case of SO(3) symmetry, the calculation is the same as above. The result is

$$[Q_k, \phi_m(x)] = \mathrm{i}\epsilon_{kmj}\phi_j(x)\ .$$

7.11 The dilatation current is

$$j^\mu = -\phi\partial^\mu\phi - x^\nu\partial^\mu\phi\partial_\nu\phi + \mathcal{L}x^\mu\ .$$

(a) The dilatation generator is

$$D = -\int \mathrm{d}^3\boldsymbol{x}\left(\phi\dot{\phi} + x^i\dot{\phi}\partial_i\phi + \frac{1}{2}x^0(\dot{\phi}^2 - \partial_i\phi\partial^i\phi)\right)\ .$$

(b) The commutator between the generator D and the field $\phi(x)$ is given by

$$\begin{aligned}
[D, \phi(y)] &= -\int \mathrm{d}^3\boldsymbol{x}[\phi(x)\pi(x) + x^i\pi(x)\partial_i\phi(x) \\
&+ \frac{1}{2}x^0\pi^2(x) - \frac{1}{2}x^0\partial_i\phi(x)\partial^i\phi(x), \phi(y)] \\
&= -\int \mathrm{d}^3\boldsymbol{x}\,(\phi(x)[\pi(x),\phi(y)] + x^0\pi(x)[\pi(x),\phi(y)] \\
&+ x^i[\pi(x),\phi(y)]\partial_i\phi(x))\ .
\end{aligned}$$

By using the commutation relations (7.C), we have

$$\rho[D,\phi(y)] = \mathrm{i}\rho(\phi(y) + y^0\pi(y) + y^i\partial_i\phi)$$
$$= \mathrm{i}\rho(\phi(y) + y^\mu\partial_\mu\phi(y)) = \mathrm{i}\delta_0\phi\ .$$

In the same way, we obtain:

$$\rho[D,\pi(x)] = \mathrm{i}\rho(2\pi + x^\mu\partial_\mu\pi) = \mathrm{i}\delta_0\pi\ .$$

(c) By applying the previous result, we easily get

$$\rho[D,\phi^2] = \rho([D,\phi]\phi + \phi[D,\phi])$$
$$= \mathrm{i}((\delta_0\phi)\phi + \phi\delta_0\phi) = \mathrm{i}\delta_0(\phi^2)\ ,$$

and generally

$$\rho[D,\phi^a] = \mathrm{i}\delta_0(\phi^a)\ .$$

Similarly, one can show that

$$\rho[D,\pi^a] = \mathrm{i}\delta_0(\pi^a)\ .$$

An arbitrary analytic function can be expanded in the following form

$$F(\phi,\pi) = \sum_{ab} c_{ab}\phi^a\pi^b\ ,$$

so that

$$\begin{aligned}\rho[D,F] &= \rho\sum_{a,b} c_{ab}[D,\phi^a\pi^b] \\ &= \rho\sum_{a,b} c_{ab}\left([D,\phi^a]\pi^b + \phi^b[D,\pi^b]\right) \\ &= \mathrm{i}\sum_{a,b} c_{ab}\left(\delta_0(\phi^a)\pi^b + \phi^a\delta_0(\pi^b)\right) \\ &= \mathrm{i}\delta_0\left(\sum_{a,b} c_{ab}\phi^a\pi^b\right) \\ &= \mathrm{i}\delta_0 F\ .\end{aligned}$$

(d) We first consider the case $\mu = i$:

$$[D,P^i] = \int \mathrm{d}^3\boldsymbol{x}[D,\pi\partial^i\phi]$$
$$= \int \mathrm{d}^3\boldsymbol{x}\left(\pi[D,\partial^i\phi] + [D,\pi]\partial^i\phi\right)\ .$$

By using part (b) of this problem, we obtain

$$[D, P^i] = \mathrm{i} \int \mathrm{d}^3\boldsymbol{x} \left[(2\pi + x_0\partial_0\pi + x^j\partial_j\pi)\partial^i\phi \right.$$
$$\left. + \pi(2\partial^i\phi + x^0\partial^i\pi + x^j\partial^i\partial_j\phi) \right] . \tag{7.21}$$

The second term in this expression is transformed in the following way

$$\int \mathrm{d}^3\boldsymbol{x} x^0 \partial_k\partial^k\phi\partial^i\phi = -\int \mathrm{d}^3\boldsymbol{x} x^0\partial^k\phi\partial_k\partial^i\phi = -\frac{1}{2}\int \mathrm{d}^3\boldsymbol{x}\partial^i(x^0\partial_k\phi\partial^k\phi) ,$$

where we used the Klein-Gordon equation, $\partial_0\pi = -\partial^i\partial_i\phi$ and then performed a partial integration. Thus, we conclude that the second term can be dropped as a surface term. The expression $\int \mathrm{d}^3\boldsymbol{x}\pi x^0\partial^i\pi$ is also a surface term. Similarly, one can show that

$$\int \mathrm{d}^3\boldsymbol{x} x^j\partial_j\pi\partial^i\phi = -3\int \mathrm{d}^3\boldsymbol{x}\pi\partial^i\phi - \int \mathrm{d}^3\boldsymbol{x} x^j\pi\partial_j\partial^i\phi .$$

Inserting these results in the formula (7.21) we obtain

$$[D, P^i] = \mathrm{i}P^i .$$

The commutator $[D, P^0] = \mathrm{i}P^0$ can be calculated in the same way.

7.12 In the expression for the vacuum expectation value, express the fields ϕ_f in terms of the creation and annihilations operators. From four terms, only one, which is proportional to $\langle 0| a(\boldsymbol{k})a^\dagger(\boldsymbol{k}') |0\rangle = \delta^{(3)}(\boldsymbol{k}-\boldsymbol{k}')$, is nonzero. Then, we have

$$\langle 0| \phi_f(t,\boldsymbol{x})\phi_f(t,\boldsymbol{x}) |0\rangle = \frac{1}{(a^2\pi)^3}\frac{1}{(2\pi)^3}\int \frac{\mathrm{d}^3\boldsymbol{k}}{2\omega_k}\left(\int \mathrm{d}^3\boldsymbol{y}\mathrm{e}^{-(\boldsymbol{x}-\boldsymbol{y})^2/a^2+\mathrm{i}\boldsymbol{k}\cdot(\boldsymbol{x}-\boldsymbol{y})}\right)^2 .$$

Calculating the Poisson integral in this formula, we obtain

$$\begin{aligned}
\langle 0| \phi_f(t)\phi_f(t) |0\rangle &= \frac{1}{2(2\pi)^3}\int \frac{\mathrm{d}^3\boldsymbol{k}}{\omega_k}\mathrm{e}^{-k^2a^2/2} \\
&= \frac{1}{(2\pi)^2}\int_0^\infty \frac{k^2\mathrm{d}k}{\sqrt{k^2+m^2}}\mathrm{e}^{-k^2a^2/2} .
\end{aligned}$$

By the change of variable $k^2 = t$, the last integral becomes

$$\begin{aligned}
\langle 0| \phi_f(t)\phi_f(t) |0\rangle &= \frac{1}{8\pi^2}\int_0^\infty \frac{\sqrt{t}\mathrm{d}t}{\sqrt{t+m^2}}\mathrm{e}^{-ta^2/2} \\
&= \frac{m^2}{16\pi^2}\mathrm{e}^{m^2a^2/4}\left[K_1(\frac{m^2a^2}{4}) - K_0(\frac{m^2a^2}{4})\right] ,
\end{aligned} \tag{7.22}$$

where $K_\nu(x)$ are modified Bessel functions of the third kind (MacDonald functions). Using the asymptotic expansions:

$$K_1(x) = \frac{1}{x} ,$$

$$K_0(x) = -(\log(x/2) + 0,5772)$$

for $x \ll 1$, we obtain in the limit $m \to 0$

$$\langle 0| \phi_f(t)\phi_f(t) |0\rangle = \frac{1}{4\pi^2 a^2} .$$

7.13 Express the operators L_m and L_n in terms of α^μ_m and use the commutation relations.

7.14 After a very simple calculation, we find that

$$\begin{aligned}\langle 0| \{\phi(x), \phi(y)\} |0\rangle = &\frac{\mathrm{i}}{2(2\pi)^2} \frac{1}{|\boldsymbol{x} - \boldsymbol{y}|} \lim_{\epsilon\to 0} \int_0^\infty \mathrm{d}k \mathrm{e}^{-\epsilon k} \Big(\mathrm{e}^{\mathrm{i}k(y^0 - x^0 - |\boldsymbol{x}-\boldsymbol{y}|)} \\ &- \mathrm{e}^{\mathrm{i}k(y^0 - x^0 + |\boldsymbol{x}-\boldsymbol{y}|)} + \mathrm{e}^{\mathrm{i}k(x^0 - y^0 - |\boldsymbol{x}-\boldsymbol{y}|)} \\ &- \mathrm{e}^{\mathrm{i}k(x^0 - y^0 + |\boldsymbol{x}-\boldsymbol{y}|)} \Big) . \end{aligned} \tag{7.23}$$

The integrals in the previous expression are regularized by introducing ϵ as a regularization parameter. At the end we have to take the limit $\epsilon \to 0$. The result is

$$\langle 0| \{\phi(x), \phi(y)\} |0\rangle = -\frac{1}{2\pi^2} \frac{1}{(x-y)^2} .$$

7.15 The vacuum expectation value $\langle \phi(x)\phi(y)\rangle$ is given by

$$\begin{aligned}\langle \phi(x)\phi(y)\rangle &= \langle \phi^+(x)\phi^-(y)\rangle \\ &= \int \frac{\mathrm{d}^3\boldsymbol{k}}{(2\pi)^{3/2}\sqrt{2\omega_k}} \frac{\mathrm{d}^3\boldsymbol{q}}{(2\pi)^{3/2}\sqrt{2\omega_q}} \mathrm{e}^{\mathrm{i}(q\cdot y - k\cdot x)} \delta^{(3)}(\boldsymbol{k} - \boldsymbol{q}) \\ &= \frac{1}{(2\pi)^3} \int \frac{\mathrm{d}^3\boldsymbol{k}}{2\omega_k} \mathrm{e}^{-\mathrm{i}k\cdot(x-y)} ,\end{aligned}$$

where we split the field ϕ into positive and negative energy parts, $\phi = \phi^+ + \phi^-$. If we do the same in the vacuum expectation value of four scalar fields, we see that only two terms remain:

$$\begin{aligned}\langle \phi(x_1)\phi(x_2)\phi(x_3)\phi(x_4)\rangle = &\langle \phi^+(x_1)\phi^+(x_2)\phi^-(x_3)\phi^-(x_4)\rangle \\ &+ \langle \phi^+(x_1)\phi^-(x_2)\phi^+(x_3)\phi^-(x_4)\rangle . \end{aligned} \tag{7.24}$$

The first term in the last expression is

$$\begin{aligned}\langle \phi^+(x_1)\phi^+(x_2)\phi^-(x_3)\phi^-(x_4)\rangle = &\prod_{i=1}^4 \int \frac{\mathrm{d}^3\boldsymbol{q}_i}{(2\pi)^{3/2}\sqrt{2\omega_i}} \left\langle a_1 a_2 a_3^\dagger a_4^\dagger \right\rangle \\ &\times \mathrm{e}^{\mathrm{i}(-q_1\cdot x_1 - q_2\cdot x_2 + q_3\cdot x_3 + q_4\cdot x_4)} ,\end{aligned}$$

where $a_i = a(\boldsymbol{q}_i)$. Using the relation

$$\begin{aligned}\left\langle a_1 a_2 a_3^\dagger a_4^\dagger \right\rangle &= \left\langle a_1(\delta^{(3)}(\boldsymbol{q}_2 - \boldsymbol{q}_3) + a_3^\dagger a_2) a_4^\dagger \right\rangle \\ &= \delta^{(3)}(\boldsymbol{q}_2 - \boldsymbol{q}_3)\delta^{(3)}(\boldsymbol{q}_1 - \boldsymbol{q}_4) + \left\langle a_1 a_3^\dagger(\delta^{(3)}(\boldsymbol{q}_2 - \boldsymbol{q}_4) - a_4^\dagger a_2) \right\rangle \\ &= \delta^{(3)}(\boldsymbol{q}_2 - \boldsymbol{q}_3)\delta^{(3)}(\boldsymbol{q}_1 - \boldsymbol{q}_4) + \delta^{(3)}(\boldsymbol{q}_1 - \boldsymbol{q}_3)\delta^{(3)}(\boldsymbol{q}_2 - \boldsymbol{q}_4) \ ,\end{aligned}$$

we obtain

$$\begin{aligned}\langle \phi^+(x_1)\phi^+(x_2)\phi^-(x_3)\phi^-(x_4)\rangle &= \frac{1}{(2\pi)^6}\int \frac{\mathrm{d}^3\boldsymbol{q_1}}{2\omega_1}\frac{\mathrm{d}^3\boldsymbol{q_2}}{2\omega_2} \mathrm{e}^{-\mathrm{i}q_2\cdot(x_2-x_3)-\mathrm{i}q_1\cdot(x_1-x_4)} \\ &+ \frac{1}{(2\pi)^6}\int \frac{\mathrm{d}^3\boldsymbol{q_1}}{2\omega_1}\frac{\mathrm{d}^3\boldsymbol{q_2}}{2\omega_2} \mathrm{e}^{-\mathrm{i}q_2\cdot(x_2-x_4)-\mathrm{i}q_1\cdot(x_1-x_3)} \\ &= \langle \phi(x_2)\phi(x_3)\rangle \langle \phi(x_1)\phi(x_4)\rangle \\ &+ \langle \phi(x_1)\phi(x_3)\rangle \langle \phi(x_2)\phi(x_4)\rangle \ .\end{aligned}$$

The following result can be derived in the same way:

$$\langle \phi^+(x_1)\phi^-(x_2)\phi^+(x_3)\phi^-(x_4)\rangle = \langle \phi(x_1)\phi(x_2)\rangle \langle \phi(x_3)\phi(x_4)\rangle \ .$$

By adding two last expressions, we get

$$\begin{aligned}\langle \phi(x_1)\phi(x_2)\phi(x_3)\phi(x_4)\rangle &= \langle \phi(x_1)\phi(x_3)\rangle \langle \phi(x_2)\phi(x_4)\rangle \\ &+ \langle \phi(x_1)\phi(x_4)\rangle \langle \phi(x_2)\phi(x_3)\rangle + \\ &+ \langle \phi(x_1)\phi(x_2)\rangle \langle \phi(x_3)\phi(x_4)\rangle \ .\end{aligned}$$

This result is a special case of Wick' s theorem.

7.16 Scalar field in two–dimensional spacetime can be represented as

$$\phi(x) = \int_{-\infty}^{\infty} \frac{\mathrm{d}k}{\sqrt{(2\pi)2\omega_k}} \left[a(k)\mathrm{e}^{-\mathrm{i}k_\mu x^\mu} + a^\dagger(k)\mathrm{e}^{\mathrm{i}k_\mu x^\mu} \right] ,$$

so that

$$\langle \phi(x)\phi(y)\rangle = \frac{1}{4\pi}\int_{-\infty}^{\infty} \frac{\mathrm{d}k}{|k|}\mathrm{e}^{\mathrm{i}|k|(y_0-x_0)-\mathrm{i}k(y-x)} \ . \tag{7.25}$$

If we introduce the notation $y_0 - x_0 = \tau$, $y - x = r$, the previous integral becomes

$$\langle \phi(x)\phi(y)\rangle = \frac{1}{4\pi}\int_0^{\infty} \frac{\mathrm{d}k}{k}\left(\mathrm{e}^{\mathrm{i}k(\tau-r)} + \mathrm{e}^{\mathrm{i}k(\tau+r)}\right) \ . \tag{7.26}$$

Denoting the integral in (7.26) by I and introducing the regularization parameter ϵ, we get:

$$\begin{aligned}\frac{\partial I}{\partial \tau} &= \frac{\mathrm{i}}{4\pi}\mathrm{lim}_{\epsilon\to 0}\int_0^{\infty} \mathrm{d}k e^{-\epsilon k}\left(\mathrm{e}^{\mathrm{i}k(\tau-r)} + \mathrm{e}^{\mathrm{i}k(\tau+r)}\right) \\ &= -\frac{1}{2\pi}\frac{\tau}{\tau^2 - r^2} \ .\end{aligned} \tag{7.27}$$

From (7.27), it follows that

$$\langle\phi(x)\phi(y)\rangle = -\frac{1}{4\pi}\log\frac{\tau^2 - r^2}{\mu^2} = -\frac{1}{4\pi}\log\frac{(x-y)^2}{\mu^2}\ ,$$

where μ is an integration constant which has the dimension of length.

7.17 By taking partial derivative of the expression $\langle 0| T(\phi(x)\phi(y)) |0\rangle$ with respect to x_0, we get:

$$\begin{aligned}\partial_{x_0}\langle 0| T(\phi(x)\phi(y)) |0\rangle &= \delta(x_0 - y_0)\langle 0| [\phi(x),\phi(y)] |0\rangle + \\ &+ \theta(x_0 - y_0)\langle 0| \partial_{x_0}\phi(x)\phi(y) |0\rangle + \theta(y_0 - x_0)\langle 0| \phi(y)\partial_{x_0}\phi(x) |0\rangle\ .\end{aligned}$$

The first term is equal to zero as a consequence of the equal–time commutation relation. By taking second order partial derivative with respect to x_0, we get:

$$\begin{aligned}\partial^2_{x^0}\langle 0| T(\phi(x)\phi(y)) |0\rangle &= \delta(x^0 - y^0)[\pi(x),\phi(y)] \\ &+ \theta(x^0 - y^0)\langle 0| \partial^2_{x^0}\phi(x)\phi(y) |0\rangle + \\ &+ \theta(y^0 - x^0)\langle 0| \phi(y)\partial^2_{x^0}\phi(x) |0\rangle\ .\end{aligned}$$

In the first term, we use the equal–time commutation relation, and finally get the result

$$\begin{aligned}\partial^2_{x^0}\langle 0| T(\phi(x)\phi(y)) |0\rangle &= -\mathrm{i}\delta^{(4)}(x-y) + \\ &+ \theta(x^0 - y^0)\langle 0| \partial^2_{x^0}\phi(x)\phi(y) |0\rangle + \\ &+ \theta(y^0 - x^0)\langle 0| \phi(y)\partial^2_{x^0}\phi(x) |0\rangle\ ,\end{aligned}$$

which implies

$$\begin{aligned}(\Box_x + m^2)\langle 0| T(\phi(x)\phi(y)) |0\rangle &= -\mathrm{i}\delta^{(4)}(x-y) + \\ &+ \theta(x_0 - y_0)\langle 0| (\Box_x + m^2)\phi(x)\phi(y) |0\rangle + \\ &+ \theta(y_0 - x_0)\langle 0| \phi(y)(\Box_x + m^2)\phi(x) |0\rangle\ .\end{aligned}$$

The last two terms vanish since the field ϕ satisfies the Klein–Gordon equation. Therefore,

$$(\Box_x + m^2)\langle 0| T(\phi(x)\phi(y)) |0\rangle = -\mathrm{i}\delta^{(4)}(x-y)\ . \tag{7.28}$$

7.18

(a) Applying the variational principle to the given action leads to the equations:

$$\begin{aligned}\mathrm{i}\frac{\partial\psi}{\partial t} &= \left(-\frac{1}{2m}\Delta + V(\boldsymbol{r})\right)\psi \\ -\mathrm{i}\frac{\partial\psi^\dagger}{\partial t} &= \left(-\frac{1}{2m}\Delta + V(\boldsymbol{r})\right)\psi^\dagger\ .\end{aligned}$$

The first of these equations is the Schrödinger equation, the second one is its conjugation equation.

(b) A particular solution of the free Schrödinger equation is a plane wave $e^{-iE_k t+i\boldsymbol{k}\cdot\boldsymbol{r}}$, where $E_k = k^2/2m$ so that the general solution is

$$\psi(t,\boldsymbol{r}) = \int \frac{d^3\boldsymbol{k}}{(2\pi)^{3/2}} a(\boldsymbol{k}) e^{-iE_k t+i\boldsymbol{k}\cdot\boldsymbol{r}} . \tag{7.29}$$

The negative energy solutions are not present in previous expression since $E_k > 0$ in nonrelativistic quantum mechanics. The field $\psi^\dagger$ is

$$\psi^\dagger(t,\boldsymbol{r}) = \int \frac{d^3\boldsymbol{k}}{(2\pi)^{3/2}} a^\dagger(\boldsymbol{k}) e^{iE_k t-i\boldsymbol{k}\cdot\boldsymbol{r}} . \tag{7.30}$$

In the quantum theory these classical fields are replaced by operators in the Hilbert space. The field conjugate to ψ is

$$\pi = \frac{\partial\mathcal{L}}{\partial\dot{\psi}} = i\psi^\dagger .$$

The equal–time commutation relations are

$$\begin{aligned}
[\psi(t,\boldsymbol{x}), \psi^\dagger(t,\boldsymbol{y})] &= \delta^{(3)}(\boldsymbol{x}-\boldsymbol{y}) ,\\
[\psi(t,\boldsymbol{x}), \psi(t,\boldsymbol{y})] = [\psi^\dagger(t,\boldsymbol{x}), \psi^\dagger(t,\boldsymbol{y})] &= 0 .
\end{aligned} \tag{7.31}$$

From the relations (7.29) and (7.30) follows

$$\begin{aligned}
a(\boldsymbol{k}) &= \frac{1}{(2\pi)^{3/2}} e^{iE_k t} \int d^3\boldsymbol{x}\, \psi(t,\boldsymbol{x}) e^{-i\boldsymbol{k}\cdot\boldsymbol{x}}\\
a^\dagger(\boldsymbol{k}) &= \frac{1}{(2\pi)^{3/2}} e^{-iE_k t} \int d^3\boldsymbol{x}\, \psi^\dagger(t,\boldsymbol{x}) e^{i\boldsymbol{k}\cdot\boldsymbol{x}} .
\end{aligned}$$

From (7.31) and previous relations one easily gets the commutation relations:

$$[a(\boldsymbol{k}), a^\dagger(\boldsymbol{p})] = \delta^{(3)}(\boldsymbol{p}-\boldsymbol{k}) , \tag{7.32}$$

$$[a(\boldsymbol{k}), a(\boldsymbol{p})] = [a^\dagger(\boldsymbol{k}), a^\dagger(\boldsymbol{p})] = 0 . \tag{7.33}$$

(c) Substituting (7.29) and (7.30) into the expression for the Green function one obtains

$$\begin{aligned}
G(x_0,\boldsymbol{x},y_0,\boldsymbol{y}) &= -i \langle 0|\, \psi(x_0,\boldsymbol{x})\psi^\dagger(y_0,\boldsymbol{y})\, |0\rangle\, \theta(x_0-y_0)\\
&= -\frac{i}{(2\pi)^3} \int d^3\boldsymbol{k} d^3\boldsymbol{p} e^{-i(E_k x_0 - \boldsymbol{k}\cdot\boldsymbol{x} - E_p y_0 + \boldsymbol{p}\cdot\boldsymbol{y})}\\
&\times \langle 0|\, a(\boldsymbol{k}) a^\dagger(\boldsymbol{p})\, |0\rangle\, \theta(x_0-y_0)\\
&= -\frac{i}{(2\pi)^3} \int d^3\boldsymbol{k} d^3\boldsymbol{p} e^{-i(E_k x_0 - \boldsymbol{k}\cdot\boldsymbol{x} - E_p y_0 + \boldsymbol{p}\cdot\boldsymbol{y})}\\
&\times \delta^{(3)}(\boldsymbol{p}-\boldsymbol{k})\theta(x_0-y_0)\\
&= -\frac{i}{(2\pi)^3} \int d^3\boldsymbol{k} e^{-i\frac{k^2}{2m}(x^0-y^0)+i\boldsymbol{k}\cdot(\boldsymbol{x}-\boldsymbol{y})} \theta(x_0-y_0)\\
&= -i \left(\frac{m}{2\pi i(x_0-y_0)}\right)^{3/2} e^{\frac{im(\boldsymbol{x}-\boldsymbol{y})^2}{2(x_0-y_0)}} \theta(x_0-y_0) .
\end{aligned}$$

(d) The eigenfunctions are

$$u_k = \sqrt{\frac{2}{\pi}} \sin(kx) \ ,$$

hence the (nonrelativistic) field operators are

$$\psi = \sqrt{\frac{2}{\pi}} \int_0^\infty \mathrm{d}k a(k) \mathrm{e}^{-\mathrm{i}\frac{k^2}{2m}t} \sin(kx) \ , \tag{7.34}$$

$$\psi^\dagger = \sqrt{\frac{2}{\pi}} \int_0^\infty \mathrm{d}k a^\dagger(k) \mathrm{e}^{\mathrm{i}\frac{k^2}{2m}t} \sin(kx) \ . \tag{7.35}$$

We shall leave to the reader to prove that

$$G(x_0, x, y_0, y) = -\mathrm{i} \left(\frac{m}{2\pi \mathrm{i}(x_0 - y_0)} \right)^{1/2} \left[\mathrm{e}^{\frac{im(x-y)^2}{2(x_0-y_0)}} - \mathrm{e}^{\frac{im(x+y)^2}{2(x_0-y_0)}} \right] \theta(x_0 - y_0) \ . \tag{7.36}$$

Generally, if the eigenfunctions of the Hamiltonian are $u_n(\boldsymbol{x})$ the Green function is

$$G(x_0, \boldsymbol{x}, y_0, \boldsymbol{y}) = -\mathrm{i} \sum_n \mathrm{e}^{-\mathrm{i}E_n(x_0-y_0)} u_n(\boldsymbol{x}) u_n^*(\boldsymbol{y}) \theta(x_0 - y_0) \ . \tag{7.37}$$

(e) The invariance of the Schrödinger equation can be proven directly. We leave that to reader.

(f) In order to find the conserved charges we should calculate only time components of the conserved currents. For the spatial translations the time component of the current is

$$\begin{aligned} j_0 &= -\frac{\partial \mathcal{L}}{\partial(\partial_0 \psi)} \partial_i \psi \epsilon^i \\ &= -\mathrm{i}\psi^\dagger \partial_i \psi \epsilon^i = -\mathrm{i}\psi^\dagger \nabla \psi \cdot \boldsymbol{\epsilon} \ . \end{aligned} \tag{7.38}$$

The conserved charge is the linear momentum

$$\boldsymbol{P} = -\int \mathrm{d}^3 \boldsymbol{x} \psi^\dagger (\mathrm{i}\nabla) \psi \ . \tag{7.39}$$

The Hamiltonian

$$H = \int \mathrm{d}^3 \boldsymbol{x} \psi^\dagger (-\frac{1}{2m}) \Delta \psi \tag{7.40}$$

is generator of time translations. The angular momentum

$$\boldsymbol{J} = -\mathrm{i} \int \mathrm{d}^3 \boldsymbol{x} \psi^\dagger (\boldsymbol{x} \times \nabla) \psi \tag{7.41}$$

is generator of rotations. Under Galilean boosts we have $\delta x_i = -v_i t$, $\delta\psi = -im\boldsymbol{v} \cdot \boldsymbol{x}\psi$ so that

$$j_0 = \boldsymbol{v} \cdot \boldsymbol{j}_0 = m\boldsymbol{v} \cdot \boldsymbol{x}\psi^\dagger\psi + \mathrm{i}\boldsymbol{v}t\psi^\dagger\nabla\psi. \tag{7.42}$$

Consequently, the boost generator is

$$\boldsymbol{G} = \int \mathrm{d}^3\boldsymbol{x}\psi^\dagger(m\boldsymbol{x} + \mathrm{i}t\nabla)\psi \ . \tag{7.43}$$

The commutation relations can be found using the commutation relations (7.31). Let us start with $[P_i, G_j]$:

$$\begin{aligned}
[P_i, G_j] &= \mathrm{i}\int \mathrm{d}^3\boldsymbol{x}\mathrm{d}^3\boldsymbol{y}[-\psi^\dagger(y)\partial_i^y\psi(y), \psi^\dagger(x)(mx_j + \mathrm{i}t\partial_j)\psi(x)] \\
&= -\mathrm{i}m\int \mathrm{d}^3\boldsymbol{x}\mathrm{d}^3\boldsymbol{y}\,(\psi^\dagger(y)[\partial_i\psi(y), \psi^\dagger(x)x_j\psi(x)] \\
&+ \ [\psi^\dagger(y), \psi^\dagger(x)x_j\psi(x)]\partial_i\psi(y)) \\
&= -\mathrm{i}m\int \mathrm{d}^3\boldsymbol{x}(-\partial_i\psi^\dagger x_j\psi(x) - x_j\psi^\dagger\partial_i\psi) \\
&= -\mathrm{i}M\delta_{ij} \ ,
\end{aligned} \tag{7.44}$$

where $M = m\int \mathrm{d}^3\boldsymbol{x}\psi^\dagger\psi$ is the mass operator. It appears since the representation is projective. We have two possibilities either to enlarge the Galilean algebra with this operator or to add a superselection rule which forbids superposition of particles of different masses.
In the similar manner the other commutation relations can be obtained:

$$\begin{aligned}
[G_i, G_j] = [H, \boldsymbol{P}] &= [H, \boldsymbol{J}] = 0 \\
[J_i, J_j] &= \mathrm{i}\epsilon_{ijk}J_k \\
[J_i, G_j] &= \mathrm{i}\epsilon_{ijk}G_k \\
[J_i, P_j] &= \mathrm{i}\epsilon_{ijk}P_k \\
[H, G_i] &= -\mathrm{i}P_i \ .
\end{aligned}$$

The Galilean algebra can also be derived from the Poincaré algebra [23].

7.19

(a) By using the first commutation relation in (7.D), we get

$$\begin{aligned}
[a(\boldsymbol{p}), a^\dagger] &= C\int \frac{\mathrm{d}^3\boldsymbol{q}}{\sqrt{2\omega_q}}[a(\boldsymbol{p}), a^\dagger(\boldsymbol{q})]\tilde{f}(q) \\
&= C\int \frac{\mathrm{d}^3\boldsymbol{q}}{\sqrt{2\omega_q}}\tilde{f}(q)\delta^{(3)}(\boldsymbol{p} - \boldsymbol{q}) \\
&= C\frac{1}{\sqrt{2\omega_p}}\tilde{f}(p) \ .
\end{aligned} \tag{7.45}$$

The second commutator can be evaluated in the same way. The result is

$$[a^\dagger(\boldsymbol{p}), a] = -C\frac{1}{\sqrt{2\omega_p}}\tilde{f}^*(p) \ . \tag{7.46}$$

(b) Using (7.45), we have

$$a(\boldsymbol{p})(a^\dagger)^n = C\frac{1}{\sqrt{2\omega_p}}\tilde{f}(p)(a^\dagger)^{n-1} + a^\dagger a(\boldsymbol{p})(a^\dagger)^{n-1} \; . \tag{7.47}$$

By repeating this procedure n times, we get

$$a(\boldsymbol{p})(a^\dagger)^n = C\frac{1}{\sqrt{2\omega_p}}n\tilde{f}(p)(a^\dagger)^{n-1} + (a^\dagger)^n a(\boldsymbol{p}) \; . \tag{7.48}$$

Hence,

$$[a(\boldsymbol{p}), (a^\dagger)^n] = C\frac{n\tilde{f}(p)}{\sqrt{2\omega_p}}(a^\dagger)^{n-1} \; . \tag{7.49}$$

(c) This calculation is straightforward:

$$\begin{aligned} a(\boldsymbol{p})\,|z\rangle &= \mathrm{e}^{-|z|^2/2}a(\boldsymbol{p})\sum_{n=0}^{\infty}\frac{z^n(a^\dagger)^n}{n!}\,|0\rangle \\ &= \mathrm{e}^{-|z|^2/2}\sum_{n=1}^{\infty}\frac{C}{\sqrt{2\omega_p}}\frac{z^n\tilde{f}(p)}{(n-1)!}(a^\dagger)^{n-1}\,|0\rangle \\ &= \frac{C}{\sqrt{2\omega_p}}\tilde{f}(p)z\,|z\rangle \; . \end{aligned} \tag{7.50}$$

(d) By using the previous relation and the property $\langle z|z\rangle = 1$, we have

$$\begin{aligned} \langle z|\,\phi\,|z\rangle &= \int\frac{\mathrm{d}^3\boldsymbol{p}}{(2\pi)^{3/2}\sqrt{2\omega_p}}\left(\langle z|\,a(\boldsymbol{p})\,|z\rangle\,\mathrm{e}^{-\mathrm{i}p\cdot x} + \langle z|\,a^\dagger(\boldsymbol{p})\,|z\rangle\,\mathrm{e}^{\mathrm{i}p\cdot x}\right) \\ &= C\int\frac{\mathrm{d}^3\boldsymbol{p}}{(2\pi)^{3/2}2\omega_p}\left(z\tilde{f}(p)\mathrm{e}^{-\mathrm{i}p\cdot x} + z^*\tilde{f}^*(p)\mathrm{e}^{\mathrm{i}p\cdot x}\right) \\ &= \frac{C}{(2\pi)^{3/2}}\left(zf(x) + z^*f^*(x)\right) \; . \end{aligned} \tag{7.51}$$

In the same manner we have

$$\begin{aligned} \langle z|:\phi^2:|z\rangle &= \int\frac{\mathrm{d}^3\boldsymbol{p}}{(2\pi)^{3/2}\sqrt{2\omega_p}}\frac{\mathrm{d}^3\boldsymbol{q}}{(2\pi)^{3/2}\sqrt{2\omega_q}}\Big(\langle z|\,a(\boldsymbol{p})a(\boldsymbol{q})\,|z\rangle\,\mathrm{e}^{-\mathrm{i}(p+q)\cdot x} \\ &+ \langle z|\,a^\dagger(\boldsymbol{q})a(\boldsymbol{p})\,|z\rangle\,\mathrm{e}^{\mathrm{i}(q-p)\cdot x} \\ &+ \langle z|\,a^\dagger(\boldsymbol{p})a(\boldsymbol{q})\,|z\rangle\,\mathrm{e}^{\mathrm{i}(p-q)\cdot x} + \langle z|\,a^\dagger(\boldsymbol{p})a^\dagger(\boldsymbol{q})\,|z\rangle\,\mathrm{e}^{\mathrm{i}(q+p)\cdot x}\Big) \\ &= C^2\int\frac{\mathrm{d}^3\boldsymbol{p}}{(2\pi)^{3/2}2\omega_p}\frac{\mathrm{d}^3\boldsymbol{q}}{(2\pi)^{3/2}2\omega_q}\Big(\tilde{f}(p)\tilde{f}(q)z^2\mathrm{e}^{-\mathrm{i}(p+q)\cdot x} \\ &+ \tilde{f}(p)\tilde{f}^*(q)|z|^2\mathrm{e}^{-\mathrm{i}(p-q)\cdot x} \\ &+ \tilde{f}^*(p)\tilde{f}(q)|z|^2\mathrm{e}^{\mathrm{i}(p-q)\cdot x} + \tilde{f}^*(p)\tilde{f}^*(q)(z^*)^2\mathrm{e}^{\mathrm{i}(p+q)\cdot x}\Big) \\ &= \frac{C^2}{(2\pi)^3}\left(zf(x) + z^*f^*(x)\right)^2 \; . \end{aligned} \tag{7.52}$$

Hence,

$$(\Delta\phi)^2 = 0 \, . \tag{7.53}$$

(e) It is easy to see that

$$\langle z|\, H\,|z\rangle = C^2|z|^2 \int \mathrm{d}^3\boldsymbol{p}|\tilde{f}(p)|^2 \, . \tag{7.54}$$

7.20

(a) By substituting the expression for ϕ in the relation

$$U(\Lambda, a)\phi(x)U^{-1}(\Lambda, a) = \phi(\Lambda x + a)$$

we obtain

$$\begin{aligned} &\int \frac{\mathrm{d}^3\boldsymbol{k}}{(2\pi)^{3/2}\sqrt{2\omega_k}} U(\Lambda, a)\left(a(\boldsymbol{k})\mathrm{e}^{-\mathrm{i}k\cdot x} + a^\dagger(\boldsymbol{k})\mathrm{e}^{\mathrm{i}k\cdot x}\right)U^{-1}(\Lambda, a) \\ &= \int \frac{\mathrm{d}^3\boldsymbol{k}'}{(2\pi)^{3/2}\sqrt{2\omega_{k'}}} \left(a(\boldsymbol{k}')\mathrm{e}^{-\mathrm{i}k'\cdot(\Lambda x + a)} + a^\dagger(\boldsymbol{k}')\mathrm{e}^{\mathrm{i}k'\cdot(\Lambda x + a)}\right) \, . \end{aligned} \tag{7.55}$$

In the integral on the right hand side we make the changing of variables $k'^\mu \Lambda_\mu^{\ \nu} = k^\nu$. In Problem 6.3, we proved that $\mathrm{d}^3\boldsymbol{k}/(2\omega_k)$ is a Lorentz invariant measure, so that

$$\frac{\mathrm{d}^3\boldsymbol{k}'}{\sqrt{2\omega_{k'}}} = \sqrt{\frac{\omega_{k'}}{2}} \frac{\mathrm{d}^3\boldsymbol{k}}{\omega_k} \, .$$

By performing the inverse Fourier transformation, we obtain the requested result.

(b) It is easy to see that

$$\begin{aligned} U(\Lambda, a)\,|k_1, \ldots, k_n\rangle &= U(\Lambda, a)a^\dagger(\boldsymbol{k}_1)U^{-1}(\Lambda, a)U(\Lambda, a)\cdots \\ &\quad \cdots U(\Lambda, a)a^\dagger(\boldsymbol{k}_n)U^{-1}(\Lambda, a)\,|0\rangle \\ &= \sqrt{\frac{\omega_{k_1'}\cdots\omega_{k_n'}}{\omega_{k_1}\cdots\omega_{k_n}}}\mathrm{e}^{\mathrm{i}a_\mu\Lambda^\mu{}_\nu(k_1^\nu + \ldots + k_n^\nu)}\,|\Lambda k_1, \ldots, \Lambda k_n\rangle \, . \end{aligned}$$

(c) From the expressions (7.6) and (7.8) and the first part of this problem, we have

$$\begin{aligned} U(\Lambda)P^\mu U^{-1}(\Lambda) &= \int \mathrm{d}^3\boldsymbol{k} k^\mu U(\Lambda)a^\dagger(\boldsymbol{k})a(\boldsymbol{k})U^{-1}(\Lambda) \\ &= \int \mathrm{d}^3\boldsymbol{k} k^\mu \frac{\omega_{k'}}{\omega_k} a^\dagger(\Lambda k)a(\Lambda k) \\ &= \Lambda_\nu^{\ \mu} \int \mathrm{d}^3\boldsymbol{k}' k'^\nu a^\dagger(\boldsymbol{k}')a(\boldsymbol{k}') \\ &= \Lambda_\nu^{\ \mu} P^\nu \, , \end{aligned}$$

where we made the change of variables $k^\mu = \Lambda_\nu^{\ \mu} k'^\nu$ in the integral.

(d) First, you should prove the following formulae:

$$U(\Lambda)[\phi(x),\phi(y)]U^{-1}(\Lambda)=[\phi(\Lambda x),\phi(\Lambda x)]\ ,$$

$$[\phi(x),\phi(y)]=\mathrm{i}\Delta(x-y)\ .$$

From the integral expression for the function $\Delta(x-y)$ (Problem 6.6), it follows that $\Delta(\Lambda x-\Lambda y)=\Delta(x-y)$, i.e. it is a relativistic covariant quantity.

7.21

(a) In Problem 7.3, we obtained the Hamiltonian

$$H=\int \mathrm{d}^3\boldsymbol{k}\omega_{\boldsymbol{k}}a^\dagger(\boldsymbol{k})a(\boldsymbol{k})\ .$$

The Backer–Hausdorff relation reads

$$PHP^{-1}=\mathrm{e}^{A}H\mathrm{e}^{-A}=H+[A,H]+\frac{1}{2}[A,[A,H]]+\dots \tag{7.56}$$

where $A=-\frac{\mathrm{i}\pi}{2}\int \mathrm{d}^3\boldsymbol{q}\left(a^\dagger(\boldsymbol{q})a(\boldsymbol{q})-\eta_p a^\dagger(\boldsymbol{q})a(-\boldsymbol{q})\right)$. The first commutator in this expression is

$$[A,H]=-\frac{\mathrm{i}\pi}{2}\eta_p\int \mathrm{d}^3\boldsymbol{k}\omega_k\left(a^\dagger(\boldsymbol{k})a(-\boldsymbol{k})-a^\dagger(-\boldsymbol{k})a(\boldsymbol{k})\right)\ .$$

By changing $\boldsymbol{k}\to-\boldsymbol{k}$ in the second term, we get $[A,H]=0$. It is clear that the other commutators in (7.56) also vanish, hence

$$[P,H]=0\ .$$

(b) Starting from Problem 7.8, we obtain the requested result.

7.22 $\tau\boldsymbol{P}\tau^{-1}=-\boldsymbol{P}$, $\tau H\tau^{-1}=H$

7.23 The first step is to show that $C\phi^\dagger C^{-1}=\eta_c^*\phi$, $C\pi C^{-1}=\eta_c\pi^\dagger$ and $C\pi^\dagger C^{-1}=\eta_c\pi$.

8

Canonical quantization of the Dirac field

8.1 If we use the anticommutation relation (8.E) the anticommutator $\mathrm{i}S_{ab}(x-y)=\{\psi_a(x),\bar{\psi}_b(y)\}$, where $a,b=1,\ldots,4$ are Dirac indices, becomes

$$\{\psi_a(x),\bar{\psi}_b(y)\}=\sum_{r,s}\frac{1}{(2\pi)^3}\int \mathrm{d}^3p\mathrm{d}^3\boldsymbol{q}\frac{m}{\sqrt{E_pE_q}}\delta_{rs}\delta^{(3)}(\boldsymbol{p}-\boldsymbol{q})$$
$$\times\Big(u_a(\boldsymbol{p},r)\bar{u}_b(\boldsymbol{q},s)\mathrm{e}^{\mathrm{i}(q\cdot y-p\cdot x)}$$
$$+\,v_a(\boldsymbol{p},r)\bar{v}_b(\boldsymbol{q},s)\mathrm{e}^{-\mathrm{i}(q\cdot y-p\cdot x)}\Big)\,.$$

Applying the solution of Problem 4.4 we have

$$\mathrm{i}S_{ab}=\frac{1}{(2\pi)^3}\int\frac{\mathrm{d}^3\boldsymbol{p}}{2E_p}\left[(\not{p}+m)_{ab}\mathrm{e}^{-\mathrm{i}p\cdot(x-y)}+(\not{p}-m)_{ab}\mathrm{e}^{\mathrm{i}p\cdot(x-y)}\right]\,. \tag{8.1}$$

The last expression can be easily transformed into the following form

$$\{\psi_a(x),\bar{\psi}_b(y)\}=(\mathrm{i}\gamma^\mu\partial^x_\mu+m)_{ab}\frac{1}{(2\pi)^3}\int\frac{\mathrm{d}^3\boldsymbol{p}}{2E_p}\left[\mathrm{e}^{-\mathrm{i}p\cdot(x-y)}-\mathrm{e}^{\mathrm{i}p\cdot(x-y)}\right]\,. \tag{8.2}$$

From (8.2) we see that $\Delta(x-y)$ is given by

$$\Delta(x-y)=-\frac{\mathrm{i}}{(2\pi)^3}\int\frac{\mathrm{d}^3\boldsymbol{p}}{2E_p}\left[\mathrm{e}^{-\mathrm{i}p\cdot(x-y)}-\mathrm{e}^{\mathrm{i}p\cdot(x-y)}\right]\,.$$

The function $\Delta(x-y)$ was defined in Problem 6.6. In the special case $x_0=y_0$ we shall make change $\boldsymbol{p}\to-\boldsymbol{p}$ in the second term of expression (8.1) and obtain

$$\{\psi_a(x),\bar{\psi}_b(y)\}|_{x_0=y_0}=(\gamma^0)_{ab}\int\frac{\mathrm{d}^3\boldsymbol{p}}{(2\pi)^3}\mathrm{e}^{\mathrm{i}\boldsymbol{p}\cdot(\boldsymbol{x}-\boldsymbol{y})}=(\gamma^0)_{ab}\delta^{(3)}(\boldsymbol{x}-\boldsymbol{y})\,. \tag{8.3}$$

8.2

(a) Substituting (8.A,B) in the expression for charge Q we obtain

$$\begin{aligned}Q &= -e\int \mathrm{d}^3\boldsymbol{x} : \psi^\dagger\psi : \\ &= -e\sum_{r,s}\int \mathrm{d}^3\boldsymbol{p}\frac{m}{E_p}\left[c_r^\dagger(\boldsymbol{p})c_s(\boldsymbol{p})u_r^\dagger(\boldsymbol{p})u_s(\boldsymbol{p})\right.\\ &+ : d_r(\boldsymbol{p})d_s^\dagger(\boldsymbol{p}) : v_r^\dagger(\boldsymbol{p})v_s(\boldsymbol{p}) + c_r^\dagger(\boldsymbol{p})d_s^\dagger(-\boldsymbol{p})u_r^\dagger(\boldsymbol{p})v_s(-\boldsymbol{p})\mathrm{e}^{2\mathrm{i}E_pt}\\ &\left.+d_r(\boldsymbol{p})c_s(-\boldsymbol{p})v_r^\dagger(\boldsymbol{p})u_s(-\boldsymbol{p})\mathrm{e}^{-2\mathrm{i}E_pt}\right] . \end{aligned} \tag{8.4}$$

From (4.52) and (8.4) we get

$$Q = -e\sum_r\int \mathrm{d}^3\boldsymbol{p}\left(c_r^\dagger(\boldsymbol{p})c_r(\boldsymbol{p}) - d_r^\dagger(\boldsymbol{p})d_r(\boldsymbol{p})\right) . \tag{8.5}$$

(b) As ψ satisfies the Dirac equation, $(-\mathrm{i}\gamma^i\partial_i+m)\psi = \mathrm{i}\gamma_0\partial_0\psi$ the Hamiltonian is

$$\begin{aligned}H &= \mathrm{i}\int \mathrm{d}^3\boldsymbol{x} : \psi^\dagger\partial_0\psi : \\ &= \sum_{r,s}\frac{1}{(2\pi)^3}\int \mathrm{d}^3\boldsymbol{x}\mathrm{d}^3\boldsymbol{p}\mathrm{d}^3\boldsymbol{q}\sqrt{\frac{m}{E_p}}\sqrt{\frac{m}{E_q}} : \left(u_r^\dagger(\boldsymbol{p})c_r^\dagger(\boldsymbol{p})\mathrm{e}^{\mathrm{i}p\cdot x}\right.\\ &\left.+v_r^\dagger(\boldsymbol{p})d_r(\boldsymbol{p})\mathrm{e}^{-\mathrm{i}p\cdot x}\right)E_q\left(u_s(\boldsymbol{q})c_s(\boldsymbol{q})\mathrm{e}^{-\mathrm{i}q\cdot x} - v_s(\boldsymbol{q})d_s^\dagger(\boldsymbol{q})\mathrm{e}^{\mathrm{i}q\cdot x}\right) : \\ &= \sum_r\int \mathrm{d}^3\boldsymbol{p}E_p\left(c_r^\dagger(\boldsymbol{p})c_r(\boldsymbol{p}) + d_r^\dagger(\boldsymbol{p})d_r(\boldsymbol{p})\right) . \end{aligned} \tag{8.6}$$

(c)

$$\boldsymbol{P} = \sum_r\int \mathrm{d}^3\boldsymbol{p}\boldsymbol{p}\left(c_r^\dagger(\boldsymbol{p})c_r(\boldsymbol{p}) + d_r^\dagger(\boldsymbol{p})d_r(\boldsymbol{p})\right) . \tag{8.7}$$

8.3

(a) It is easy to see that

$$\begin{aligned}[H,\psi] &= \sum_{r,s}\frac{1}{(2\pi)^{3/2}}\int \mathrm{d}^3\boldsymbol{p}\mathrm{d}^3\boldsymbol{q}E_p\sqrt{\frac{m}{E_q}}\\ &\times\left[c_r^\dagger(\boldsymbol{p})c_r(\boldsymbol{p}) + d_r^\dagger(\boldsymbol{p})d_r(\boldsymbol{p}), c_s(\boldsymbol{q})u_s(\boldsymbol{q})\mathrm{e}^{-\mathrm{i}q\cdot x} + d_s^\dagger(\boldsymbol{q})v_s(\boldsymbol{q})\mathrm{e}^{\mathrm{i}q\cdot x}\right]\\ &= \sum_{r,s}\frac{1}{(2\pi)^{3/2}}\int \mathrm{d}^3\boldsymbol{p}\mathrm{d}^3\boldsymbol{q}E_p\sqrt{\frac{m}{E_q}}\delta_{rs}\delta^{(3)}(\boldsymbol{p}-\boldsymbol{q})\\ &\times\left(-c_r(\boldsymbol{p})u_s(\boldsymbol{q})\mathrm{e}^{-\mathrm{i}q\cdot x} + d_r^\dagger(\boldsymbol{p})v_s(\boldsymbol{q})\mathrm{e}^{\mathrm{i}q\cdot x}\right)\\ &= \sum_r\int\frac{\mathrm{d}^3\boldsymbol{p}}{(2\pi)^{3/2}}\sqrt{mE_p}\left(-c_r(\boldsymbol{p})u_r(\boldsymbol{p})\mathrm{e}^{-\mathrm{i}p\cdot x} + d_r^\dagger(\boldsymbol{p})v_r(\boldsymbol{p})\mathrm{e}^{\mathrm{i}p\cdot x}\right)\\ &= -\mathrm{i}\frac{\partial\psi}{\partial t} ,\end{aligned}$$

where we have used:

$$[c_r^\dagger(\boldsymbol{p})c_r(\boldsymbol{p}), c_s(\boldsymbol{q})] = -\{c_r^\dagger(\boldsymbol{p}), c_s(\boldsymbol{q})\}c_r(\boldsymbol{p})$$
$$= -\delta_{rs}\delta^{(3)}(\boldsymbol{p}-\boldsymbol{q})c_r(\boldsymbol{p}) ,$$

and the similar expression for $d-$operators.

(b) If we had used commutation relations instead of anticommutation relations in the quantization process we would have obtained:

$$H = \sum_r \int \mathrm{d}^3\boldsymbol{p}E_p \left(c_r^\dagger(\boldsymbol{p})c_r(\boldsymbol{p}) - d_r^\dagger(\boldsymbol{p})d_r(\boldsymbol{p})\right) .$$

From here we conclude that the energy spectrum would have been unbounded from below, which is physically unacceptable.

8.4

$$[H, c_r^\dagger(\boldsymbol{p})c_r(\boldsymbol{p})] = \sum_s \int \mathrm{d}^3\boldsymbol{q}E_q[c_s^\dagger(\boldsymbol{q})c_s(\boldsymbol{q}) + d_s^\dagger(\boldsymbol{q})d_s(\boldsymbol{q}), c_r^\dagger(\boldsymbol{p})c_r(\boldsymbol{p})]$$
$$= \sum_s \int \mathrm{d}^3\boldsymbol{q}E_q \left([c_s^\dagger(\boldsymbol{q})c_s(\boldsymbol{q}), c_r^\dagger(\boldsymbol{p})]c_r(\boldsymbol{p})\right.$$
$$\left.+ c_r^\dagger(\boldsymbol{p})[c_s^\dagger(\boldsymbol{q})c_s(\boldsymbol{q}), c_r(\boldsymbol{p})]\right)$$
$$= \sum_s \int \mathrm{d}^3\boldsymbol{q}E_q \left(c_s^\dagger(\boldsymbol{q})\{c_s(\boldsymbol{q}), c_r^\dagger(\boldsymbol{p})\}c_r(\boldsymbol{p})\right.$$
$$- \{c_s^\dagger(\boldsymbol{q}), c_r^\dagger(\boldsymbol{p})\}c_s(\boldsymbol{q})c_r(\boldsymbol{p})$$
$$\left.+ c_r^\dagger(\boldsymbol{p})(c_s^\dagger(\boldsymbol{q})\{c_s(\boldsymbol{q}), c_r(\boldsymbol{p})\} - \{c_s^\dagger(\boldsymbol{q}), c_r(\boldsymbol{p})\}c_s(\boldsymbol{q}))\right)$$
$$= E_p\left(c_r^\dagger(\boldsymbol{p})c_r(\boldsymbol{p}) - c_r^\dagger(\boldsymbol{p})c_r(\boldsymbol{p})\right) = 0$$

8.5 The form variation of a spinor field is

$$\delta_0\psi = \delta\psi - \delta x^\mu \partial_\mu \psi =$$
$$= -\frac{\mathrm{i}}{4}\omega^{\mu\nu}\sigma_{\mu\nu}\psi - \omega^{\mu\nu}x_\nu\partial_\mu\psi$$
$$= \frac{1}{2}\omega^{\mu\nu}\left(x_\mu\partial_\nu - x_\nu\partial_\mu - \frac{\mathrm{i}}{2}\sigma_{\mu\nu}\right)\psi .$$

On the other hand we have $\delta_0\psi = -\frac{\mathrm{i}}{2}\omega^{\mu\nu}M_{\mu\nu}\psi$. Comparing these results we conclude that the generators are given by

$$M_{\mu\nu} = \mathrm{i}(x_\mu\partial_\nu - x_\nu\partial_\mu) + \frac{1}{2}\sigma_{\mu\nu} .$$

8.6

(a) Applying the formula $[AB, C] = A\{B, C\} - \{A, C\}B$ we obtain

$$
\begin{aligned}
[M_{\mu\nu}, \psi_a(x)] &= \int \mathrm{d}^3\boldsymbol{y} \left[\psi_b^\dagger(y) \left(\mathrm{i}(y_\mu \partial_\nu - y_\nu \partial_\mu) + \frac{1}{2}\sigma_{\mu\nu} \right)_{bc} \psi_c(y), \psi_a(x) \right] \\
&= -\int \mathrm{d}^3\boldsymbol{y} \{\psi_b^\dagger(y), \psi_a(x)\} \left(\mathrm{i}(y_\mu \partial_\nu - y_\nu \partial_\mu) + \frac{1}{2}\sigma_{\mu\nu} \right)_{bc} \psi_c(y) \\
&= -[\mathrm{i}(x_\mu \partial_\nu - x_\nu \partial_\mu) + \frac{1}{2}\sigma_{\mu\nu}]_{ac} \psi_c(x) \, ,
\end{aligned}
$$

where we have used anticommutation relations (8.C,D). This result is a consequence of Lorentz symmetry.

(b) Substituting the expressions for angular momentum and momentum of the Dirac field we get

$$
\begin{aligned}
[M_{\mu\nu}, P_\rho] &= \mathrm{i} \int \mathrm{d}^3\boldsymbol{x} \mathrm{d}^3\boldsymbol{y} \\
&\times \left[\psi_a^\dagger(x) \left(\mathrm{i}(x_\mu \partial_\nu - x_\nu \partial_\mu) + \frac{1}{2}\sigma_{\mu\nu} \right)_{ab} \psi_b(x), \psi_c^\dagger(y) \partial_\rho \psi_c(y) \right] .
\end{aligned}
$$

First we suppose that all indices are the spatial: $\mu = i, \nu = j, \ \rho = k$. Then,

$$
\begin{aligned}
[M_{ij}, P_k] &= \mathrm{i} \int \mathrm{d}^3\boldsymbol{x} \mathrm{d}^3\boldsymbol{y} \\
&\times \Bigg(\psi_a^\dagger(x) \left\{ \left(\mathrm{i}(x_i \partial_j - x_j \partial_i) + \frac{1}{2}\sigma_{ij} \right)_{ab} \psi_b(x), \psi_c^\dagger(y) \right\} \partial_k \psi_c(y) \\
&- \psi_c^\dagger(y) \{\psi_a^\dagger(x), \partial_k \psi_c(y)\} \left(\mathrm{i}(x_i \partial_j - x_j \partial_i) + \frac{1}{2}\sigma_{ij} \right)_{ab} \psi_b(x) \Bigg) \\
&= \mathrm{i} \int \mathrm{d}^3\boldsymbol{x} \mathrm{d}^3\boldsymbol{y} \\
&\times \Bigg(\psi_a^\dagger(x) \left(\mathrm{i}(x_i \partial_j - x_j \partial_i) + \frac{1}{2}\sigma_{ij} \right)_{ab} \delta^{(3)}(\boldsymbol{x} - \boldsymbol{y}) \partial_k \psi_b(y) \\
&- \psi_c^\dagger(y) \partial_k^y \delta^{(3)}(\boldsymbol{x} - \boldsymbol{y}) \delta_{ac} \left(\mathrm{i}(x_i \partial_j - x_j \partial_i) + \frac{1}{2}\sigma_{ij} \right)_{ab} \psi_b(x) \Bigg) \, ,
\end{aligned}
$$

where we used the equal-time anticommutation relations (8.C,D). The integration over $\boldsymbol{y}$ leads to

$$
[M_{ij}, P_k] = \mathrm{i} \int \mathrm{d}^3\boldsymbol{x} \left(\mathrm{i} g_{jk} \psi^\dagger \partial_i \psi - \mathrm{i} g_{ik} \psi^\dagger \partial_j \psi \right) \, ,
$$

or

$$
[M_{ij}, P_k] = \mathrm{i}(g_{jk} P_i - g_{ik} P_j).
$$

Now we take $\mu = 0, \ \nu = i$, and $\rho = k$, i.e. we calculate the commutator $[M_{0i}, P_k]$. In order to do it we first compute anticommutator

$$
\{\partial_{x^0} \psi(x), \bar{\psi}(y)\}|_{x^0 = y^0} \, .
$$

Taking partial derivative of (8.1) with respect to x_0 and substituting $x^0 = y^0$ we get

$$\begin{aligned}\{\partial_{x^0}\psi_a(x), \bar{\psi}_b(y)\}|_{x^0=y^0} &= \frac{\mathrm{i}}{2(2\pi)^3}\int \mathrm{d}^3\boldsymbol{p}\left[(-E_p\gamma^0 + \boldsymbol{p}\cdot\boldsymbol{\gamma} - m)_{ab}\mathrm{e}^{\mathrm{i}\boldsymbol{p}\cdot(\boldsymbol{x}-\boldsymbol{y})}\right. \\ &\quad \left. + (E_p\gamma^0 - \boldsymbol{p}\cdot\boldsymbol{\gamma} - m)_{ab}\mathrm{e}^{-\mathrm{i}\boldsymbol{p}\cdot(\boldsymbol{x}-\boldsymbol{y})}\right] \\ &= \frac{\mathrm{i}}{(2\pi)^3}\int \mathrm{d}^3\boldsymbol{p}(\boldsymbol{p}\cdot\boldsymbol{\gamma} - m)_{ab}\mathrm{e}^{\mathrm{i}\boldsymbol{p}\cdot(\boldsymbol{x}-\boldsymbol{y})} \\ &= \boldsymbol{\gamma}_{ab}\nabla_x\delta^{(3)}(\boldsymbol{x}-\boldsymbol{y}) - \mathrm{i}m\delta_{ab}\delta^{(3)}(\boldsymbol{x}-\boldsymbol{y})\ .\end{aligned}$$

Then

$$\begin{aligned}[M_{0i}, P_k] &= \mathrm{i}\int \mathrm{d}^3\boldsymbol{x}\mathrm{d}^3\boldsymbol{y} \\ &\times \left(\psi_a^\dagger(x)\left\{\left(\mathrm{i}(x_0\partial_i - x_i\partial_0) + \frac{1}{2}\sigma_{0i}\right)_{ab}\psi_b(x), \psi_c^\dagger(y)\right\}\partial_k\psi_c(y)\right. \\ &- \left.\psi_c^\dagger(y)\{\psi_a^\dagger(x), \partial_k\psi_c(y)\}\left(\mathrm{i}(x_0\partial_i - x_i\partial_0) + \frac{1}{2}\sigma_{0i}\right)_{ab}\psi_b(x)\right) \\ &= \mathrm{i}\int \mathrm{d}^3\boldsymbol{x}\mathrm{d}^3\boldsymbol{y}\left(\mathrm{i}x_0\psi^\dagger(x)\partial_i^x\delta^{(3)}(\boldsymbol{x}-\boldsymbol{y})\partial_k\psi(y)\right. \\ &- \mathrm{i}x_i\psi_a^\dagger(x)(\boldsymbol{\gamma}\gamma_0\nabla_x - im\gamma_0)_{ac}\delta^{(3)}(\boldsymbol{x}-\boldsymbol{y})\partial_k\psi_c(y) \\ &+ \psi_a^\dagger(x)\frac{1}{2}(\sigma_{0i})_{ab}\delta^{(3)}(\boldsymbol{x}-\boldsymbol{y})\partial_k\psi_b(y) \\ &- \mathrm{i}x_0\psi^\dagger(y)\partial_k^y\delta^{(3)}(\boldsymbol{x}-\boldsymbol{y})\partial_i\psi(x) \\ &+ \mathrm{i}x_i\psi^\dagger(y)\partial_k^y\delta^{(3)}(\boldsymbol{x}-\boldsymbol{y})\partial_0\psi(x) \\ &- \left.\psi_a^\dagger(y)\frac{1}{2}(\sigma_{0i})_{ab}\partial_k^y\delta^{(3)}(\boldsymbol{x}-\boldsymbol{y})\psi_b(x)\right) \\ &= \mathrm{i}\int \mathrm{d}^3\boldsymbol{x}\left(-\mathrm{i}x_i\psi^\dagger\boldsymbol{\gamma}\gamma^0\partial_k\nabla\psi - mx_i\psi^\dagger\gamma_0\partial_k\psi - \mathrm{i}x_i\partial_k\psi^\dagger\partial_0\psi\right) \\ &= \mathrm{i}\int \mathrm{d}^3\boldsymbol{x}\left(\mathrm{i}g_{ik}\psi^\dagger\partial_0\psi + x_i\bar{\psi}(\mathrm{i}\gamma^0\partial_0 + \mathrm{i}\boldsymbol{\gamma}\nabla - m)\partial_k\psi\right)\ .\end{aligned}$$

The second term in the last line vanishes since ψ satisfies the Dirac equation. Then we get

$$[M_{0i}, P_k] = \mathrm{i}g_{ik}P_0\ .$$

The remaining commutators $[M_{0i}, P_0]$ and $[M_{ij}, P_0]$ can be computed in the same way.

8.7 The helicity operator is

$$S_p = \frac{1}{2}\int \mathrm{d}^3\boldsymbol{x} : \psi^\dagger\frac{\boldsymbol{\Sigma}\cdot\boldsymbol{p}}{|\boldsymbol{p}|}\psi :\ . \tag{8.8}$$

Inserting expressions for fields ψ and $\psi^\dagger$ in the previous formula and using the fact that $u_r(\boldsymbol{p})$ and $v_r(\boldsymbol{p})$ are eigenspinors of $\boldsymbol{\Sigma}\cdot\boldsymbol{p}/|\boldsymbol{p}|$ with eigenvalues $(-1)^{r+1}$ and $(-1)^r$, respectively (see Problem 4.7) we get

$$
\begin{aligned}
S_p = \frac{1}{2(2\pi)^3}\int \mathrm{d}^3\boldsymbol{x}\sum_{r,s=1}^{2}\int \mathrm{d}^3\boldsymbol{p}\mathrm{d}^3\boldsymbol{q}\frac{m}{\sqrt{E_pE_q}} \\
\times\Big[c_r^\dagger(\boldsymbol{q})c_s(\boldsymbol{p})(-1)^{s+1}u_r^\dagger(\boldsymbol{q})u_s(\boldsymbol{p})\mathrm{e}^{\mathrm{i}(q-p)\cdot x} \\
+ c_r^\dagger(\boldsymbol{q})d_s^\dagger(\boldsymbol{p})(-1)^{s}u_r^\dagger(\boldsymbol{q})v_s(\boldsymbol{p})\mathrm{e}^{\mathrm{i}(q+p)\cdot x} \\
+ d_r(\boldsymbol{q})c_s(\boldsymbol{p})(-1)^{s+1}v_r^\dagger(\boldsymbol{q})u_s(\boldsymbol{p})\mathrm{e}^{-\mathrm{i}(q+p)\cdot x} \\
- d_s^\dagger(\boldsymbol{p})d_r(\boldsymbol{q})(-1)^{s}v_r^\dagger(\boldsymbol{q})v_s(\boldsymbol{p})\mathrm{e}^{\mathrm{i}(p-q)\cdot x}\Big]\ . \qquad (8.9)
\end{aligned}
$$

Performing the $\boldsymbol{x}$ integration and applying orthogonality relations (4.52) one gets that the second and the third term in the expression (8.9) vanish. Finally, integration over the momentum $\boldsymbol{q}$ gives

$$
S_p = \frac{1}{2}\sum_{r=1}^{2}\int \mathrm{d}^3\boldsymbol{p}(-1)^{r+1}\left(c_r^\dagger(\boldsymbol{p})c_r(\boldsymbol{p}) + d_r^\dagger(\boldsymbol{p})d_r(\boldsymbol{p})\right)\ . \qquad (8.10)
$$

Let us emphasize that we have used the expansion of the fields with respect to helicity basis.

8.8 The two–particle state given in the problem is eigenstate of the operators H, Q, and S_p. Using the explicit form of the Hamiltonian from Problem 8.2 we have

$$
\begin{aligned}
Hc_{r_1}^\dagger(\boldsymbol{p}_1)c_{r_2}^\dagger(\boldsymbol{p}_2)\left|0\right\rangle = \sum_r\int \mathrm{d}^3\boldsymbol{p}E_p\left(c_r^\dagger(\boldsymbol{p})c_r(\boldsymbol{p})\right. \\
\left.+ d_r^\dagger(\boldsymbol{p})d_r(\boldsymbol{p})\right)c_{r_1}^\dagger(\boldsymbol{p}_1)c_{r_2}^\dagger(\boldsymbol{p}_2)\left|0\right\rangle\ . \qquad (8.11)
\end{aligned}
$$

Let us calculate the first term in the previous expression. Commuting $c_r(\boldsymbol{p})$ to the right we get

$$
\begin{aligned}
c_r^\dagger(\boldsymbol{p})c_r(\boldsymbol{p})c_{r_1}^\dagger(\boldsymbol{p}_1)c_{r_2}^\dagger(\boldsymbol{p}_2)\left|0\right\rangle = \delta_{r_1r}\delta^{(3)}(\boldsymbol{p}-\boldsymbol{p}_1)c_r^\dagger(\boldsymbol{p})c_{r_2}^\dagger(\boldsymbol{p}_2)\left|0\right\rangle \\
- c_r^\dagger(\boldsymbol{p})c_{r_1}^\dagger(\boldsymbol{p}_1)c_r(\boldsymbol{p})c_{r_2}^\dagger(\boldsymbol{p}_2)\left|0\right\rangle\ . \qquad (8.12)
\end{aligned}
$$

Repeating once more we get

$$
\begin{aligned}
c_r^\dagger(\boldsymbol{p})c_r(\boldsymbol{p})c_{r_1}^\dagger(\boldsymbol{p}_1)c_{r_2}^\dagger(\boldsymbol{p}_2)\left|0\right\rangle = \delta_{r_1r}\delta^{(3)}(\boldsymbol{p}-\boldsymbol{p}_1)c_r^\dagger(\boldsymbol{p})c_{r_2}^\dagger(\boldsymbol{p}_2)\left|0\right\rangle \\
- c_r^\dagger(\boldsymbol{p})c_{r_1}^\dagger(\boldsymbol{p}_1)\delta_{rr_2}\delta^{(3)}(\boldsymbol{p}-\boldsymbol{p}_2)\left|0\right\rangle\ . \qquad (8.13)
\end{aligned}
$$

It is easy to see that

$$
d_r^\dagger(\boldsymbol{p})d_r(\boldsymbol{p})c_{r_1}^\dagger(\boldsymbol{p}_1)c_{r_2}^\dagger(\boldsymbol{p}_2)\left|0\right\rangle = 0\ . \qquad (8.14)
$$

Inserting (8.13) and (8.14) in (8.11) and integrating over momentum $\boldsymbol{p}$ we obtain

$$H c^\dagger_{r_1}(\boldsymbol{p}_1) c^\dagger_{r_2}(\boldsymbol{p}_2) |0\rangle = (E_{p_1} + E_{p_2}) c^\dagger_{r_1}(\boldsymbol{p}_1) c^\dagger_{r_2}(\boldsymbol{p}_2) |0\rangle \ . \tag{8.15}$$

Similar as before we have:

$$Q c^\dagger_{r_1}(\boldsymbol{p}_1) c^\dagger_{r_2}(\boldsymbol{p}_2) |0\rangle = -2e c^\dagger_{r_1}(\boldsymbol{p}_1) c^\dagger_{r_2}(\boldsymbol{p}_2) |0\rangle \ , \tag{8.16}$$

for charge and

$$\begin{aligned} &S_p \, c^\dagger_{r_1}(\boldsymbol{p}_1) c^\dagger_{r_2}(\boldsymbol{p}_2) |0\rangle \\ &= \frac{1}{2}\left((-1)^{r_1+1} + (-1)^{r_2+1}\right) c^\dagger_{r_1}(\boldsymbol{p}_1) c^\dagger_{r_2}(\boldsymbol{p}_2) |0\rangle \end{aligned} \tag{8.17}$$

for helicity. To summarize: energy, charge and helicity of the two–particle state $|\boldsymbol{p}_1, r_1; \boldsymbol{p}_2, r_2\rangle$ are

$$E_{p_1} + E_{p_2}, \quad -2e, \quad \frac{1}{2}\left((-1)^{r_1+1} + (-1)^{r_2+1}\right) \ , \tag{8.18}$$

respectively.

8.9 The commutator is

$$\begin{aligned} [Q^a, Q^b] &= \frac{1}{4}\int \mathrm{d}^3\boldsymbol{x}\mathrm{d}^3\boldsymbol{y} \tau^a_{ij}\tau^b_{kl}[\psi^\dagger_i(x)\psi_j(x), \psi^\dagger_k(y)\psi_l(y)] \\ &= \frac{1}{4}\int \mathrm{d}^3\boldsymbol{x}\mathrm{d}^3\boldsymbol{y} \tau^a_{ij}\tau^b_{kl}(\psi^\dagger_i(x)\psi_l(y)\delta_{jk} - \psi^\dagger_k(y)\psi_j(x)\delta_{il})\delta^{(3)}(\boldsymbol{x}-\boldsymbol{y}) \\ &= \frac{1}{4}\int \mathrm{d}^3\boldsymbol{x}(\psi^\dagger_i\tau^a_{ij}\tau^b_{jl}\psi_l - \psi^\dagger_k\tau^b_{kl}\tau^a_{lj}\psi_j) \\ &= \frac{1}{4}\int \mathrm{d}^3\boldsymbol{x}\psi^\dagger[\tau^a, \tau^b]\psi \\ &= \frac{\mathrm{i}}{2}\epsilon^{abc}\int \mathrm{d}^3\boldsymbol{x}\psi^\dagger\tau^c\psi = \mathrm{i}\epsilon^{abc}Q^c \ . \end{aligned}$$

The generators Q^a satisfy the commutation relations of SU(2) algebra as we expected.

8.10 The charges are

$$Q^b = \int \mathrm{d}^3\boldsymbol{x} j^b_0 = \int \mathrm{d}^3\boldsymbol{x}(\epsilon^{abc}\dot{\pi}^a\pi^c + \frac{1}{2}\Psi^\dagger_i\tau^b_{ij}\Psi_j) \ . \tag{8.19}$$

(a) The commutator is

$$\begin{aligned} [Q^b, Q^e] = \int \mathrm{d}^3\boldsymbol{x}\mathrm{d}^3\boldsymbol{y} \, \Big(&\epsilon^{abc}\epsilon^{def}[\dot{\pi}^a(x)\pi^c(x), \dot{\pi}^d(y)\pi^f(y)] \\ &+ \frac{\tau^b_{ij}}{2}\frac{\tau^e_{mn}}{2}[\Psi^\dagger_i(x)\Psi_j(x), \Psi^\dagger_m(y)\Psi_n(y)]\Big) \end{aligned}$$

$$= \int \mathrm{d}^3\boldsymbol{x}\mathrm{d}^3\boldsymbol{y} \left(\epsilon^{abc}\epsilon^{def} (\mathrm{i}\delta^{(3)}(\boldsymbol{x}-\boldsymbol{y})\delta^{cd}\dot{\pi}^a(x)\pi^f(y) \right.$$
$$- \mathrm{i}\delta^{(3)}(\boldsymbol{x}-\boldsymbol{y})\delta^{af}\dot{\pi}^d(y)\pi^c(x))$$
$$\left. + \frac{\tau^b_{ij}}{2}\frac{\tau^e_{mn}}{2}\delta^{(3)}(\boldsymbol{x}-\boldsymbol{y})(\delta_{jm}\Psi^\dagger_i(x)\Psi_n(y) - \delta_{in}\Psi^\dagger_m(y)\Psi_j(x)) \right)$$
$$= \int \mathrm{d}^3\boldsymbol{x} \left(\mathrm{i}(\dot{\pi}^e\pi^b - \dot{\pi}^b\pi^e) + \frac{\mathrm{i}}{2}\epsilon^{bed}\Psi^\dagger\tau^d\Psi \right)$$
$$= \mathrm{i}\epsilon^{bed} \int \mathrm{d}^3\boldsymbol{x} \left(\epsilon^{adc}\dot{\pi}^a\pi^c + \frac{1}{2}\Psi^\dagger\tau^d\Psi \right)$$
$$= \mathrm{i}\epsilon^{bed}Q^d .$$

(b) The results are

$$[Q^b, \pi^a(x)] = -\mathrm{i}\epsilon^{abc}\pi^c(x) ,$$
$$[Q^b, \psi_i(x)] = -\frac{\tau^b_{in}}{2}\psi_n(x) ,$$
$$[Q^b, \bar{\psi}_i(x)] = \bar{\psi}_n(x)\frac{\tau^b_{ni}}{2} .$$

8.11 The conserved charge for dilatation is

$$D = \int \mathrm{d}^3\boldsymbol{x} j^0 = -\mathrm{i}\int \mathrm{d}^3\boldsymbol{x} \left(\frac{3}{2}\psi^\dagger\psi + x^j\psi^\dagger\partial_j\psi - x^0\bar{\psi}\gamma^j\partial_j\psi \right) . \tag{8.20}$$

Let us find the commutator between the operator D and momentum P^i

$$[D, P^i] = \int \mathrm{d}^3\boldsymbol{x}\mathrm{d}^3\boldsymbol{y} \left([\frac{3}{2}\psi^\dagger(x)\psi(x) + x^j\psi^\dagger(x)\partial_j\psi(x), \psi^\dagger(y)\partial^i\psi(y)] \right.$$
$$\left. - [x^0\bar{\psi}(x)\gamma^j\partial_j\psi(x), \psi^\dagger(y)\partial^i\psi(y)] \right) .$$

We decompose the previous expression on three commutators. The first one is

$$[\psi^\dagger(x)\psi(x), \psi^\dagger(y)\partial^i\psi(y)] = [\psi^\dagger_a(x)\psi_a(x), \psi^\dagger_b(y)]\partial^i\psi_b(y)$$
$$+ \psi^\dagger_b(y)[\psi^\dagger_a(x)\psi_a(x), \partial^i\psi_b(y)]$$
$$= \psi^\dagger_a(x)\{\psi_a(x), \psi^\dagger_b(y)\}\partial^i\psi_b(y)$$
$$- \psi^\dagger_b(y)\{\psi^\dagger_a(x), \partial^i\psi_b(y)\}\psi_a(x) ,$$

where we have dropped the vanishing terms. The anticommutation relations (8.C–D) give the following result

$$[\psi^\dagger(x)\psi(x), \psi^\dagger(y)\partial^i\psi(y)] = \psi^\dagger(x)\partial^i\psi(y)\delta^{(3)}(\boldsymbol{x}-\boldsymbol{y})$$
$$- \psi^\dagger(y)\psi(x)\partial^i_y\delta^{(3)}(\boldsymbol{x}-\boldsymbol{y}) . \tag{8.21}$$

The remaining commutators can be calculated in the same way. The result is:

$$[\psi^\dagger(x)\partial_j\psi(x), \psi^\dagger(y)\partial^i\psi(y)] = \psi^\dagger(x)\partial^i\psi(y)\partial_j^x\delta^{(3)}(\boldsymbol{x}-\boldsymbol{y})$$
$$- \psi^\dagger(y)\partial_j\psi(x)\partial_y^i\delta^{(3)}(\boldsymbol{x}-\boldsymbol{y}) , \qquad (8.22)$$

$$[\bar{\psi}(x)\gamma^j\partial_j\psi(x), \psi^\dagger(y)\partial^i\psi(y)] = \bar{\psi}(x)\gamma^j\partial^i\psi(y)\partial_j^x\delta^{(3)}(\boldsymbol{x}-\boldsymbol{y})$$
$$- \bar{\psi}(y)\gamma^j\partial_j\psi(x)\partial_y^i\delta^{(3)}(\boldsymbol{x}-\boldsymbol{y}) . \qquad (8.23)$$

Inserting (8.21), (8.22) and (8.23) in the expression for commutator and applying

$$\partial_x^k\delta^{(3)}(\boldsymbol{x}-\boldsymbol{y}) = -\partial_y^k\delta^{(3)}(\boldsymbol{x}-\boldsymbol{y}) , \qquad (8.24)$$

we get

$$[D, P^i] = -\int \mathrm{d}^3\boldsymbol{x}\psi^\dagger\partial^i\psi = \mathrm{i}P^i . \qquad (8.25)$$

Similarly one can show that

$$[D, P^0] = \mathrm{i}P^0 . \qquad (8.26)$$

8.12

(a) Using the expression (5.G) the energy–momentum tensor is

$$T_{\alpha\beta} = \mathrm{i}\bar{\psi}\gamma_\alpha\partial_\beta\psi - g_{\alpha\beta}(\mathrm{i}\bar{\psi}\partial\!\!\!/\psi - gx^2\bar{\psi}\psi) .$$

Taking derivative of the previous expression we get

$$\partial^\alpha T_{\alpha\beta} = 2gx_\beta\bar{\psi}\psi ,$$

where we have used the equations of motion:

$$\mathrm{i}\partial\!\!\!/\psi - gx^2\psi = 0 ,$$

$$\mathrm{i}\partial_\mu\bar{\psi}\gamma^\mu + gx^2\bar{\psi} = 0 .$$

The result $\partial^\alpha T_{\alpha\beta} \neq 0$ shows that there is no translation symmetry in the theory. As a consequence, the energy and momentum are not conserved in this theory.

(b) From the expression for the four-momentum we have

$$P^0(t) = \int \mathrm{d}^3\boldsymbol{x}(-\mathrm{i}\bar{\psi}\gamma^j\partial_j\psi + gx^2\bar{\psi}\psi) ,$$

$$P^i(t) = \mathrm{i}\int \mathrm{d}^3\boldsymbol{x}\psi^\dagger\partial^i\psi ,$$

so

$$
\begin{aligned}
[P^0(t), P^i(t)] &= \int\!\!\int \mathrm{d}^3\boldsymbol{x}\mathrm{d}^3\boldsymbol{y} \\
&\times \Big([\bar{\psi}(t,\boldsymbol{x})\gamma^j\partial_j\psi(t,\boldsymbol{x}), \psi^\dagger(t,\boldsymbol{y})\partial^i\psi(t,\boldsymbol{y})] \\
&+ \mathrm{i}gx^2[\bar{\psi}(t,\boldsymbol{x})\psi(t,\boldsymbol{x}), \psi^\dagger(t,\boldsymbol{y})\partial^i\psi(t,\boldsymbol{y})] \Big) \\
&= \int\!\!\int \mathrm{d}^3\boldsymbol{x}\mathrm{d}^3\boldsymbol{y} \\
&\times \Big((\gamma^0\gamma^j)_{ab}[\psi_a^\dagger(t,\boldsymbol{x})\partial_j\psi_b(t,\boldsymbol{x}), \psi_c^\dagger(t,\boldsymbol{y})\partial^i\psi_c(t,\boldsymbol{y})] \\
&+ \mathrm{i}gx^2\gamma^0_{ab}[\psi_a^\dagger(t,\boldsymbol{x})\psi_b(t,\boldsymbol{x}), \psi_c^\dagger(t,\boldsymbol{y})\partial^i\psi_c(t,\boldsymbol{y})] \Big) \; .
\end{aligned}
$$

The commutators in the previous expression can be found in the same way as in the previous problem

$$
\begin{aligned}
[P^0(t), P^i(t)] &= \int \mathrm{d}^3\boldsymbol{x} \, (-\partial_j\bar{\psi}\gamma^j\partial^i\psi - \bar{\psi}\gamma^j\partial_j\partial_i\psi \\
&+ \mathrm{i}gx^2(\bar{\psi}\partial^i\psi + (\partial^i\bar{\psi})\psi)) \\
&= \int \mathrm{d}^3\boldsymbol{x} \, (-\partial_j(\bar{\psi}\gamma^j\partial^i\psi) + igx^2\partial^i(\bar{\psi}\psi)) \\
&= -2\mathrm{i}g \int \mathrm{d}^3\boldsymbol{x} x^i \bar{\psi}\psi \; ,
\end{aligned}
$$

where we dropped the surface terms.

(c) It is easy to show that $\partial_\mu M^{\mu\nu\rho} = 0$, which is a consequence of the Lorentz symmetry of the Lagrangian density.

8.13

(a) Under the Lorentz transformation the commutator $[J^\mu(x), J^\nu(y)]$ transforms in the following way

$$
\begin{aligned}
&U(\Lambda)[J^\mu(x), J^\nu(y)]U^{-1}(\Lambda) \\
&= U(\Lambda)[\bar{\psi}_a(x)\gamma^\mu_{ab}\psi_b(x), \bar{\psi}_c(y)\gamma^\nu_{cd}\psi_d(y)]U^{-1}(\Lambda) \\
&= [U\bar{\psi}_a(x)U^{-1}\gamma^\mu_{ab}U\psi_b(x)U^{-1}, U\bar{\psi}_c(y)U^{-1}\gamma^\nu_{cd}U\psi_d(y)U^{-1}] \; .
\end{aligned} \tag{8.27}
$$

Taking the adjoint of (8.G) and multiplying by γ^0 we obtain

$$
U(\Lambda)\bar{\psi}(x)U^{-1}(\Lambda) = \bar{\psi}(\Lambda x)S(\Lambda) \; . \tag{8.28}
$$

By using (8.G), last expression and $S^{-1}\gamma^\mu S = \Lambda^\mu{}_\nu\gamma^\nu$ in (8.27) we get

$$
U(\Lambda)[J^\mu(x), J^\nu(y)]U^{-1}(\Lambda) = \Lambda_\rho{}^\mu\Lambda_\sigma{}^\nu[J^\rho(\Lambda x), J^\sigma(\Lambda y)] \; . \tag{8.29}
$$

From the last result we see that the commutator $[J^\mu(x), J^\nu(y)]$ is a covariant quantity.

(b) Using the fact that the commutator is a Lorentz tensor we calculate it in the frame where $x^0 = y^0 = t$, $\boldsymbol{x} \neq \boldsymbol{y}$. We get

$$\begin{aligned}
&[J_\mu(t,\boldsymbol{x}), J_\nu(t,\boldsymbol{y})] \\
&= (\gamma_0\gamma_\mu)_{ab}(\gamma_0\gamma_\nu)_{cd}[\psi_a^\dagger(t,\boldsymbol{x})\psi_b(t,\boldsymbol{x}), \psi_c^\dagger(t,\boldsymbol{y})\psi_d(t,\boldsymbol{y})] \\
&= (\gamma_0\gamma_\mu)_{ab}(\gamma_0\gamma_\nu)_{cd}\left(\psi_a^\dagger(t,\boldsymbol{x})\{\psi_b(t,\boldsymbol{x}), \psi_c^\dagger(t,\boldsymbol{y})\}\psi_d(t,\boldsymbol{y})\right. \\
&- \left.\psi_c^\dagger(t,\boldsymbol{y})\{\psi_a^\dagger(t,\boldsymbol{x}), \psi_d(t,\boldsymbol{y})\}\psi_b(t,\boldsymbol{x})\right) . \qquad (8.30)
\end{aligned}$$

Using the anticommutation relation (8.D) in (8.30) gives

$$\begin{aligned}
&[J_\mu(t,\boldsymbol{x}), J_\nu(t,\boldsymbol{y})] \\
&= \left(\bar\psi(t,\boldsymbol{x})\gamma_\mu\gamma_0\gamma_\nu\psi(t,\boldsymbol{y}) - \bar\psi(t,\boldsymbol{y})\gamma_\nu\gamma_0\gamma_\mu\psi(t,\boldsymbol{x})\right)\delta^{(3)}(\boldsymbol{x}-\boldsymbol{y}) .
\end{aligned}$$

Since $\boldsymbol{x} \neq \boldsymbol{y}$ then $\delta^{(3)}(\boldsymbol{x}-\boldsymbol{y}) = 0$ and the commutator is equal to zero in the special frame we have chosen. Because of the covariance it follows that it is equal to zero for $(x-y)^2 < 0$. Therefore, microcausality principle is valid.

8.14 First show that

$$\langle\psi_a(x)\bar\psi_b(y)\rangle = \frac{1}{(2\pi)^3}\int\frac{\mathrm{d}^3\boldsymbol{p}}{2E_p}(\not{p}+m)_{ab}\mathrm{e}^{-\mathrm{i}p\cdot(x-y)} , \qquad (8.31)$$

$$\langle\bar\psi_a(x)\psi_b(y)\rangle = \frac{1}{(2\pi)^3}\int\frac{\mathrm{d}^3\boldsymbol{p}}{2E_p}(\not{p}-m)_{ba}\mathrm{e}^{-\mathrm{i}p\cdot(x-y)} . \qquad (8.32)$$

If in the expression $\langle\bar\psi_a(x_1)\psi_b(x_2)\psi_c(x_3)\bar\psi_d(x_4)\rangle$, we substitute the expansions (8.A–B), we obtain

$$\begin{aligned}
&\langle\bar\psi_a(x_1)\psi_b(x_2)\psi_c(x_3)\bar\psi_d(x_4)\rangle \\
&= \sum_{r_1,\dots,r_4}\frac{m^2}{(2\pi)^6}\left(\prod_{i=1}^4\int\frac{\mathrm{d}^3\boldsymbol{p}_i}{\sqrt{E_{p_i}}}\right) \\
&\times\left(\left\langle d_1c_2d_3^\dagger c_4^\dagger\right\rangle\bar v_{1a}u_{2b}v_{3c}\bar u_{4d}\mathrm{e}^{\mathrm{i}(-p_1\cdot x_1-p_2\cdot x_2+p_3\cdot x_3+p_4\cdot x_4)}\right. \\
&+ \left.\left\langle d_1d_2^\dagger c_3c_4^\dagger\right\rangle\bar v_{1a}v_{2b}u_{3c}\bar u_{4d}\mathrm{e}^{\mathrm{i}(-p_1\cdot x_1+p_2\cdot x_2-p_3\cdot x_3+p_4\cdot x_4)}\right) ,
\end{aligned}$$

where the vanishing terms are discarded. Also, we use the abbreviations:

$$d_1 = d_{r_1}(\boldsymbol{p}_1), u_1 = u_{r_1}(\boldsymbol{p}_1), \text{ etc.}$$

Applying the expressions for projectors to positive and negative energy solutions from Problem 4.4 and using

$$\left\langle d_1c_2d_3^\dagger c_4^\dagger\right\rangle = -\delta_{r_1r_3}\delta_{r_2r_4}\delta^{(3)}(\boldsymbol{p}_1-\boldsymbol{p}_3)\delta^{(3)}(\boldsymbol{p}_2-\boldsymbol{p}_4) ,$$

$$\left\langle d_1 d_2^\dagger c_3 c_4^\dagger \right\rangle = \delta_{r_1 r_2}\delta_{r_3 r_4}\delta^{(3)}(\boldsymbol{p}_1 - \boldsymbol{p}_2)\delta^{(3)}(\boldsymbol{p}_3 - \boldsymbol{p}_4)$$

we have

$$\begin{aligned}
&\langle \bar{\psi}_a(x_1)\psi_b(x_2)\psi_c(x_3)\bar{\psi}_d(x_4)\rangle \\
&= -\frac{1}{(2\pi)^6}\int \frac{\mathrm{d}^3\boldsymbol{p}_1\mathrm{d}^3\boldsymbol{p}_2}{4E_{p_1}E_{p_2}}(\not{p}_1 - m)_{ca}(\not{p}_2 + m)_{bd}\mathrm{e}^{-\mathrm{i}p_1\cdot(x_1-x_3)-\mathrm{i}p_2\cdot(x_2-x_4)} \\
&+ \frac{1}{(2\pi)^6}\int \frac{\mathrm{d}^3\boldsymbol{p}_1\mathrm{d}^3\boldsymbol{p}_3}{4E_{p_1}E_{p_3}}(\not{p}_1 - m)_{ba}(\not{p}_3 + m)_{cd}\mathrm{e}^{-\mathrm{i}p_1\cdot(x_1-x_2)-\mathrm{i}p_3\cdot(x_3-x_4)} .
\end{aligned}$$

By using (8.31) and (8.32) the last expression takes the form

$$\begin{aligned}
\langle \bar{\psi}_a(x_1)\psi_b(x_2)\psi_c(x_3)\bar{\psi}_d(x_4)\rangle = &-\langle \bar{\psi}_a(x_1)\psi_c(x_3)\rangle \langle \psi_b(x_2)\bar{\psi}_d(x_4)\rangle \\
&+ \langle \bar{\psi}_a(x_1)\psi_b(x_2)\rangle \langle \psi_c(x_3)\bar{\psi}_d(x_4)\rangle .
\end{aligned}$$

The previous formula is special case of the Wick theorem.

8.15 Substituting (8.A-B) in the commutator we obtain

$$\begin{aligned}
\frac{1}{2}[\bar{\psi}, \gamma^\mu\psi] = &\frac{1}{2(2\pi)^3}\sum_{r,s}\int \mathrm{d}^3\boldsymbol{p}\mathrm{d}^3\boldsymbol{q}\frac{m}{\sqrt{E_pE_q}}\Big[\bar{u}_r(\boldsymbol{p})\gamma^\mu u_s(\boldsymbol{q}) \\
&\times (c_r^\dagger(\boldsymbol{p})c_s(\boldsymbol{q}) - c_s(\boldsymbol{q})c_r^\dagger(\boldsymbol{p}))\mathrm{e}^{\mathrm{i}(p-q)\cdot x} \\
&+ \bar{u}_r(\boldsymbol{p})\gamma^\mu v_s(\boldsymbol{q})(c_r^\dagger(\boldsymbol{p})d_s^\dagger(\boldsymbol{q}) - d_s^\dagger(\boldsymbol{q})c_r^\dagger(\boldsymbol{p}))\mathrm{e}^{\mathrm{i}(p+q)\cdot x} \\
&+ \bar{v}_r(\boldsymbol{p})\gamma^\mu u_s(\boldsymbol{q})(d_r(\boldsymbol{p})c_s(\boldsymbol{q}) - c_s(\boldsymbol{q})d_r(\boldsymbol{p}))\mathrm{e}^{-\mathrm{i}(p+q)\cdot x} \\
&+ \bar{v}_r(\boldsymbol{p})\gamma^\mu v_s(\boldsymbol{q})(d_r(\boldsymbol{p})d_s^\dagger(\boldsymbol{q}) - d_s^\dagger(\boldsymbol{q})d_r(\boldsymbol{p}))\mathrm{e}^{\mathrm{i}(q-p)\cdot x}\Big] .
\end{aligned} \tag{8.33}$$

Using the anticommutation relations (8.E) we obtain

$$\begin{aligned}
\frac{1}{2}[\bar{\psi}, \gamma^\mu\psi] = &: \bar{\psi}\gamma^\mu\psi : - \\
&- \frac{1}{2(2\pi)^3}\int \mathrm{d}^3\boldsymbol{p}\frac{p^\mu}{E_p}\sum_r(\bar{u}_r(\boldsymbol{p})u_r(\boldsymbol{p}) + \bar{v}_r(\boldsymbol{p})v_r(\boldsymbol{p})) ,
\end{aligned}$$

where we have used the Gordon identities (Problem 4.21) in addition. The requested result follows after applying the orthogonality relations (4.D).

8.16 Let us first prove that

$$\langle 0|\, T(\bar{\psi}_a(x)\psi_b(y))\, |0\rangle = -\mathrm{i}S_{Fba}(y-x) .$$

Using the definition of time ordering and the expressions (8.31) and (8.32) we obtain

$$\langle 0| T(\bar{\psi}_a(x)\psi_b(y)) |0\rangle = \frac{1}{(2\pi)^3}\int \frac{\mathrm{d}^3\boldsymbol{p}}{2E_p}\Big[(\not{p}-m)_{ba}\mathrm{e}^{\mathrm{i}p\cdot(y-x)}\theta(x_0-y_0) - (\not{p}+m)_{ba}\mathrm{e}^{\mathrm{i}p\cdot(x-y)}\theta(y_0-x_0)\Big] . \tag{8.34}$$

With a help of Problem 6.13 we see that right hand side of the expression (8.34) is $-\mathrm{i}S_{Fba}(y-x)$ and we have

$$\begin{aligned}\langle 0| T(\bar{\psi}(x)\Gamma\psi(y)) |0\rangle &= \Gamma_{ab}\langle 0| T(\bar{\psi}_a(x)\psi_b(y)) |0\rangle \\ &= -\mathrm{i}\Gamma_{ab}S_{Fba}(y-x) \\ &= -\mathrm{i}\ \mathrm{tr}\left[\Gamma S_F(y-x)\right] \\ &= -\mathrm{i}\int \frac{\mathrm{d}^4p}{(2\pi)^4}\frac{\mathrm{e}^{-\mathrm{i}p\cdot(y-x)}}{p^2-m^2+\mathrm{i}\epsilon}\mathrm{tr}\left[(\not{p}+m)\Gamma\right] .\end{aligned}$$

Using the identities from the Problems 3.6(b),(d),(e) and (i) we obtain

$$\mathrm{tr}\left[(\not{p}+m)\gamma_5\right] = \mathrm{tr}\left[(\not{p}+m)\gamma_5\gamma_\mu\right] = 0,\ \mathrm{tr}\left[(\not{p}+m)\gamma_\mu\gamma_\nu\right] = 4mg_{\mu\nu} .$$

From here the requested result follows.

8.17

(a) In the Weyl representation for γ–matrices the charge conjugate spinor is

$$\begin{aligned}\psi_c &= C\bar{\psi}^T \\ &= \mathrm{i}\begin{pmatrix}\sigma_2 & 0\\ 0 & -\sigma_2\end{pmatrix}\begin{pmatrix}0 & 1\\ 1 & 0\end{pmatrix}\begin{pmatrix}\varphi^*\\ -\mathrm{i}\sigma_2\chi\end{pmatrix} \\ &= \begin{pmatrix}\chi\\ -\mathrm{i}\sigma_2\varphi^*\end{pmatrix} .\end{aligned}$$

The condition $\psi_M = \psi_M^c$ gives $\varphi = \chi$.

(b) If

$$\psi_M = \begin{pmatrix}\chi\\ -\mathrm{i}\sigma_2\chi^*\end{pmatrix} \text{ and } \phi_M = \begin{pmatrix}\varphi\\ -\mathrm{i}\sigma_2\varphi^*\end{pmatrix} ,$$

then

$$\begin{aligned}\bar{\psi}_M\phi_M &= -\mathrm{i}\chi^\dagger\sigma_2\varphi^* + \mathrm{i}\chi^T\sigma_2\varphi \\ &= -\mathrm{i}\sigma_{2ab}\chi_a^*\varphi_b^* + \mathrm{i}\sigma_{2ab}\chi_a\varphi_b \\ &= -\mathrm{i}\sigma_{2ba}\varphi_b^*\chi_a^* + \mathrm{i}\sigma_{2ba}\varphi_b\chi_a \\ &= -\mathrm{i}\varphi^\dagger\sigma_2\chi^* + \mathrm{i}\varphi^T\sigma_2\chi = \bar{\phi}_M\psi_M .\end{aligned}$$

In the last expression we used that φ and χ are Grassmann variables. The other identities can be proved in the same way. For the second one the following identity is useful: $\sigma_2\sigma^\mu\sigma_2 = \bar{\sigma}^{\mu T}$.

(c) The Majorana field operator is

$$\begin{aligned}\psi_M &= \frac{1}{\sqrt{2}}(\psi + \psi_c) \\ &= \int \frac{\mathrm{d}^3\boldsymbol{p}}{(2\pi)^3}\sqrt{\frac{m}{E_p}}\sum_r \left(\frac{c_r(\boldsymbol{p}) + d_r(\boldsymbol{p})}{\sqrt{2}} u_r(\boldsymbol{p})\mathrm{e}^{-\mathrm{i}p\cdot x}\right. \\ &+ \left.\frac{c_r^\dagger(\boldsymbol{p}) + d_r^\dagger(\boldsymbol{p})}{\sqrt{2}} v_r(\boldsymbol{p})\mathrm{e}^{\mathrm{i}p\cdot x}\right) .\end{aligned}$$

The annihilation and creation operators can easily be read off:

$$b_M(\boldsymbol{p}, r) = \frac{c_r(\boldsymbol{p}) + d_r(\boldsymbol{p})}{\sqrt{2}} , \quad b_M^\dagger(\boldsymbol{p}, r) = \frac{c_r^\dagger(\boldsymbol{p}) + d_r^\dagger(\boldsymbol{p})}{\sqrt{2}} .$$

The anticommutation relations are derived from (8.E):

$$\{b_M(\boldsymbol{p}, r), b_M^\dagger(\boldsymbol{q}, s)\} = \delta_{rs}\delta^{(3)}(\boldsymbol{p} - \boldsymbol{q}) ,$$

$$\{b_M(\boldsymbol{p}, r), b_M(\boldsymbol{q}, s)\} = \{b_M^\dagger(\boldsymbol{p}, r), b_M^\dagger(\boldsymbol{q}, s)\} = 0 .$$

(d) The Dirac spinor is $\psi_D = \psi_1 + \mathrm{i}\psi_2$ where $\psi_{1,2}$ are Majorana spinors. The Lagrangian density is

$$\mathcal{L} = \mathrm{i}\bar{\psi}_1 \partial\!\!\!/ \psi_1 + \mathrm{i}\bar{\psi}_2 \partial\!\!\!/ \psi_2 - m(\bar{\psi}_1\psi_1 + \bar{\psi}_2\psi_2) + \mathrm{i}e(\bar{\psi}_1 A\!\!\!/ \psi_2 - \bar{\psi}_2 A\!\!\!/ \psi_1) .$$

8.18 Under Lorentz transformations the operator $V_\mu(x) = \bar{\psi}(x)\gamma_\mu\psi(x)$ transforms in the following way:

$$\begin{aligned}U(\Lambda)V_\mu(x)U^{-1}(\Lambda) &= U(\Lambda)\bar{\psi}(x)U^{-1}(\Lambda)\gamma_\mu U(\Lambda)\psi(x)U^{-1}(\Lambda) \\ &= \bar{\psi}(\Lambda x)S(\Lambda)\gamma_\mu S^{-1}(\Lambda)\psi(\Lambda x) = \Lambda^\nu{}_\mu V_\nu(\Lambda x) ,\end{aligned} \tag{8.35}$$

since $S\gamma_\mu S^{-1} = \Lambda^\nu{}_\mu\gamma_\nu$. The other operator $A_\mu(x) = \bar{\psi}(x)\gamma_5\partial_\mu\psi(x)$ transforms as

$$\begin{aligned}U(\Lambda)A_\mu(x)U^{-1}(\Lambda) &= U(\Lambda)\bar{\psi}(x)U^{-1}(\Lambda)\gamma_5\partial_\mu U(\Lambda)\psi(x)U^{-1}(\Lambda) \\ &= \bar{\psi}(\Lambda x)\gamma_5\partial_\mu\psi(\Lambda x) ,\end{aligned}$$

where we used well known relation $S\gamma_5 S^{-1} = \gamma_5$ (see Problem 4.38). Since $\partial_\mu = \Lambda^\rho{}_\mu\partial'_\rho$ we have

$$U(\Lambda)A_\mu(x)U^{-1}(\Lambda) = \Lambda^\rho{}_\mu A_\rho(\Lambda x) . \tag{8.36}$$

Under parity vector V_μ transforms as follows:

$$\begin{aligned}V_\mu(x) \to PV_\mu(x)P^{-1} &= \psi^\dagger(t, -\boldsymbol{x})\gamma_\mu\gamma_0\psi(t, -\boldsymbol{x}) \\ &= \begin{cases} V_0(t, -\boldsymbol{x}), & \text{for } \mu = 0 \\ -V_i(t, -\boldsymbol{x}), & \text{for } \mu = i \end{cases} \\ &= V^\mu(t, -\boldsymbol{x}) ,\end{aligned}$$

since

$$P\bar{\psi}(x)P^{-1} = (P\psi(x)P^{-1})^{\dagger}\gamma_0 = (\gamma_0\psi(t,-\boldsymbol{x}))^{\dagger}\gamma_0 = \psi^{\dagger}(t,-\boldsymbol{x}) .$$

In the similar way we get

$$\begin{aligned} PA_{\mu}(x)P^{-1} &= -\bar{\psi}(t,-\boldsymbol{x})\gamma_5\partial_{\mu}\psi(t,-\boldsymbol{x}) \\ &= \begin{cases} -\bar{\psi}(t,-\boldsymbol{x})\gamma_5\partial'_0\psi(t,-\boldsymbol{x}), & \text{for } \mu=0 \\ \bar{\psi}(t,-\boldsymbol{x})\gamma_5\partial'_i\psi(t,-\boldsymbol{x}), & \text{for } \mu=i \end{cases} \\ &= -A^{\mu}(t,-\boldsymbol{x}) . \end{aligned}$$

From $\tau\psi(t,\boldsymbol{x})\tau^{-1} = T\psi(-t,\boldsymbol{x})$, where τ is an antiunitary operator of time reversal follows

$$\tau\bar{\psi}(t,\boldsymbol{x})\tau^{-1} = \tau\psi^{\dagger}(t,\boldsymbol{x})\tau^{-1}\gamma_0^{*} = \psi^{\dagger}(-t,\boldsymbol{x})T^{\dagger}\gamma_0^{*} .$$

From the previous expressions we get

$$\tau V_{\mu}(t,\boldsymbol{x})\tau^{-1} = \psi^{\dagger}(-t,\boldsymbol{x})T^{\dagger}(\gamma_0\gamma_{\mu})^{*}T\psi(-t,\boldsymbol{x}) . \tag{8.37}$$

With a help of $T\gamma_{\mu}T^{-1} = \gamma^{\mu *}$ we get

$$\tau V_{\mu}(x)\tau^{-1} = \bar{\psi}(-t,\boldsymbol{x})\gamma^{\mu}\psi(-t,\boldsymbol{x}) = V^{\mu}(-t,\boldsymbol{x}) . \tag{8.38}$$

We would suggest to reader to prove the previous result by taking $T = \mathrm{i}\gamma^1\gamma^3$. The identity

$$(\mathrm{i}\gamma^1\gamma^3)^{\dagger}\gamma_0^{*}\gamma_{\mu}^{*}\mathrm{i}\gamma^1\gamma^3 = \gamma^0\gamma^{\mu} , \tag{8.39}$$

has to be shown. Under time reversal the operator $A_{\mu}(x)$ transforms as

$$\tau A_{\mu}(x)\tau^{-1} = -\bar{\psi}(-t,\boldsymbol{x})\gamma_5\partial'^{\mu}\psi(-t,\boldsymbol{x}) = -A^{\mu}(-t,\boldsymbol{x}) . \tag{8.40}$$

From $\mathcal{C}\psi_a(x)\mathcal{C}^{-1} = (C\gamma_0^T)_{ab}\psi_b^{\dagger}(x)$ follows $\mathcal{C}\bar{\psi}_a\mathcal{C}^{-1} = -\psi_b C_{ba}^{-1}$, where $\mathcal{C}$ is a unitary charge conjugation operator while C is a matrix. It is easy to see

$$\begin{aligned} \mathcal{C}V^{\mu}\mathcal{C}^{-1} &= -\psi_c C_{ca}^{-1}\gamma_{ab}^{\mu}C_{bd}\bar{\psi}_d \\ &= \psi_c(\gamma^{\mu})_{cd}^T\bar{\psi}_d \\ &= \psi_c(\gamma^{\mu})_{dc}\bar{\psi}_d \\ &= -\bar{\psi}_d\gamma_{dc}^{\mu}\psi_c \\ &= -V^{\mu} . \end{aligned}$$

The minus sign in the forth line of the previous calculation appears since the fields ψ and $\bar{\psi}$ anticommute. An infinity constant is ignored. Compare this result with result of Problem 4.37. In the similar way result $\mathcal{C}A_{\mu}\mathcal{C}^{-1} = \partial_{\mu}\bar{\psi}\gamma_5\psi$ is derived.

8.19 The Dirac Lagrangian density transforms as

$$U(\Lambda)\ldots U^{-1}(\Lambda) ,$$

with respect to Lorentz transformations. Therefore, we have:

$$\begin{aligned}
&U(\Lambda)\mathcal{L}(x)U^{-1}(\Lambda)\\
&= \mathrm{i}U(\Lambda)\bar{\psi}(x)U^{-1}(\Lambda)\gamma^\mu\partial_\mu U(\Lambda)\psi(x)U^{-1}(\Lambda) - mU(\Lambda)\bar{\psi}(x)\psi(x)U^{-1}(\Lambda)\\
&= \mathrm{i}\bar{\psi}(\Lambda x)S\gamma^\mu\partial_\mu S^{-1}\psi(\Lambda x) - m\bar{\psi}(\Lambda x)SS^{-1}\psi(\Lambda x)\\
&= \mathrm{i}(\Lambda^{-1})^\mu{}_\nu\bar{\psi}(\Lambda x)\gamma^\nu\Lambda^\rho{}_\mu\partial'_\rho\psi(\Lambda x) - \bar{\psi}(\Lambda x)\psi(\Lambda x)\\
&= \mathrm{i}\bar{\psi}(\Lambda x)\gamma^\mu\partial'_\mu\psi(\Lambda x) - m\bar{\psi}(\Lambda x)\psi(\Lambda x)\\
&= \mathcal{L}(\Lambda x) .
\end{aligned}$$

Under the parity $\mathcal{L}$ transforms as follows

$$\begin{aligned}
P\mathcal{L}P^{-1} &= \mathrm{i}\psi^\dagger(t,-\boldsymbol{x})\gamma^\mu\partial_\mu\gamma^0\psi(t,-\boldsymbol{x}) -\\
&\quad - m\bar{\psi}(t,-\boldsymbol{x})\psi(t,-\boldsymbol{x}) .
\end{aligned}$$

From

$$\gamma^\mu\gamma^0\partial_\mu = \gamma^0\gamma^0\partial'_0 + \gamma^0\gamma^i\partial'_i = \gamma^0\gamma^\mu\partial'_\mu ,$$

we get

$$P\mathcal{L}(t,\boldsymbol{x})P^{-1} = \mathcal{L}(t,-\boldsymbol{x}) .$$

The transformation rules under time reversal and charge conjugation in the previous problem were found using the general properties of matrices T and C. Here, we use explicit expressions for them. Starting from

$$\tau\psi(t,\boldsymbol{x})\tau^{-1} = \mathrm{i}\gamma^1\gamma^3\psi(-t,\boldsymbol{x}) , \tag{8.41}$$

we obtain

$$\begin{aligned}
\tau\bar{\psi}(t,\boldsymbol{x})\tau^{-1} &= \tau\psi^\dagger(t,\boldsymbol{x})\tau^{-1}\gamma_0^*\\
&= -\mathrm{i}\psi^\dagger(-t,\boldsymbol{x})(\gamma^3)^\dagger(\gamma^1)^\dagger(\gamma^0)^*\\
&= -\mathrm{i}\bar{\psi}(-t,\boldsymbol{x})\gamma^3\gamma^1 .
\end{aligned}$$

Further,

$$\begin{aligned}
\tau\mathcal{L}\tau^{-1} &= -\mathrm{i}\bar{\psi}(-t,\boldsymbol{x})\gamma^3\gamma^1(\gamma^\mu)^*\gamma^1\gamma^3\partial_\mu\psi(-t,\boldsymbol{x})\\
&\quad - m\bar{\psi}(-t,\boldsymbol{x})\gamma^3\gamma^1\gamma^1\gamma^3\psi(-t,\boldsymbol{x}) .
\end{aligned}$$

Applying

$$(\gamma^0)^* = \gamma^0,\ (\gamma^1)^* = \gamma^1,\ (\gamma^2)^* = -\gamma^2,\ (\gamma^3)^* = \gamma^3 ,$$

the anticommutation relation among γ–matrices and introducing derivatives with respect to new coordinates $t' = -t$, $\boldsymbol{x}' = \boldsymbol{x}$ instead of the old ones gives

$$\begin{aligned}\mathcal{T}\mathcal{L}\mathcal{T}^{-1} &= \mathrm{i}\bar{\psi}(-t,\boldsymbol{x})\gamma^\mu\partial'_\mu\psi(-t,\boldsymbol{x}) - m\bar{\psi}(-t,\boldsymbol{x})\psi(-t,\boldsymbol{x}) \\ &= \mathcal{L}(-t,\boldsymbol{x})\ .\end{aligned}$$

The transformation law for field ψ under charge conjugation

$$\mathcal{C}\psi_a\mathcal{C}^{-1} = \mathrm{i}(\gamma^2)_{ab}\psi_b^\dagger$$

induces

$$\mathcal{C}\bar{\psi}_a\mathcal{C}^{-1} = \mathrm{i}\psi_b(\gamma^2\gamma^0)_{ba}\ .$$

Then Lagrangian density transforms as

$$\mathcal{C}\mathcal{L}\mathcal{C}^{-1} = -\mathrm{i}\psi_c(\gamma^2\gamma^0\gamma^\mu\gamma^2)_{ca}\partial_\mu\psi_a^\dagger + m\psi_b(\gamma^2\gamma^0\gamma^2)_{ba}\psi_a^\dagger\ .$$

Since

$$\gamma^2\gamma^0\gamma^\mu\gamma^2\partial_\mu = (-\gamma^0\partial_0 + \gamma^1\partial_1 - \gamma^2\partial_2 + \gamma^3\partial_3)\gamma_0\ ,$$

then the kinetic term becomes

$$-\mathrm{i}\psi_c\left[-\gamma^0\partial_0 + \gamma^1\partial_1 - \gamma^2\partial_2 + \gamma^3\partial_3)\right]_{cd}\bar{\psi}_d\ .$$

In the Dirac representation of γ–matrices the following relations are satisfied:

$$(\gamma^0)^T = \gamma^0\ ,\quad (\gamma^1)^T = -\gamma^1\ ,\quad (\gamma^2)^T = \gamma^2\ ,\quad (\gamma^3)^T = -\gamma^3\ ,$$

and the kinetic term is

$$\mathrm{i}\psi_c(\gamma^{\mu T})_{cd}\partial_\mu\bar{\psi}_d = -\mathrm{i}\partial_\mu\bar{\psi}_d\gamma^\mu_{dc}\psi_c\ .$$

As in the previous problem we anticommute the fields $\bar{\psi}$ and ψ, and ignore the infinity constant $\delta^{(3)}(0)$. At the end we obtain

$$\mathcal{C}\mathcal{L}\mathcal{C}^{-1} = -\mathrm{i}\partial_\mu\bar{\psi}\gamma^\mu\psi - m\bar{\psi}\psi\ ,$$

which is the starting Lagrangian density up to four divergence.

8.20 From

$$S(\Lambda)\sigma_{\mu\nu}S^{-1}(\Lambda) = \Lambda^\rho{}_\mu\Lambda^\sigma{}_\nu\sigma_{\rho\sigma}\ , \tag{8.42}$$

follows

$$U(\Lambda)T_{\mu\nu}U^{-1}(\Lambda) = \Lambda^\rho{}_\mu\Lambda^\sigma{}_\nu T_{\rho\sigma}(\Lambda x)\ , \tag{8.43}$$

and therefore $T_{\mu\nu}$ is a second rank tensor. Under parity the transformation rule is:

$$\begin{aligned}PT_{0i}(t,\boldsymbol{x})P^{-1} &= -T_{0i}(t,-\boldsymbol{x})\ ,\\ PT_{ij}(t,\boldsymbol{x})P^{-1} &= T_{ij}(t,-\boldsymbol{x})\ .\end{aligned}$$

Charge conjugation act on a $T_{\mu\nu}$ tensor according to

$$\mathcal{C}T_{\mu\nu}(x)\mathcal{C}^{-1} = -T_{\mu\nu}(x)\ . \tag{8.44}$$

In order to confirm the previous result you should to prove that

$$C^{-1}\sigma_{\mu\nu}C = -(\sigma_{\mu\nu})^T \ . \tag{8.45}$$

The identity

$$T\sigma_{\mu\nu}T^{-1} = -(\sigma^{\mu\nu})^* \ , \tag{8.46}$$

can be derived easily. Consequently,

$$\tau T_{0i}(t, \boldsymbol{x})\tau^{-1} = T_{0i}(-t, \boldsymbol{x}) \ ,$$

$$\tau T_{ij}(t, \boldsymbol{x})\tau^{-1} = -T_{ij}(-t, \boldsymbol{x}) \ .$$

9

Canonical quantization of the electromagnetic field

9.1 The commutator is

$$[A^\mu(t,\boldsymbol{x}),\dot{A}^\nu(t,\boldsymbol{y})]=\sum_{\lambda,\lambda'}\frac{\mathrm{i}}{(2\pi)^3}\int\frac{\mathrm{d}^3\boldsymbol{k}\mathrm{d}^3\boldsymbol{q}}{2\sqrt{\omega_k\omega_q}}\omega_q\epsilon^\mu_\lambda(\boldsymbol{k})\epsilon^\nu_{\lambda'}(\boldsymbol{q})$$
$$\times\left([a_\lambda(\boldsymbol{k}),a^\dagger_{\lambda'}(\boldsymbol{q})]\mathrm{e}^{\mathrm{i}(\boldsymbol{k}\cdot\boldsymbol{x}-\boldsymbol{q}\cdot\boldsymbol{y})}\right.$$
$$\left.-\;[a^\dagger_\lambda(\boldsymbol{k}),a_{\lambda'}(\boldsymbol{q})]\mathrm{e}^{-\mathrm{i}(\boldsymbol{k}\cdot\boldsymbol{x}-\boldsymbol{q}\cdot\boldsymbol{y})}\right)\;.$$

Using the commutation relations (9.G) as well as completeness relations (9.D) we obtain

$$\begin{aligned}[A^\mu(t,\boldsymbol{x}),\dot{A}^\nu(t,\boldsymbol{y})]&=-\frac{\mathrm{i}}{2(2\pi)^3}g^{\mu\nu}\int\mathrm{d}^3\boldsymbol{k}\left(\mathrm{e}^{\mathrm{i}\boldsymbol{k}\cdot(\boldsymbol{x}-\boldsymbol{y})}+\mathrm{e}^{\mathrm{i}\boldsymbol{k}\cdot(\boldsymbol{y}-\boldsymbol{x})}\right)\\&=-\mathrm{i}g^{\mu\nu}\delta^{(3)}(\boldsymbol{x}-\boldsymbol{y})\;.\end{aligned}$$

9.2 Using the commutation relations (9.G) and the completeness relation (9.D) we get

$$\mathrm{i}D^{\mu\nu}=[A^\mu(x),A^\nu(y)]=-g^{\mu\nu}\frac{1}{(2\pi)^3}\int\frac{\mathrm{d}^3\boldsymbol{k}}{2|\boldsymbol{k}|}\left(\mathrm{e}^{-\mathrm{i}k\cdot(x-y)}-\mathrm{e}^{\mathrm{i}k\cdot(x-y)}\right)\;.\qquad(9.1)$$

In order to calculate the integral (9.1) we shall use spherical coordinates (using notation $x_0-y_0=t$, $|\boldsymbol{x}-\boldsymbol{y}|=r$)

$$\begin{aligned}\mathrm{i}D^{\mu\nu}(x-y)&=-g^{\mu\nu}\frac{1}{2(2\pi)^2}\int_0^\infty k\mathrm{d}k\int_0^\pi\mathrm{d}\theta\sin\theta\\&\times\left(\mathrm{e}^{-\mathrm{i}(kt-kr\cos\theta)}-\mathrm{e}^{\mathrm{i}(kt-kr\cos\theta)}\right)\\&=-g^{\mu\nu}\frac{1}{2(2\pi)^2}\frac{1}{\mathrm{i}r}\int_0^\infty\mathrm{d}k\left(\mathrm{e}^{-\mathrm{i}kt}(\mathrm{e}^{\mathrm{i}kr}-\mathrm{e}^{-\mathrm{i}kr})+\mathrm{e}^{\mathrm{i}kt}(\mathrm{e}^{-\mathrm{i}kr}-\mathrm{e}^{\mathrm{i}kr})\right)\end{aligned}$$

$$= -g^{\mu\nu}\frac{1}{2(2\pi)^2}\frac{1}{\mathrm{i}r}\int_{-\infty}^{\infty}\mathrm{d}k\left(\mathrm{e}^{-\mathrm{i}kt+\mathrm{i}kr}-\mathrm{e}^{-\mathrm{i}kt-\mathrm{i}kr}\right)$$
$$= -g^{\mu\nu}\frac{1}{4\pi\mathrm{i}r}\left(\delta(t-r)-\delta(t+r)\right)$$
$$= \mathrm{i}g^{\mu\nu}\frac{1}{2\pi}\epsilon(t)\delta(t^2-r^2)\ , \qquad (9.2)$$

where

$$\epsilon(t) = \begin{cases} 1, & t > 0 \\ -1, & t < 0 \\ 0, & t = 0 \end{cases} .$$

The previous result in terms of x and y coordinates has the form

$$\begin{aligned} \mathrm{i}D^{\mu\nu}(x-y) &= -\mathrm{i}g^{\mu\nu}D(x-y) \\ &= g^{\mu\nu}\frac{\mathrm{i}}{4\pi|\boldsymbol{x}-\boldsymbol{y}|}\left(\delta(x_0-y_0-|\boldsymbol{x}-\boldsymbol{y}|)-\delta(x_0-y_0+|\boldsymbol{x}-\boldsymbol{y}|)\right) \\ &= \frac{\mathrm{i}}{2\pi}g^{\mu\nu}\epsilon(x_0-y_0)\delta^{(4)}((x-y)^2)\ . \end{aligned}$$

9.3 Both the electric and magnetic fields are gauge invariants. The simplest way to calculate the commutators is in the Lorentz gauge. The first commutator is

$$[E^i(x), E^j(y)] = \partial_x^i\partial_y^j[A^0(x), A^0(y)] + \partial_x^0\partial_y^0[A^i(x), A^j(y)]\ , \qquad (9.3)$$

where we used relation between the electric field and the electromagnetic potential:

$$\boldsymbol{E} = -\nabla A^0 - \frac{\partial \boldsymbol{A}}{\partial t}\ .$$

Using Problem 9.2 we get

$$[E^i(x), E^j(y)] = \mathrm{i}(\partial_x^i\partial_x^j - \delta_{ij}\partial_x^0\partial_x^0)D(x-y)\ .$$

The commutator between the components of the magnetic field is:

$$\begin{aligned} [B^i(x), B^j(y)] &= \epsilon^{ikl}\epsilon^{jmn}\partial_k^x\partial_m^y[A^l(x), A^n(y)] \\ &= \mathrm{i}\epsilon^{ikl}\epsilon^{jml}\partial_k^x\partial_m^y D(x-y) \\ &= \mathrm{i}(\delta^{ij}\delta^{km} - \delta^{im}\delta^{kj})\partial_k^x\partial_m^y D(x-y) \\ &= \mathrm{i}(-\delta^{ij}\Delta + \partial_i^x\partial_j^x)D(x-y)\ . \end{aligned}$$

In the similar way one can get

$$[E^i(x), B^j(y)] = \mathrm{i}\epsilon^{jki}\partial_0^x\partial_k^x D(x-y)\ .$$

Now, consider the equal–time commutators i.e. take that $x^0 = y^0$. First show that

$$\partial_{x^0}D(x-y)|_{x^0=y^0} = -\delta^{(3)}(\boldsymbol{x}-\boldsymbol{y})\ ,$$

$$\begin{aligned}
\partial^2_{x^0} D(x-y)|_{x^0=y^0} &= 0 \ , \\
\partial^i_x D(x-y)|_{x^0=y^0} &= 0 \ , \\
\partial^i_x \partial^j_x D(x-y)|_{x^0=y^0} &= 0 \ , \\
\partial^x_i \partial^x_0 D(x-y)|_{x^0=y^0} &= -\partial^x_i \delta^{(3)}(\boldsymbol{x}-\boldsymbol{y}) \ .
\end{aligned}$$

The easiest way to prove the previous formulae is to start with the integral expression for D–function:

$$D(x-y) = -\frac{\mathrm{i}}{(2\pi)^3}\int \frac{\mathrm{d}^3\boldsymbol{k}}{2|\boldsymbol{k}|}\left(\mathrm{e}^{-\mathrm{i}k\cdot(x-y)} - \mathrm{e}^{\mathrm{i}k\cdot(x-y)}\right) \ .$$

The results for the equal–time commutators are:

$$\begin{aligned}
[E^i(x), E^j(y)]|_{x^0=y^0} &= 0 \ , \\
[B^i(x), B^j(y)]|_{x^0=y^0} &= 0 \ , \\
[E^i(x), B^j(y)]|_{x^0=y^0} &= -\mathrm{i}\epsilon^{ijk}\partial^x_k \delta^{(3)}(\boldsymbol{x}-\boldsymbol{y}) \ .
\end{aligned}$$

9.4 We shall first calculate the commutator between the Hamiltonian and the electromagnetic potential $A^\nu(x)$:

$$\begin{aligned}
[H, A^\nu(x)] &= -\frac{1}{2}\int \mathrm{d}^3\boldsymbol{y}[\pi^\mu\pi_\mu + \nabla A^\mu \nabla A_\mu, A^\nu(x)] \\
&= -\frac{1}{2}\int \mathrm{d}^3\boldsymbol{y}\left(\pi^\mu(y)[\pi_\mu(y), A^\nu(x)] + [\pi^\mu(y), A^\nu(x)]\pi_\mu(y)\right) \\
&= -\frac{1}{2}\int \mathrm{d}^3\boldsymbol{y}\delta^{(3)}(\boldsymbol{x}-\boldsymbol{y})\left(\pi^\mu(y)(-\mathrm{i})g^\nu_\mu - \mathrm{i}g^{\mu\nu}\pi_\mu(y)\right) \\
&= \mathrm{i}\pi^\nu(x) \\
&= -\mathrm{i}\partial^0 A^\nu \ .
\end{aligned}$$

The commutator between three–momentum of electromagnetic field and electromagnetic potential can be calculated in the similar manner

$$\begin{aligned}
[P^i, A^\nu(x)] &= -\int \mathrm{d}^3\boldsymbol{y}[\dot{A}^\rho(y)\partial^i A_\rho(y), A^\nu(x)] \\
&= -\mathrm{i}g^{\rho\nu}\int \mathrm{d}^3\boldsymbol{y}\delta^{(3)}(\boldsymbol{x}-\boldsymbol{y})\partial^i A_\rho(y) \\
&= -\mathrm{i}\partial^i A^\nu(x) \ .
\end{aligned}$$

9.5 The helicity of the state $\epsilon^\mu_{(\pm)}(\boldsymbol{k})$ is determined under the rotation for angle θ about $\boldsymbol{k}/|\boldsymbol{k}| = \boldsymbol{e}_z$–axis. Namely,

$$
\begin{aligned}
\epsilon'_{\pm} &= \Lambda(\theta)\epsilon_{\pm} \\
&= \begin{pmatrix} 1 & 0 & 0 & 0 \\ 0 & \cos\theta & \sin\theta & 0 \\ 0 & -\sin\theta & \cos\theta & 0 \\ 0 & 0 & 0 & 1 \end{pmatrix} \begin{pmatrix} 0 \\ 1/\sqrt{2} \\ \pm\mathrm{i}/\sqrt{2} \\ 0 \end{pmatrix} \\
&= \mathrm{e}^{\pm\mathrm{i}\theta} \begin{pmatrix} 0 \\ 1/\sqrt{2} \\ \pm\mathrm{i}/\sqrt{2} \\ 0 \end{pmatrix} \\
&= \mathrm{e}^{\pm\mathrm{i}\theta}\epsilon_{\pm} \, .
\end{aligned}
$$

From the last line we can read off that helicity is $\lambda = \pm 1$. Polarization of these photons is circular.

9.6 The four–momentum of the photon for observer S' is

$$
k'^{\mu} = \Lambda^{\mu}{}_{\nu} k^{\nu} = \begin{pmatrix} \gamma & -\beta\gamma & 0 & 0 \\ -\beta\gamma & \gamma & 0 & 0 \\ 0 & 0 & 1 & 0 \\ 0 & 0 & 0 & 1 \end{pmatrix} \begin{pmatrix} k \\ 0 \\ 0 \\ k \end{pmatrix} = \begin{pmatrix} k\gamma \\ -k\beta\gamma \\ 0 \\ k \end{pmatrix} .
$$

Under the Lorentz transformation the polarization vector $\epsilon^{\mu}(k)$ transforms as

$$
\epsilon'^{\mu}(k') = \Lambda^{\mu}{}_{\nu}\epsilon^{\nu}(k) - \mathrm{i}\alpha(k')k'^{\mu} \, .
$$

The second term comes from the gauge transformation of the electromagnetic potential; $\alpha(\kappa')$ is an arbitrary function of the momentum. This term can be easily obtained by substituting

$$
A'^{\mu} = \epsilon'^{\mu}(k')\mathrm{e}^{-\mathrm{i}k'\cdot x'} \, ,
$$

and

$$
\Lambda(x') = \alpha \mathrm{e}^{-\mathrm{i}k'\cdot x'}
$$

in the gauge transformation rule

$$
\tilde{A}^{\mu} = A'^{\mu} + \partial'^{\mu}\Lambda(x') \, .
$$

If we choose the function $\alpha = \mathrm{i}\beta/k$ we get

$$
\epsilon'^{\mu}(k') = \begin{pmatrix} 0 \\ \gamma^{-1} \\ 0 \\ \beta \end{pmatrix} .
$$

Note that the vector $\boldsymbol{\epsilon}'$ is orthogonal to the photon direction of motion $\boldsymbol{k}'/k'$. This was a condition to determine the function $\alpha(k')$. Thus, the polarization of photon is transversal for both observers.

9.7

(a) In the first step use the commutation relations (9.G) to derive the expression:

$$[a_3(\boldsymbol{k}) - a_0(\boldsymbol{k}), a_3^\dagger(\boldsymbol{q}) - a_0^\dagger(\boldsymbol{q})] = 0\ .$$

From the previous result it is not hard to show that $\langle \Phi_n | \Phi_n \rangle = \delta_{n0}$.

(b) There are only two terms in the expression $\langle \Phi | A^\mu | \Phi \rangle$ which are not equal to zero:

$$\langle \Phi | A^\mu | \Phi \rangle = C_0^* C_1 \langle \Phi_0 | A^\mu | \Phi_1 \rangle + C_0 C_1^* \langle \Phi_1 | A^\mu | \Phi_0 \rangle\ .$$

It is easy to see that

$$\langle \Phi_0 | A^\mu | \Phi_1 \rangle = -\frac{1}{(2\pi)^{3/2}} \int \frac{\mathrm{d}^3\boldsymbol{k}}{\sqrt{2|\boldsymbol{k}|}} f(\boldsymbol{k}) \mathrm{e}^{-\mathrm{i}k\cdot x} \left(\epsilon^\mu_{(0)}(\boldsymbol{k}) + \epsilon^\mu_{(3)}(\boldsymbol{k}) \right)\ .$$

By applying the relation

$$\epsilon^\mu_{(0)}(\boldsymbol{k}) + \epsilon^\mu_{(3)}(\boldsymbol{k}) = \frac{k^\mu}{|\boldsymbol{k}|}\ ,$$

we get

$$\langle \Phi | A^\mu | \Phi \rangle = \partial^\mu \Lambda\ ,$$

where Λ is given by

$$\Lambda = -\frac{\mathrm{i}}{(2\pi)^{3/2}} \int \frac{\mathrm{d}^3\boldsymbol{k}}{\sqrt{2|\boldsymbol{k}|}|\boldsymbol{k}|} \left(C_0^* C_1 f(\boldsymbol{k}) \mathrm{e}^{-\mathrm{i}k\cdot x} - C_0 C_1^* f^*(\boldsymbol{k}) \mathrm{e}^{\mathrm{i}k\cdot x} \right)\ .$$

9.8 The quantities defined in this problem are projectors on massless states with the helicities ± 1 and 0. Let us first calculate $P_\perp^{\mu\nu} P_{\nu\sigma\perp}$:

$$\begin{aligned} P_\perp^{\mu\nu} P_{\nu\sigma\perp} &= \frac{k^\mu \bar{k}^\nu + k^\nu \bar{k}^\mu}{k\cdot\bar{k}} \frac{k_\nu \bar{k}_\sigma + k_\sigma \bar{k}_\nu}{k\cdot\bar{k}} \\ &= \frac{k^\mu \bar{k}_\sigma + k_\sigma \bar{k}^\mu}{k\cdot\bar{k}} \\ &= P^\mu_{\sigma\perp}\ , \end{aligned}$$

since $\bar{k}\cdot\bar{k} = 0$. The other expressions can be evaluated in the same way. The results are:

$$P^{\mu\nu} P_{\nu\sigma} = P^\mu_\sigma\ , \quad P^{\mu\nu} + P_\perp^{\mu\nu} = g^{\mu\nu}\ ,$$
$$g^{\mu\nu} P_{\mu\nu} = 2\ , \quad g^{\mu\nu} P^\perp_{\mu\nu} = 2\ , \quad P_{\mu\nu} P_\perp^{\nu\sigma} = 0\ .$$

9.9

(a) The components of the angular momentum M^{ij} were calculated in Problem 5.18 using the Nether technique. It follows that (in the Coulomb gauge)

$$J^l = \epsilon^{lij} \int \mathrm{d}^3\boldsymbol{x} \left(\dot{A}^j A^i + x^i \dot{A}^k \partial^j A^k \right)\ .$$

(b) The spin part of the angular momentum is

$$S^l = \epsilon^{lij} \int \mathrm{d}^3\boldsymbol{x} \dot{A}^j A^i \ .$$

By substituting the explicit expression for the electromagnetic potential we get

$$\begin{aligned} S^l = \frac{\mathrm{i}}{2}\epsilon^{lij} \sum_{\lambda,\lambda'} \int \mathrm{d}^3\boldsymbol{k} \Big(& -\epsilon^j_\lambda(\boldsymbol{k})\epsilon^i_{\lambda'}(-\boldsymbol{k}) a_\lambda(\boldsymbol{k}) a_{\lambda'}(-\boldsymbol{k}) \mathrm{e}^{-2\mathrm{i}\omega_k t} - \\ & - \epsilon^j_\lambda(\boldsymbol{k})\epsilon^i_{\lambda'}(\boldsymbol{k}) : a_\lambda(\boldsymbol{k}) a^\dagger_{\lambda'}(\boldsymbol{k}) : + \epsilon^j_\lambda(\boldsymbol{k})\epsilon^i_{\lambda'}(\boldsymbol{k}) a^\dagger_\lambda(\boldsymbol{k}) a_{\lambda'}(\boldsymbol{k}) + \\ & + \epsilon^j_\lambda(\boldsymbol{k})\epsilon^i_{\lambda'}(-\boldsymbol{k}) a^\dagger_\lambda(\boldsymbol{k}) a^\dagger_{\lambda'}(-\boldsymbol{k}) \mathrm{e}^{2\mathrm{i}\omega_k t} \Big) \ . \end{aligned}$$

The first and the last term are symmetric under the change of indices i and j, so that the multiplication by the antisymmetric ϵ symbol give vanishing contribution. Then:

$$\boldsymbol{S} = \frac{\mathrm{i}}{2} \sum_{\lambda,\lambda'} \int \mathrm{d}^3\boldsymbol{k} \left(\boldsymbol{\epsilon}_{\lambda'}(\boldsymbol{k}) \times \boldsymbol{\epsilon}_\lambda(\boldsymbol{k}) \right) \left(a^\dagger_\lambda(\boldsymbol{k}) a_{\lambda'}(\boldsymbol{k}) - a^\dagger_{\lambda'}(\boldsymbol{k}) a_\lambda(\boldsymbol{k}) \right) \ .$$

By using $\boldsymbol{\epsilon}_1(\boldsymbol{k}) \times \boldsymbol{\epsilon}_2(\boldsymbol{k}) = \boldsymbol{k}/|\boldsymbol{k}|$ we get

$$\boldsymbol{S} = \mathrm{i} \int \mathrm{d}^3\boldsymbol{k} \frac{\boldsymbol{k}}{|\boldsymbol{k}|} \left(a^\dagger_2(\boldsymbol{k}) a_1(\boldsymbol{k}) - a^\dagger_1(\boldsymbol{k}) a_2(\boldsymbol{k}) \right) \ .$$

By using the operators $a_\pm(\boldsymbol{k})$ which were defined in the problem, the spin $\boldsymbol{S}$ becomes diagonal

$$\boldsymbol{S} = \int \mathrm{d}^3\boldsymbol{k} \frac{\boldsymbol{k}}{|\boldsymbol{k}|} \left(a^\dagger_+(\boldsymbol{k}) a_+(\boldsymbol{k}) - a^\dagger_-(\boldsymbol{k}) a_-(\boldsymbol{k}) \right) \ .$$

From the previous result we conclude that the operator

$$\Lambda = \int \mathrm{d}^3\boldsymbol{k} \left(a^\dagger_+(\boldsymbol{k}) a_+(\boldsymbol{k}) - a^\dagger_-(\boldsymbol{k}) a_-(\boldsymbol{k}) \right) \ ,$$

is the helicity.

(c) By applying the commutation relations (9.J) we get

$$[a^\dagger_\pm(\boldsymbol{k}), a_\pm(\boldsymbol{q})] = -\delta^{(3)}(\boldsymbol{k} - \boldsymbol{q}) \ ,$$

from which we have

$$\begin{aligned} \Lambda a^\dagger_\pm(\boldsymbol{q}) |0\rangle &= [\Lambda, a^\dagger_\pm(\boldsymbol{q})] |0\rangle \\ &= \pm \int \mathrm{d}^3\boldsymbol{k} \delta^{(3)}(\boldsymbol{k} - \boldsymbol{q}) a^\dagger_\pm(\boldsymbol{k}) |0\rangle \\ &= \pm a^\dagger_\pm(\boldsymbol{q}) |0\rangle \ . \end{aligned}$$

(d) The commutator between the angular momentum and the electromagnetic potential is:

$$
\begin{aligned}
[J^l, A^m(t,\boldsymbol{x})] &= \epsilon^{lij}\int \mathrm{d}^3\boldsymbol{y}\left[\dot{A}^j(t,\boldsymbol{y}), A^m(t,\boldsymbol{x})\right] A^i(t,\boldsymbol{y}) + \\
&+ y^i[\dot{A}^n(t,\boldsymbol{y}), A^m(t,\boldsymbol{x})]\partial^j A^n(t,\boldsymbol{y}) \\
&= -\mathrm{i}\epsilon^{lij}\int \mathrm{d}^3\boldsymbol{y}\delta^{(3)}_{\perp nm}(\boldsymbol{x}-\boldsymbol{y})\left(\delta_{nj}A^i(t,\boldsymbol{y}) + y^i\partial^j A^n(t,\boldsymbol{y})\right) \\
&= -\mathrm{i}\epsilon^{lij}\frac{1}{(2\pi)^3}\int \mathrm{d}^3\boldsymbol{y}\int \mathrm{d}^3\boldsymbol{k}\mathrm{e}^{\mathrm{i}\boldsymbol{k}\cdot(\boldsymbol{x}-\boldsymbol{y})}\left(\delta_{nm} - \frac{k^n k^m}{\boldsymbol{k}^2}\right) \\
&\times \left(\delta_{jn}A^i(t,\boldsymbol{y}) + y^i\partial^j A^n(t,\boldsymbol{y})\right) \; . \qquad (9.4)
\end{aligned}
$$

The term which contains $k^n k^m/\boldsymbol{k}^2$ is equal to zero:

$$
\begin{aligned}
&\int \mathrm{d}^3\boldsymbol{y}\int \mathrm{d}^3\boldsymbol{k}\frac{k^n k^m}{\boldsymbol{k}^2}\mathrm{e}^{\mathrm{i}\boldsymbol{k}\cdot(\boldsymbol{x}-\boldsymbol{y})}\left(A^i\delta_{nj} + y^i\partial^j A^n\right) \\
&= \int \mathrm{d}^3\boldsymbol{y}\int \mathrm{d}^3\boldsymbol{k}\left(A^i\delta_{nj} + y^i\partial^j A^n\right)\frac{k^m}{\boldsymbol{k}^2}(\mathrm{i}\frac{\partial}{\partial y^n}\mathrm{e}^{\mathrm{i}\boldsymbol{k}\cdot(\boldsymbol{x}-\boldsymbol{y})}) \; . \qquad (9.5)
\end{aligned}
$$

Integrating by parts in (9.5) we get that it vanishes. Then from (9.4) follows

$$
[J^l, A^m(t,\boldsymbol{x})] = \mathrm{i}\epsilon^{lmi}A^i + \mathrm{i}(\boldsymbol{r}\times\nabla)^l A^m \; .
$$

9.10 The electric field is

$$
\boldsymbol{E} = \int \frac{\mathrm{d}^3\boldsymbol{k}}{\sqrt{2(2\pi)^3\omega_k}}\sum_{\lambda=1}^{2}\mathrm{i}\omega_k\boldsymbol{\epsilon}_\lambda(\boldsymbol{k})\left(a_\lambda(\boldsymbol{k})\mathrm{e}^{-\mathrm{i}k\cdot x} - a^\dagger_\lambda(\boldsymbol{k})\mathrm{e}^{\mathrm{i}k\cdot x}\right) \; ,
$$

while the magnetic field is given by

$$
\boldsymbol{B} = \int \frac{\mathrm{d}^3\boldsymbol{k}}{\sqrt{2(2\pi)^3\omega_k}}\sum_{\lambda=1}^{2}\mathrm{i}(\boldsymbol{k}\times\boldsymbol{\epsilon}_\lambda(\boldsymbol{k}))\left(a_\lambda(\boldsymbol{k})\mathrm{e}^{-\mathrm{i}k\cdot x} - a^\dagger_\lambda(\boldsymbol{k})\mathrm{e}^{\mathrm{i}k\cdot x}\right) \; .
$$

(a) The vacuum expectation value of the anticommutator between the electric and the magnetic field is

$$
\begin{aligned}
&\langle 0|\{E^m(x), B^n(y)\}|0\rangle = \langle 0|E^m(x)B^n(y)|0\rangle + \langle 0|B^n(y)E^m(x)|0\rangle \\
&= \int \frac{\mathrm{d}^3\boldsymbol{k}\mathrm{d}^3\boldsymbol{q}}{2(2\pi)^3\sqrt{\omega_k\omega_q}}\sum_{\lambda=1}^{2}\sum_{\lambda'=1}^{2}\omega_k\epsilon^m_\lambda(\boldsymbol{k})(\boldsymbol{q}\times\boldsymbol{\epsilon}_{\lambda'}(\boldsymbol{q}))^n \\
&\times \left(\langle 0|a_\lambda(\boldsymbol{k})a^\dagger_{\lambda'}(\boldsymbol{q})|0\rangle\,\mathrm{e}^{-\mathrm{i}k\cdot x+\mathrm{i}q\cdot y} + \langle 0|a_{\lambda'}(\boldsymbol{q})a^\dagger_\lambda(\boldsymbol{k})|0\rangle\,\mathrm{e}^{\mathrm{i}k\cdot x-\mathrm{i}q\cdot y}\right) \\
&= \int \frac{\mathrm{d}^3\boldsymbol{k}}{2(2\pi)^3}\sum_{\lambda=1}^{2}\epsilon^m_\lambda(\boldsymbol{k})(\boldsymbol{k}\times\boldsymbol{\epsilon}_\lambda(\boldsymbol{k}))^n\left(\mathrm{e}^{-\mathrm{i}k\cdot(x-y)} + \mathrm{e}^{\mathrm{i}k\cdot(x-y)}\right) \; . \qquad (9.6)
\end{aligned}
$$

By using

$$\sum_{\lambda=1}^{2} \epsilon^{nij} k^i \epsilon_\lambda^j(\boldsymbol{k}) \epsilon_\lambda^m(\boldsymbol{k}) = \epsilon^{nim} k^i ,$$

the formula (9.6) becomes

$$\langle 0| \{E^m(x), B^n(y)\} |0\rangle = \int \frac{d^3\boldsymbol{k}}{2(2\pi)^3} \epsilon^{njm} k^j \left(e^{-ik\cdot(x-y)} + e^{ik\cdot(x-y)} \right) .$$

The result can be rewritten in the following form:

$$\begin{aligned} \langle 0| \{E^m(x), B^n(y)\} |0\rangle &= \epsilon^{njm} \frac{\partial^2}{\partial x^0 \partial x^j} \int \frac{d^3\boldsymbol{k}}{2(2\pi)^3 \omega_k} \\ &\quad \times \left(e^{-ik\cdot(x-y)} + e^{ik\cdot(x-y)} \right) \\ &= -\frac{1}{2\pi^2} \epsilon^{njm} \frac{\partial^2}{\partial x^o \partial x^j} \frac{1}{(x-y)^2} . \end{aligned} \tag{9.7}$$

The integral in the first line was calculated in Problem 7.14.

(b) As before,

$$\begin{aligned} \langle 0| \{B^i(x), B^j(y)\} |0\rangle = \int \frac{d^3\boldsymbol{k}}{2(2\pi)^3 \omega_k} &\sum_{\lambda=1}^{2} (\boldsymbol{k} \times \epsilon_\lambda(\boldsymbol{k}))^i (\boldsymbol{k} \times \epsilon_\lambda(\boldsymbol{k}))^j \\ \times &\left(e^{-ik\cdot(x-y)} + e^{ik\cdot(x-y)} \right) . \end{aligned}$$

Since

$$\begin{aligned} (\boldsymbol{k} \times \epsilon_\lambda(\boldsymbol{k}))^i (\boldsymbol{k} \times \epsilon_\lambda(\boldsymbol{k}))^j &= \sum_{\lambda=1}^{2} \epsilon^{imn} \epsilon^{jpq} k^m k^p \epsilon_\lambda^n(\boldsymbol{k}) \epsilon_\lambda^q(\boldsymbol{k}) \\ &= \epsilon^{imn} \epsilon^{jpn} k^m k^p \\ &= (\boldsymbol{k}^2 \delta^{ij} - k^i k^j) . \end{aligned}$$

we have

$$\begin{aligned} \langle 0| \{B^i(x), B^j(y)\} |0\rangle &= \int \frac{d^3\boldsymbol{k}}{2(2\pi)^3 \omega_k} (\boldsymbol{k}^2 \delta^{ij} - k^i k^j) \\ &\quad \times \left(e^{-ik\cdot(x-y)} + e^{ik\cdot(x-y)} \right) \\ &= -\frac{1}{2\pi^2} \left(\frac{\partial^2}{\partial x^i \partial x^j} - \Delta \delta^{ij} \right) \frac{1}{(x-y)^2} . \end{aligned}$$

(c) This expectation value can be obtained in the same way as the previous ones. The result is

$$\langle 0| \{E^i(x), E^j(y)\} |0\rangle = -\frac{1}{2\pi^2} \left(-\frac{\partial^2}{\partial (x^0)^2} \delta_{ij} + \frac{\partial^2}{\partial x^i \partial x^j} \right) \frac{1}{(x-y)^2} . \tag{9.8}$$

9.11

(a) The vector potential $\boldsymbol{A}$ can be decomposed into parallel and normal components:

$$\boldsymbol{A} = \boldsymbol{A}_\perp + \boldsymbol{A}_\parallel .$$

The normal component of the vector potential is along the z–axis, while $\boldsymbol{A}_\parallel$ is parallel to the plates. In the Coulomb gauge ($A^0 = 0$, $\mathrm{div}\boldsymbol{A} = 0$) the electric field is

$$\boldsymbol{E} = -\frac{\partial \boldsymbol{A}}{\partial t} .$$

Since the plates are ideal conductors, the parallel component of the electric field and the normal component of magnetic field vanish on the plates, i.e.

$$\left.\frac{\partial \boldsymbol{A}_\parallel}{\partial t}\right|_{z=0} = \left.\frac{\partial \boldsymbol{A}_\parallel}{\partial t}\right|_{z=a} = 0 , \tag{9.9}$$

$$B_z|_{z=0} = B_z|_{z=a} = 0 . \tag{9.10}$$

The vector potential $\boldsymbol{A}$ satisfies the equation

$$\left(\frac{\partial^2}{\partial t^2} - \Delta\right)\boldsymbol{A} = 0 .$$

If we assume that a particular solution of this equation has the following form

$$\boldsymbol{A} = F(t,x,y)(Z_1(z)\boldsymbol{e_1} + Z_2(z)\boldsymbol{e_2} + Z_3(z)\boldsymbol{e_3}) , \tag{9.11}$$

then we get:

$$\frac{d^2 Z_i}{dz^2} + k_3^2 Z_i = 0 \tag{9.12}$$

and

$$\left(\frac{\partial^2}{\partial t^2} - \frac{\partial^2}{\partial x^2} - \frac{\partial^2}{\partial y^2} + k_3^2\right) F = 0 . \tag{9.13}$$

The solution of the first equation is

$$Z_i = a_i \sin(k_3 z) + b_i \cos(k_3 z) .$$

The boundary conditions (9.9–9.10) give $b_1 = b_2 = 0$ and $k_3 = n\pi/a$ ($n = 0, 1, 2, \ldots$). A particular solution for the function F is $F = \mathrm{e}^{-\mathrm{i}\omega t + \mathrm{i}k_1 x + \mathrm{i}k_2 y}$. Inserting it into (9.13) we obtain

$$\omega = \pm\omega_{k,n} = \pm\sqrt{k_1^2 + k_2^2 + \left(\frac{n\pi}{a}\right)^2} .$$

From the Coulomb gauge condition follows that $a_3 = 0$ and

$$\mathrm{i}a_1 k_1 + \mathrm{i}a_2 k_2 - \frac{n\pi}{a} b_3 = 0$$

for $n \neq 0$; obviously there are two independent states of polarization, unless $n = 0$. For $n = 0$ polarization vector is $\boldsymbol{e}_3$, and there is only one mode. Thus, a particular solution is

$$\boldsymbol{A} = F\left(\boldsymbol{\epsilon}_{\|}\sin(n\pi z/a) + b_3\boldsymbol{e}_3\cos(n\pi z/a)\right) ,$$

where $\boldsymbol{\epsilon}_{\|}$ belongs to the xy–plane. Then, the general solution reads:

$$\begin{aligned}\boldsymbol{A} = & \sum_{n=1}^{\infty}\int\frac{d^2k}{2\pi}\frac{1}{\sqrt{2\omega_{k,n}}}\sum_{\lambda=1}^{2}[a_\lambda(k_1,k_2,n)\mathrm{e}^{-\mathrm{i}\omega_{k,n}t+\mathrm{i}k_1x+\mathrm{i}k_2y}\\ & \times(\sin(n\pi z/a)\boldsymbol{\epsilon}_{\|}(\boldsymbol{k},n,\lambda)+\cos(n\pi z/a)\boldsymbol{e}_3)+\\ & + a_\lambda^\dagger(k_1,k_2,n)\mathrm{e}^{\mathrm{i}\omega_{k,n}t-\mathrm{i}k_1x-\mathrm{i}k_2y}\\ & \times(\sin(n\pi z/a)\boldsymbol{\epsilon}_{\|}^*(\boldsymbol{k},n,\lambda)+\cos(n\pi z/a)\boldsymbol{e}_z)]+\\ & +\int\frac{d^2k}{2\pi}\frac{1}{\sqrt{2\omega_k}}[a(k_1,k_2)\mathrm{e}^{-\mathrm{i}\omega_k t+\mathrm{i}k_1x+\mathrm{i}k_2y}+\\ & + a^\dagger(k_1,k_2)\mathrm{e}^{\mathrm{i}\omega_k t-\mathrm{i}k_1x-\mathrm{i}k_2y}]\boldsymbol{e}_3 ,\end{aligned} \tag{9.14}$$

where $\omega_k = \sqrt{k_1^2+k_2^2}$.

(b) The canonical commutation relations have the following form

$$[a_\lambda(k_1,k_2,n),a_{\lambda'}^\dagger(k_1',k_2',m)] = \delta_{nm}\delta_{\lambda\lambda'}\delta(k_1-k_1')\delta(k_2-k_2') ,$$

$$[a(k_1,k_2),a^\dagger(k_1',k_2')] = \delta(k_1-k_1')\delta(k_2-k_2') ,$$

while the other commutators vanish. The Hamiltonian is given by

$$\begin{aligned}H = & \int\mathrm{d}^2k\sum_{n=1}^{\infty}\frac{1}{2}\omega_{k,n}\sum_{\lambda=1}^{2}[a_\lambda^\dagger(k_1,k_2,n)a_\lambda(k_1,k_2,n)\\ & + a_\lambda(k_1,k_2,n)a_\lambda^\dagger(k_1,k_2,n)]\\ & + \frac{1}{2}\int\mathrm{d}^2k\omega_k[a^\dagger(k_1,k_2)a(k_1,k_2)+a(k_1,k_2)a^\dagger(k_1,k_2)] .\end{aligned} \tag{9.15}$$

(c) The energy of the ground state $|0\rangle$ is

$$\begin{aligned}\langle 0|H|0\rangle = & \sum_{n=1}^{\infty}\sum_{\lambda=1}^{2}\int\mathrm{d}^2k\frac{1}{2}\omega_{k,n}\langle 0|a_\lambda(k_1,k_2,n)a_\lambda^\dagger(k_1,k_2,n)|0\rangle\\ & +\int\mathrm{d}^2k\frac{1}{2}\omega_k\langle 0|a(k_1,k_2)a^\dagger(k_1,k_2)|0\rangle\\ = & \sum_{n=1}^{\infty}\frac{1}{2}\int\mathrm{d}^2k\omega_{k,n}2\delta^{(2)}(0)+\frac{1}{2}\int\mathrm{d}^2k\omega_k\delta^{(2)}(0) .\end{aligned}$$

Since

$$\delta^{(2)}(0) = \int \frac{\mathrm{d}x\mathrm{d}y}{(2\pi)^2} \mathrm{e}^{\mathrm{i}k_1 x + \mathrm{i}k_2 y}\Big|_{\boldsymbol{k}_\parallel = 0} = \frac{L^2}{(2\pi)^2}$$

we have

$$E = \frac{L^2}{2(2\pi)^2} \int \mathrm{d}^2 k \left(2\sum_{n=1}^{\infty} \sqrt{k_1^2 + k_2^2 + \left(\frac{n\pi}{a}\right)^2} + \sqrt{k_1^2 + k_2^2} \right) . \tag{9.16}$$

(d) The vacuum energy of the same part of space in the absence of the plates is given by

$$\begin{aligned} E_0 &= \frac{1}{2} \int \frac{L^2 \mathrm{d}^2 k}{(2\pi)^2} \int \frac{a \mathrm{d}k_3}{2\pi} 2\sqrt{k_1^2 + k_2^2 + k_3^2} \\ &= \int \frac{L^2 \mathrm{d}^2 k}{(2\pi)^2} \int_0^\infty dn \sqrt{k_1^2 + k_2^2 + \left(\frac{n\pi}{a}\right)^2} . \end{aligned}$$

Then ϵ is

$$\epsilon = \frac{1}{2} \int_0^\infty \frac{k \mathrm{d}k}{2\pi} \left[k + 2\sum_{n=1}^{\infty} \sqrt{k^2 + \left(\frac{n\pi}{a}\right)^2} - 2\int_0^\infty \mathrm{d}n \sqrt{k^2 + \left(\frac{n\pi}{a}\right)^2} \right] . \tag{9.17}$$

The last integral can be rewritten as follows

$$\epsilon = \frac{\pi^2}{8a^3} \int_0^\infty \mathrm{d}u \left(\sqrt{u} + 2\sum_{n=1}^{\infty} \sqrt{u + n^2} - 2\int_0^\infty \mathrm{d}n \sqrt{u + n^2} \right) , \tag{9.18}$$

where a new variable $u = a^2 k^2 / \pi^2$ was introduced. After the regularization ϵ takes the form

$$\begin{aligned} \epsilon = \frac{\pi^2}{8a^3} \int_0^\infty \mathrm{d}u \Bigg(&\sqrt{u} f\Big(\frac{\pi\sqrt{u}}{a}\Big) + 2\sum_{n=1}^{\infty} \sqrt{u + n^2} f\Big(\frac{\pi\sqrt{u + n^2}}{a}\Big) - \\ &- 2\int_0^\infty \mathrm{d}n \sqrt{u + n^2} f\Big(\frac{\pi\sqrt{u + n^2}}{a}\Big) \Bigg) , \end{aligned} \tag{9.19}$$

and becomes finite. If we define a new function

$$F(n) = \int_0^\infty \mathrm{d}u \sqrt{u + n^2} f\Big(\frac{\pi\sqrt{u + n^2}}{a}\Big) ,$$

ϵ becomes

$$\epsilon = \frac{\pi^2}{8a^3} \left(F(0) + 2\sum_{n=1}^{\infty} F(n) - 2\int_0^\infty \mathrm{d}n F(n) \right) . \tag{9.20}$$

To calculate the previous expression we will use the Euler-Maclaurin formula:

$$\sum_{n=1}^{\infty} F(n) - \int_0^\infty dn F(n) + \frac{1}{2}F(0) = -\frac{1}{2!}B_2 F'(0) - \frac{1}{4!}B_4 F'''(0) + \dots .$$

$B_2, B_4, \dots$ are Bernouli numbers and they are defined by

$$\frac{y}{e^y - 1} = \sum_{\nu=0}^{\infty} B_\nu \frac{y^\nu}{\nu!} .$$

Consequently,

$$\epsilon = \frac{\pi^2}{4a^3}\left(-\frac{1}{2!}B_2 F'(0) - \frac{1}{4!}B_4 F'''(0) + \dots\right) . \tag{9.21}$$

It is easy to get $F'(0) = 0, F'''(0) = -4$. Then the vacuum energy per unit surface is

$$\epsilon = -\frac{\pi^2}{720a^3} .$$

From the expression for the energy we can derive the force:

$$f = -\frac{\partial \epsilon}{\partial a} = -\frac{\pi^2}{240a^4} .$$

If $a = 1\mu m$ and $L = 1cm$ the force is $10^{-8}N$. The vacuum energy of the electromagnetic field between the two conducting plates produces a weak attractive force between them. This effect was measured in 1958.

(e) The integral I can be found in [9]:

$$\begin{aligned} I &= 2\pi \int_0^\infty \frac{k dk}{(k^2 + m^2)^\alpha} \\ &= \pi \frac{\Gamma(\alpha - 1)}{\Gamma(\alpha)} \frac{1}{(m^2)^{\alpha-1}} . \end{aligned} \tag{9.22}$$

Then

$$\begin{aligned} \frac{E}{L^2} &= \frac{1}{2}\int \frac{d^2k}{(2\pi)^2}\left(\lim_{\mu\to 0}\frac{1}{(k^2+\mu^2)^{-1/2}} + 2\sum_{n=1}^{\infty}\frac{1}{\sqrt{k^2 + \left(\frac{n\pi}{a}\right)^2}}\right) \\ &= -\frac{1}{12\pi}\left(\lim_{\mu\to 0}(\mu^2)^{3/2} + 2\frac{\pi^3}{a^3}\sum_{n=1}^{\infty} n^3\right) \\ &= -\frac{\pi^2}{6a^3}\sum_{n=1}^{\infty} n^3 . \end{aligned} \tag{9.23}$$

From

$$\zeta(1-n) = \frac{(-1)^{1+n}B_n}{n} ,$$

follows that $\zeta(-3) = 1/120$ since $B_4 = -1/30$. Finally, we get the same result as before

$$\frac{E}{L^2} = -\frac{\pi^2}{720a^3} .$$

10

Processes in the lowest order of the perturbation theory

10.1 The transition probability is

$$|S_{\mathrm{fi}}|^2 = (2\pi)^8[\delta^{(4)}(p_1' + p_2' - p_1 - p_2)]^2 \frac{m_A m_B m_C m_D}{V^4 E_1 E_2 E_1' E_2'}|\mathcal{M}|^2 \ . \tag{10.1}$$

The square of the four-dimensional delta function is

$$\begin{aligned}
[\delta^{(4)}(p_{\mathrm{f}} - p_{\mathrm{i}})]^2 &= \delta^{(4)}(p_{\mathrm{f}} - p_{\mathrm{i}})\delta^{(4)}(0) \\
&= \frac{1}{(2\pi)^4}\delta^{(4)}(p_{\mathrm{f}} - p_{\mathrm{i}}) \int_V \mathrm{d}^3\boldsymbol{x} \int_{-\frac{T}{2}}^{\frac{T}{2}} \mathrm{d}t \\
&= \frac{TV}{(2\pi)^4}\delta^{(4)}(p_{\mathrm{f}} - p_{\mathrm{i}}) \ ,
\end{aligned} \tag{10.2}$$

where: $p_{\mathrm{i}} = p_1 + p_2$ and $p_{\mathrm{f}} = p_1' + p_2'$ are initial and final four–momentum respectively. The differential cross section (10.D) is

$$d\sigma = \frac{|S_{\mathrm{fi}}|^2}{T}\frac{1}{|\boldsymbol{J}_{\mathrm{in}}|}\frac{V^2 \mathrm{d}^3\boldsymbol{p}_1' \mathrm{d}^3\boldsymbol{p}_2'}{(2\pi)^6} \ . \tag{10.3}$$

The current density flux, in the center–of–mass frame is

$$|\boldsymbol{J}_{\mathrm{in}}| = |\bar{\psi}\gamma\psi| = \frac{|\boldsymbol{p}_1|(E_1 + E_2)}{VE_1E_2} \ . \tag{10.4}$$

By substituting (10.1), (10.2) and (10.4) into (10.3) the following formula is obtained

$$\begin{aligned}
\mathrm{d}\sigma = &\frac{1}{(2\pi)^2}\delta(E_1' + E_2' - E_1 - E_2)\delta^{(3)}(\boldsymbol{p}_1' + \boldsymbol{p}_2' - \boldsymbol{p}_1 - \boldsymbol{p}_2)|\mathcal{M}|^2 \\
\times &\frac{m_A m_B m_C m_D}{(E_1 + E_2)E_1'E_2'|\boldsymbol{p}_1|}\mathrm{d}^3\boldsymbol{p}_1' \mathrm{d}^3\boldsymbol{p}_2' \ .
\end{aligned} \tag{10.5}$$

By integrating over $\boldsymbol{p}_2'$ we get

$$\frac{\mathrm{d}\sigma}{\mathrm{d}\Omega} = \frac{1}{(2\pi)^2}\int \delta(\sqrt{p_1'^2+m_C^2}+\sqrt{p_1'^2+m_D^2}-E_1-E_2)|\mathcal{M}|^2$$
$$\times \frac{m_A m_B m_C m_D}{(E_1+E_2)E_1'E_2'}\frac{p_1'^2\mathrm{d}p_1'}{p_1} ,$$

where the fact that we are doing calculations in the center–of–mass frame have been used. By applying formula

$$\int \mathrm{d}x g(x)\delta(f(x)) = \left.\frac{g(x)}{|f'(x)|}\right|_{f(x)=0} \tag{10.6}$$

the requested result is obtained.

10.2 Four–dimensional delta function and integration measure are Lorentz invariant quantities (Problem 6.3) so is the given integral. In the inertial frame in which $\boldsymbol{P}=0$ the integral becomes

$$I = \frac{1}{4}\int \frac{\mathrm{d}^3\boldsymbol{p}}{\sqrt{p^2+m^2}}\frac{\mathrm{d}^3\boldsymbol{q}}{\sqrt{q^2+m'^2}}\delta^{(3)}(\boldsymbol{p}+\boldsymbol{q})\delta(E_p+E_q-P^0) . \tag{10.7}$$

By integrating over $\boldsymbol{q}$ in (10.7) and introducing spherical coordinates we obtain

$$I = \pi\int_0^\infty p^2\mathrm{d}p\frac{1}{\sqrt{p^2+m^2}\sqrt{p^2+m'^2}}\delta(\sqrt{p^2+m^2}+\sqrt{p^2+m'^2}-P^0) .$$

By applying the formula (10.6) one gets

$$I = \frac{\pi}{P_0}\sqrt{\frac{(m^2-m'^2-P_0^2)^2}{4P_0^2}-m'^2} .$$

10.3 The Feynman amplitude, $\mathrm{i}\mathcal{M}$ is a complex number so that

$$\begin{aligned}(\mathrm{i}\mathcal{M})^* = (\mathrm{i}\mathcal{M})^\dagger &= (\bar{u}(\boldsymbol{p},r)\gamma_\mu(1-\gamma_5)u(\boldsymbol{q},s))^\dagger\,\epsilon^{\mu*}(\boldsymbol{k},\lambda)\\ &= u^\dagger(\boldsymbol{q},s)(1-\gamma_5)\gamma^0\gamma_\mu\gamma^0\gamma^0u(\boldsymbol{p},r)\epsilon^{\mu*}(\boldsymbol{k},\lambda)\\ &= \bar{u}(\boldsymbol{q},s)(1+\gamma_5)\gamma_\mu u(\boldsymbol{p},r)\epsilon^{\mu*}(\boldsymbol{k},\lambda) ,\end{aligned}$$

where identities from Problems 3.1 and 3.3 are used. The average value of the squared amplitude is ($a, b, \ldots$ are Dirac's indices)

$$\begin{aligned}\sum_{\lambda=1}^{2}\sum_{r,s=1}^{2}|\mathcal{M}|^2 &= \sum_{\lambda=1}^{2}\sum_{r,s=1}^{2}\bar{u}_a(\boldsymbol{p},r)[\gamma_\mu(1-\gamma_5)]_{ab}u_b(\boldsymbol{q},s)\\ &\quad\times\bar{u}_c(\boldsymbol{q},s)[(1+\gamma_5)\gamma_\nu]_{cd}u_d(\boldsymbol{p},r)\epsilon^\mu(\boldsymbol{k},\lambda)\epsilon^{\nu*}(\boldsymbol{k},\lambda)\\ &= \left(\sum_{r=1}^{2}u_d(\boldsymbol{p},r)\bar{u}_a(\boldsymbol{p},r)\right)[\gamma_\mu(1-\gamma_5)]_{ab}\\ &\quad\times\left(\sum_{s=1}^{2}u_b(\boldsymbol{q},s)\bar{u}_c(\boldsymbol{q},s)\right)[(1+\gamma_5)\gamma_\nu]_{cd}\sum_{\lambda=1}^{2}\epsilon^{\mu*}(\boldsymbol{k},\lambda)\epsilon^\nu(\boldsymbol{k},\lambda) .\end{aligned}$$

By applying expression for the projection operator into positive–energy solutions (Problem 4.4) we have

$$\begin{aligned}\sum_{\lambda=1}^{2}\sum_{r,s=1}^{2}|\mathcal{M}|^2 &= \left(\frac{\not p+m}{2m}\right)_{da}[\gamma_\mu(1-\gamma_5)]_{ab}\\ &\times\left(\frac{\not q+m}{2m}\right)_{bc}[(1+\gamma_5)\gamma_\nu]_{cd}\sum_{\lambda=1}^{2}\epsilon^\mu(\boldsymbol{k},\lambda)\epsilon^{\nu *}(\boldsymbol{k},\lambda)\\ &= \frac{1}{4m^2}\sum_{\lambda=1}^{2}\epsilon^\mu(\boldsymbol{k},\lambda)\epsilon^{\nu *}(\boldsymbol{k},\lambda)\\ &\times \mathrm{tr}\left[(\not p+m)\gamma_\mu(1-\gamma_5)(\not q+m)(1+\gamma_5)\gamma_\nu\right] .\end{aligned}$$

Using the facts that γ_5 anticommutes with γ^μ matrices and that $(\gamma_5)^2=1$, the last expression becomes

$$\sum_{\lambda=1}^{2}\sum_{r,s=1}^{2}|\mathcal{M}|^2 = \frac{1}{2m^2}\mathrm{tr}\left[(\not p+m)\gamma_\mu(1-\gamma_5)\not q\gamma_\nu\right]\sum_{\lambda=1}^{2}\epsilon^\mu(\boldsymbol{k},\lambda)\epsilon^{\nu *}(\boldsymbol{k},\lambda) .$$

By applying the corresponding traces form Problem 3.6 one obtains

$$\begin{aligned}\sum_{\lambda=1}^{2}\sum_{r,s=1}^{2}|\mathcal{M}|^2 &= \frac{2}{m^2}\left[p_\mu q_\nu + p_\nu q_\mu - (p\cdot q)g_{\mu\nu} + \mathrm{i}\epsilon_{\alpha\nu\beta\mu}q^\alpha p^\beta\right]\\ &\times\sum_{\lambda=1}^{2}\epsilon^\mu(\boldsymbol{k},\lambda)\epsilon^{\nu *}(\boldsymbol{k},\lambda) .\end{aligned}\tag{10.8}$$

To sum over the photon polarizations is reduced to replacement

$$\sum_{\lambda=1}^{2}\epsilon^\mu(\boldsymbol{k},\lambda)\epsilon^{\nu *}(\boldsymbol{k},\lambda)\to -g^{\mu\nu}\tag{10.9}$$

Because the other two terms in (9.E) do not give any contribution, the result is $4p\cdot q/m^2$.

10.4 In the first part of the Problem we shall apply Wick's theorem for bosons and in the second part we shall make use of the Wick's theorem for fermions.

(a) It is clear that all normal–ordered terms fall off, because their vacuum expectation value is equal to zero. Thus the only remaining terms are those with four contractions. If we contract one $\phi(x)$ with one $\phi(y)$ four times we shall get $(\langle 0|\,T(\phi(x)\phi(y))\,|0\rangle)^4$. This can be done in $4! = 24$ ways. The next possibility is to make two contractions between fields $\phi(x)$ and $\phi(y)$. One field $\phi(x)$ can be contracted in 4 ways with one of the $\phi(y)$'s. The next $\phi(x)$ can be contracted in three ways with one of the remaining

$\phi(y)'$s . The obtained result has to be multiplied by 6, because this is the number of ways in which two fields $\phi(x)$ can be chosen from the four possible. Thus, there are $4 \cdot 3 \cdot 6 = 72$ possible contractions of this type. There are three mutual contractions between two fields $\phi(x)$, the similar is obtained for fields $\phi(y)$, so the corresponding coefficient is 9. Thus,

$$\begin{aligned}
&\langle 0| T(\phi^4(x)\phi^4(y)) |0\rangle) = 24(\langle 0| T(\phi(x)\phi(y)) |0\rangle)^4 \\
&+ 72 \langle 0| T(\phi(x)\phi(x)) |0\rangle \langle 0| T(\phi(y)\phi(y)) |0\rangle (\langle 0| T(\phi(x)\phi(y)) |0\rangle)^2 \\
&+ 9(\langle 0| T(\phi(x)\phi(x)) |0\rangle)^2(\langle 0| T(\phi(y)\phi(y)) |0\rangle)^2 \\
&= 24(\mathrm{i}\Delta_{\mathrm{F}}(x-y))^4 + 72(\mathrm{i}\Delta_{\mathrm{F}}(x-x))\mathrm{i}\Delta_{\mathrm{F}}(y-y)(\mathrm{i}\Delta_{\mathrm{F}}(x-y))^2 \\
&+ 9(\mathrm{i}\Delta_{\mathrm{F}}(x-x))^2(\mathrm{i}\Delta_{\mathrm{F}}(y-y))^2 \ .
\end{aligned}$$

The last expression can be represented by the following diagram:

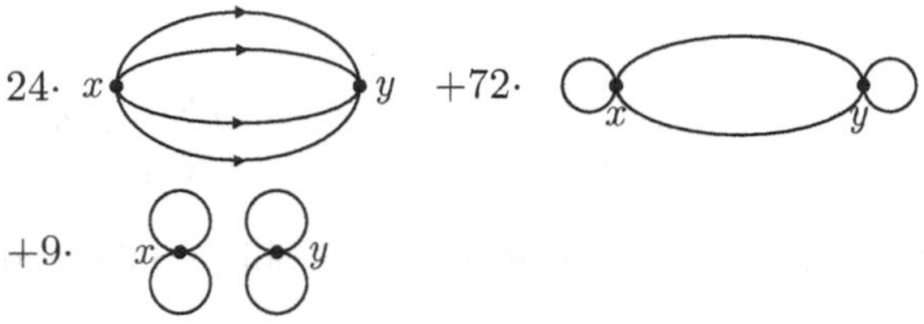

(b) Here, the equal–time contractions are forbidden. The result is

$$\begin{aligned}
T(:\phi^4(x) :: \phi^4(y):) &= 16 : \phi^3(x)\phi^3(y) : \mathrm{i}\Delta_{\mathrm{F}}(x-y) \\
&+ 72 : \phi^2(x)\phi^2(y) : (\mathrm{i}\Delta_{\mathrm{F}}(x-y))^2 \\
&+ 96 : \phi(x)\phi(y) : (\mathrm{i}\Delta_{\mathrm{F}}(x-y))^3 \\
&+ 24(\mathrm{i}\Delta_{\mathrm{F}}(x-y))^4 \ . \qquad (10.10)
\end{aligned}$$

(c) By applying Wick's theorem for fermions one obtains

$$\begin{aligned}
&\langle 0| T(\bar{\psi}(x)\psi(x)\bar{\psi}(y)\psi(y)) |0\rangle \\
&= \mathrm{i}S_{\mathrm{F}}(x-x)\mathrm{i}S_{\mathrm{F}}(y-y) - \mathrm{i}S_{\mathrm{F}}(x-y)\mathrm{i}S_{\mathrm{F}}(y-x) \ .
\end{aligned}$$

10.5

(a) The given diagram is obtained from the expression

$$-\frac{\mathrm{i}\lambda}{4!}\int \mathrm{d}^4y \,\langle 0| T(\phi(x_1)\phi(x_2)\phi^4(y)) |0\rangle \ ,$$

where $\phi(x_1)$ is to be contracted with one $\phi(y)$ (there are four ways to do this) and $\phi(x_2)$ with one of the remaining three $\phi(y)'$s. The symmetry factor is $\frac{1}{4!}4 \cdot 3 = \frac{1}{2}$. This result can be easily checked by using the formula given in the problem, where $g = 1, \alpha = 0$ and $\beta = 1$.

(b) This diagram is one of the terms in

$$\frac{1}{2!}\left(-\frac{i\lambda}{4!}\right)^2 \int d^4y_1 d^4y_2 \langle 0| T(\phi(x_1)\phi(x_2)\phi^4(y_1)\phi^4(y_2)) |0\rangle \ ,$$

where $\phi(x_1)$ is contracted with one of the four $\phi(y_1)$'s (there are four ways to do this); $\phi(x_2)$ with one of the remaining $\phi(y_1)$ fields (there are three ways to do this). It is necessary to make two more contractions between $\phi(y_1)$ and $\phi(y_2)$ which can be done in $4 \cdot 3 = 12$ ways. Thus we have:

$$S^{-1} = 2!\frac{1}{2!}\left(\frac{1}{4!}\right)^2 4 \cdot 3 \cdot 4 \cdot 3 = \frac{1}{4} \ ,$$

so the symmetry factor is $S = 4$. The same result is obtained by plugging $g = 1, \alpha_2 = 1$ and $\beta = 1$ into the formula given in the problem.

(c) In order to get this diagram it is necessary to make the following contractions in this third–order expression:

$$\frac{1}{3!}\left(-\frac{i\lambda}{4!}\right)^3 \int d^4y_1 d^4y_2 d^4y_3 \langle 0| T(\phi(x_1)\phi(x_2)\phi^4(y_1)\phi^4(y_2)\phi^4(y_3)) |0\rangle \ , \tag{10.11}$$

$\phi(x_1)$ with one of the four $\phi(y_1)'$s (four ways); $\phi(x_2)$ with one of the remaining $\phi(y_1)$ fields (three ways); two $\phi(y_1)$ fields with four $\phi(y_2)$ fields ($4 \cdot 2 = 8$ ways); the remaining $\phi(y_1)$ field with $\phi(y_3)$ fields (4 ways); three contractions between three $\phi(y_2)'$s and three $\phi(y_3)$ fields ($3 \cdot 2 = 6$ ways). Finally, one has to divide the obtained expression by two, because of the symmetry $y_2 \leftrightarrow y_3$. By combining all the factors we have:

$$S^{-1} = 3!\frac{1}{3!}\left(\frac{1}{4!}\right)^3 4 \cdot 3 \cdot 4 \cdot 2 \cdot 4 \cdot 3 \cdot 2 \cdot \frac{1}{2} = \frac{1}{12} \ , \tag{10.12}$$

so $S = 12$. This result can be checked by applying the formula given in the problem: $g = 2,\ n = 3,\ \alpha_3 = 1,\ \beta = 0$.

10.6 The result is

$$\begin{aligned}
&\frac{1}{2}\left(\frac{-i\lambda}{3!}\right)^2 \int d^4y_1 d^4y_2 \langle 0| T(\phi(x_1)\phi(x_2)\phi^3(y_1)\phi^3(y_2)) |0\rangle = \\
&= \int d^4y_1 d^4y_2 (-i\lambda)^2 \Big[\frac{1}{2} i\Delta_F(x_1 - y_1) i\Delta_F(x_2 - y_2)(i\Delta_F(y_1 - y_2))^2 \\
&+ \frac{1}{12} i\Delta_F(x_1 - x_2)(i\Delta_F(y_1 - y_2))^3 \\
&+ \frac{1}{8} i\Delta_F(x_1 - x_2) i\Delta_F(y_1 - y_1) i\Delta_F(y_2 - y_2) i\Delta_F(y_1 - y_2) \\
&+ \frac{1}{2} i\Delta_F(x_1 - y_1) i\Delta_F(x_2 - y_1) i\Delta_F(y_1 - y_2) i\Delta_F(y_2 - y_2) \\
&+ \frac{1}{4} i\Delta_F(x_1 - y_1) i\Delta_F(x_2 - y_2) i\Delta_F(y_1 - y_1) i\Delta_F(y_2 - y_2)\Big]
\end{aligned} \tag{10.13}$$

which can be represented by the following diagram:

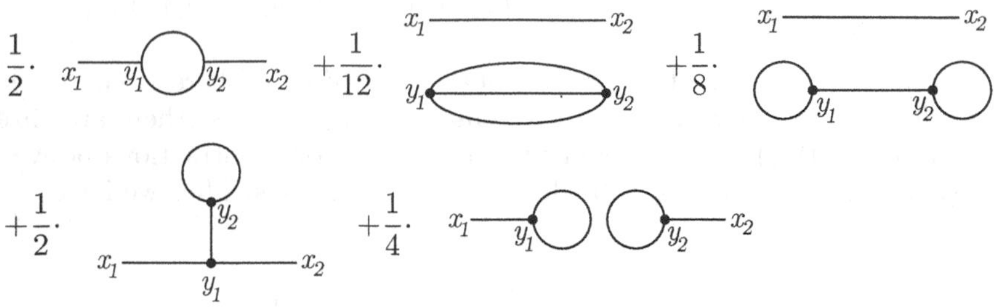

The coefficient $\frac{1}{2}$ in the first term (10.13) can be obtained in the following way: contraction $\phi(x_1)$ with $\phi(y_1)$ can be done in three ways, as well as the contraction $\phi(x_2)$ with $\phi(y_2)$. Two contractions $\phi(y_1)$ with $\phi(y_2)$ can be done in two ways. The obtained result has to be multiplied by 2! which comes from the interchange y_1–vertex with y_2–vertex, because, for instance, we could contract $\phi(x_1)$ with $\phi(y_2)$ instead of $\phi(y_1)$. Thus, the overall coefficient is

$$\frac{1}{2}\frac{3 \cdot 3 \cdot 2}{3! \cdot 3!} \cdot 2 = \frac{1}{2} \, . \tag{10.14}$$

In the second and third term there is no additional multiplying by 2 which comes from the $y_1 \leftrightarrow y_2$ interchange!

10.7

(a) Diagram for this process is represented in Fig. 10.1.

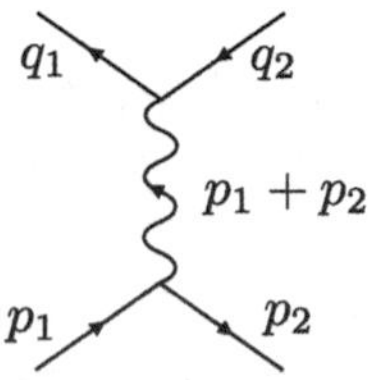

Fig. 10.1. The tree–level Feynman diagram for the scattering $\mu^-(p_1) + \mu^+(p_2) \to e^-(q_1) + e^+(q_2)$

The Feynman amplitude is given by the following expression

$$\mathrm{i}\mathcal{M} = \frac{\mathrm{i}e^2}{(p_1+p_2)^2 + \mathrm{i}\epsilon}\bar{v}(\boldsymbol{p}_2, s)\gamma^\mu u(\boldsymbol{p}_1, r)\bar{u}(\boldsymbol{q}_1, r')\gamma_\mu v(\boldsymbol{q}_2, s') \, ,$$

hence

$$
\begin{aligned}
\langle|\mathcal{M}|^2\rangle &= \frac{e^4}{4}\frac{1}{(p_1+p_2)^4}\sum_{r,s=1}^{2}\sum_{r',s'=1}^{2}\bar{v}_a(\boldsymbol{p}_2,s)\gamma^{\mu}_{ab}u_b(\boldsymbol{p}_1,r)\\
&\times \bar{u}_c(\boldsymbol{q}_1,r')(\gamma_\mu)_{cd}v_d(\boldsymbol{q}_2,s')\bar{u}_e(\boldsymbol{p}_1,r)\gamma^{\nu}_{ef}v_f(\boldsymbol{p}_2,s)\\
&\times \bar{v}_g(\boldsymbol{q}_2,s')(\gamma_\nu)_{gh}u_h(\boldsymbol{q}_1,r')\\
&= \frac{e^4}{4(p_1+p_2)^4}\sum_s\left(v_f(\boldsymbol{p}_2,s)\bar{v}_a(\boldsymbol{p}_2,s)\right)\gamma^{\mu}_{ab}\\
&\times \sum_r\left(u_b(\boldsymbol{p}_1,r)\bar{u}_e(\boldsymbol{p}_1,r)\right)(\gamma^\nu)_{ef}\\
&\times \sum_{r'}\left(u_h(\boldsymbol{q}_1,r')\bar{u}_c(\boldsymbol{q}_1,r')\right)(\gamma_\mu)_{cd}\\
&\times \sum_{s'}\left(v_d(\boldsymbol{q}_2,s')\bar{v}_g(\boldsymbol{q}_2,s')\right)(\gamma_\nu)_{gh}\ .
\end{aligned}
$$

By performing matrix multiplying in the preceding expression we obtain two traces (Problem 4.4)

$$
\begin{aligned}
\langle|\mathcal{M}|^2\rangle &= \frac{e^4}{4(p_1+p_2)^4}\frac{1}{16m_e^2m_\mu^2}\mathrm{tr}[(\not{q}_1+m_e)\gamma_\mu(\not{q}_2-m_e)\gamma_\nu]\\
&\times \mathrm{tr}[(\not{p}_2-m_\mu)\gamma^\mu(\not{p}_1+m_\mu)\gamma^\nu]\ .
\end{aligned}
$$

By applying corresponding identities from Problem 3.6 we get

$$
\begin{aligned}
\langle|\mathcal{M}|^2\rangle &= \frac{e^4}{4(p_1+p_2)^4}\frac{1}{m_e^2m_\mu^2}\left[q_{1\mu}q_{2\nu}+q_{2\mu}q_{1\nu}-(q_1\cdot q_2)g_{\mu\nu}-m_e^2g_{\mu\nu}\right]\\
&\times \left[p_1^\mu p_2^\nu+p_2^\mu p_1^\nu-(p_1\cdot p_2)g^{\mu\nu}-m_\mu^2g^{\mu\nu}\right]\ .
\end{aligned}
$$

After multiplying and reducing the preceding expression one obtains

$$
\begin{aligned}
\langle|\mathcal{M}|^2\rangle &= \frac{e^4}{4(p_1+p_2)^4m_e^2m_\mu^2}\left[2(p_2\cdot q_1)(p_1\cdot q_2)+2(p_2\cdot q_2)(p_1\cdot q_1)\right.\\
&\left.+\ 2m_e^2(p_1\cdot p_2)+2m_\mu^2(q_1\cdot q_2)+4m_e^2m_\mu^2\right]\ .
\end{aligned}
\tag{10.15}
$$

In the center–of–mass frame the four–momenta are

$$
\begin{aligned}
p_1 &= (E,0,0,p)\ ,\\
p_2 &= (E,0,0,-p)\ ,\\
q_1 &= (E,q\sin\theta,0,q\cos\theta)\ ,\\
q_2 &= (E,-q\sin\theta,0,-q\cos\theta)\ ,
\end{aligned}
$$

where p and q are intensities of the corresponding three–momenta vectors. After simple scalar product computations in (10.15) one gets:

$$
\begin{aligned}
\langle|\mathcal{M}|^2\rangle &= \frac{e^4}{16E^4m_e^2m_\mu^2}\left[(E^4+m_e^2m_\mu^2)(1+\cos^2\theta)\right.\\
&\left.+\ E^2(m_e^2+m_\mu^2)\sin^2\theta\right]\ .
\end{aligned}
\tag{10.16}
$$

In the high energy limit ($p \approx E$) expression (10.16) becomes

$$\langle|\mathcal{M}|^2\rangle = \frac{e^4}{16m_e^2 m_\mu^2}(1+\cos^2\theta) \ . \tag{10.17}$$

Using the previous expression and Problem 10.1 the differential cross section is

$$\frac{\mathrm{d}\sigma}{\mathrm{d}\Omega} = \frac{e^4}{256\pi^2 E^2}(1+\cos^2\theta) \ .$$

(b) We shall discuss just the main results. From the diagram

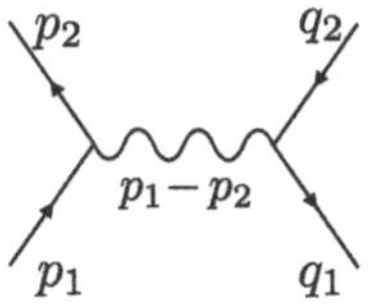

Fig. 10.2. The Feynman diagram for the scattering $e^-(p_1) + \mu^+(q_1) \to e^-(p_2) + \mu^+(q_2)$ in the lowest order

the amplitude is

$$\mathrm{i}\mathcal{M} = \bar{u}(\boldsymbol{p}_2, r_2)(\mathrm{i}e\gamma^\mu)u(\boldsymbol{p}_1, r_1)\frac{-\mathrm{i}g_{\mu\nu}}{(p_1-p_2)^2+\mathrm{i}\epsilon}\bar{v}(\boldsymbol{q}_1, s_1)(\mathrm{i}e\gamma^\nu)v(\boldsymbol{q}_2, s_2) \ .$$

The squared Feynman amplitude module (averaged over spin states of the initial particles and summed over spin states of the final particles) is:

$$\begin{aligned}\langle|\mathcal{M}|^2\rangle &= \frac{e^4}{4(p_1-p_2)^4}\frac{1}{16m_e^2 m_\mu^2}\mathrm{tr}\left[(\not{p}_2+m_e)\gamma^\mu(\not{p}_1+m_e)\gamma^\nu\right] \\ &\times \mathrm{tr}\left[(\not{q}_1-m_\mu)\gamma_\mu(\not{q}_2-m_\mu)\gamma_\nu\right] \\ &= \frac{e^4}{2(p_1-p_2)^4 m_e^2 m_\mu^2}\left[(p_2\cdot q_1)(p_1\cdot q_2)+(p_1\cdot q_1)(p_2\cdot q_2)\right. \\ &\left.- m_\mu^2(p_1\cdot p_2) - m_e^2(q_1\cdot q_2) + 2m_e^2 m_\mu^2\right] \ .\end{aligned}$$

Finally in the center–of–mass frame (in the high energy limit) we have:

$$\langle|\mathcal{M}|^2\rangle = \frac{e^4}{8m_e^2 m_\mu^2}\frac{4+(1+\cos\theta)^2}{(1-\cos\theta)^2} \ . \tag{10.18}$$

The differential cross section in the center–of–mass frame is:

$$\frac{\mathrm{d}\sigma}{\mathrm{d}\Omega} = \frac{e^4}{128\pi^2 E^2}\frac{4+(1+\cos\theta)^2}{(1-\cos\theta)^2} \ . \tag{10.19}$$

Note that for $\theta \approx 0$ differential cross section diverges. This is a consequence of the fact that for these angles the prevailing contribution in the expression for $\mathrm{i}\mathcal{M}$ comes from the virtual photon (this contribution is actually divergent because $k^2 = (p_1-p_2)^2 \approx 0$).

10.8 The Compton scattering is the process $e^-\gamma \to e^-\gamma$. In the lowest order contribution to this scattering is given by the following two diagrams:

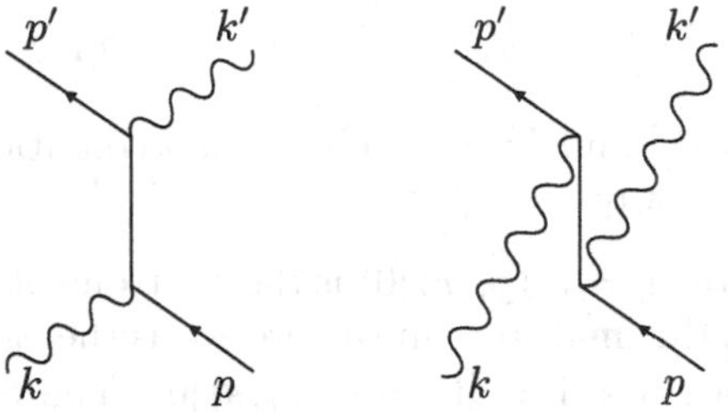

so that the Feynman amplitude is

$$\begin{aligned} i\mathcal{M} &= \bar{u}(\boldsymbol{p}', s')(ie\gamma^\mu)\epsilon^*_\mu(\boldsymbol{k}', \lambda')\frac{i(\not{p} + \not{k} + m)}{(p+k)^2 - m^2}(ie\gamma^\nu)\epsilon_\nu(\boldsymbol{k}, \lambda)u(\boldsymbol{p}, s) + \\ &+ \bar{u}(\boldsymbol{p}', s')(ie\gamma^\nu)\epsilon_\nu(\boldsymbol{k}, \lambda)\frac{i(\not{p} - \not{k}' + m)}{(p-k')^2 - m^2}(ie\gamma^\mu)\epsilon^*_\mu(\boldsymbol{k}', \lambda')u(\boldsymbol{p}, s) \\ &= -ie^2\epsilon^*_\mu(\boldsymbol{k}', \lambda')\epsilon_\nu(\boldsymbol{k}, \lambda)\bar{u}(\boldsymbol{p}', s')\left[\frac{\gamma^\mu(\not{p} + \not{k} + m)\gamma^\nu}{(p+k)^2 - m^2} + \right. \\ &\left. + \frac{\gamma^\nu(\not{p} - \not{k}' + m)\gamma^\mu}{(p-k')^2 - m^2}\right]u(\boldsymbol{p}, s) . \end{aligned} \tag{10.20}$$

As we see the Feynman amplitude has the following form

$$i\mathcal{M} = i\mathcal{M}^{\mu\nu}\epsilon^*_\mu(\boldsymbol{k}', \lambda')\epsilon_\nu(\boldsymbol{k}, \lambda) .$$

In order to prove the gauge invariance of $\mathcal{M}$ it is enough to show that

$$i\mathcal{M}^{\mu\nu}k_\nu = i\mathcal{M}^{\mu\nu}k'_\mu = 0 . \tag{10.21}$$

First we prove that $i\mathcal{M}^{\mu\nu}k_\nu = 0$. In the second term in (10.20) we will use $p - k' = p' - k$. Hence

$$i\mathcal{M}^{\mu\nu} = -ie^2\bar{u}(\boldsymbol{p}', s')\left[\frac{\gamma^\mu(\not{p} + \not{k} + m)\gamma^\nu}{(p+k)^2 - m^2} + \frac{\gamma^\nu(\not{p}' - \not{k} + m)\gamma^\mu}{(p'-k)^2 - m^2}\right]u(\boldsymbol{p}, s) . \tag{10.22}$$

The numerators can be also simplified using:

$$\begin{aligned} (\not{p} + m)\gamma^\nu u(p) &= (\gamma^\mu p_\mu + m)\gamma^\nu u(p) = (2g^{\mu\nu} - \gamma^\nu\gamma^\mu)p_\mu u(p) + m\gamma^\nu u(p) \\ &= 2p^\nu u(p) - \gamma^\nu(\not{p} - m)u(p) = 2p^\nu u(p), \end{aligned}$$

and similarly

$$\bar{u}(\boldsymbol{p}')\gamma^\nu(\not{p}' + m) = 2p'^\nu\bar{u}(\boldsymbol{p}') . \tag{10.23}$$

After performing these two simplifications $i\mathcal{M}^{\mu\nu}k_\nu$ becomes

$$
\begin{aligned}
\mathrm{i}\mathcal{M}^{\mu\nu}k_\nu &= -\mathrm{i}e^2 k_\nu \bar{u}(\boldsymbol{p}', s')\left[\frac{\gamma^\mu \not{k}\gamma^\nu + 2\gamma^\mu p^\nu}{2p\cdot k} + \frac{-\gamma^\nu \not{k}\gamma^\mu + 2\gamma^\mu p'^\nu}{-2p\cdot k'}\right] u(\boldsymbol{p}, s) \\
&= -\mathrm{i}e^2 \bar{u}(p', s')\left[\frac{\gamma^\mu k^2 + 2\gamma^\mu p\cdot k}{2p\cdot k} + \frac{-k^2\gamma^\mu + 2\gamma^\mu p'\cdot k}{-2p\cdot k'}\right] u(\boldsymbol{p}, s) = 0 ,
\end{aligned}
$$

where we used $p^2 = m^2$ and $k^2 = 0$. The second condition $\mathrm{i}\mathcal{M}^{\mu\nu}k'_\mu = 0$ can be proved in the same way.

10.9 The initial state, $|\mathrm{i}\rangle = c^\dagger(\boldsymbol{p}_\mathrm{i}, r)\,|0\rangle$ is the electron with momentum p_i and polarization r, while the final state in the process is the electron with momentum $\boldsymbol{p}_\mathrm{f}$ and polarization s, i. e. $|\mathrm{f}\rangle = c^\dagger(\boldsymbol{p}_\mathrm{f}, s)\,|0\rangle$. The transition amplitude matrix element is:

$$
S_{\mathrm{fi}} = \mathrm{i}e \int \mathrm{d}^4x\, \langle \mathrm{f}|\, \bar{\psi}(x)\gamma_\mu \psi(x)\, |\mathrm{i}\rangle\, A^\mu(x) , \tag{10.24}
$$

where ψ and $\bar{\psi}$ are field operators and A^μ is a classical electromagnetic field.

(a) From (10.24) one obtains

$$
S_{\mathrm{fi}} = \mathrm{i}ea\sqrt{\frac{m}{E_\mathrm{i}V}}\sqrt{\frac{m}{E_\mathrm{f}V}} \int \mathrm{d}^4x \bar{u}(\boldsymbol{p}_\mathrm{f}, s)\gamma_0 u(\boldsymbol{p}_\mathrm{i}, r)\mathrm{e}^{-\mathrm{i}p_\mathrm{i}\cdot x + \mathrm{i}p_\mathrm{f}\cdot x}\mathrm{e}^{-k^2\boldsymbol{x}^2} . \tag{10.25}
$$

Because of

$$
\int \mathrm{d}^3\boldsymbol{x}\mathrm{e}^{-k^2\boldsymbol{x}^2 + \mathrm{i}(\boldsymbol{p}_\mathrm{i} - \boldsymbol{p}_\mathrm{f})\cdot\boldsymbol{x}} = \left(\frac{\pi}{k^2}\right)^{3/2} \mathrm{e}^{-(\boldsymbol{p}_\mathrm{i} - \boldsymbol{p}_\mathrm{f})^2/4k^2} ,
$$

we have

$$
\begin{aligned}
S_{\mathrm{fi}} &= \mathrm{i}ea\sqrt{\frac{m}{E_\mathrm{i}V}}\sqrt{\frac{m}{E_\mathrm{f}V}}\left(\frac{\pi}{k^2}\right)^{3/2} 2\pi\delta(E_\mathrm{i} - E_\mathrm{f}) \\
&\quad \times \mathrm{e}^{-\frac{(\boldsymbol{p}_\mathrm{i} - \boldsymbol{p}_\mathrm{f})^2}{4k^2}}\bar{u}(\boldsymbol{p}_\mathrm{f}, s)\gamma_0 u(\boldsymbol{p}_\mathrm{i}, r) .
\end{aligned} \tag{10.26}
$$

Delta function which appears in the transition amplitude (10.26) indicates on the energy conservation law, which is satisfied because potential A^μ does not depend on time. As three–space is inhomogeneous (the potential depends on $\boldsymbol{x}$), the three-momentum is not conserved. The average value of the squared transition amplitude is obtained from (10.26)

$$
\begin{aligned}
\langle |S_{\mathrm{fi}}|^2 \rangle &= \frac{1}{2}\frac{e^2 m^2 a^2}{V^2 E_\mathrm{i} E_\mathrm{f}} 2\pi T\delta(E_\mathrm{i} - E_\mathrm{f})\left(\frac{\pi}{k^2}\right)^3 \\
&\quad \times \mathrm{e}^{-\frac{(\boldsymbol{p}_\mathrm{i} - \boldsymbol{p}_\mathrm{f})^2}{2k^2}} \sum_{r,s=1}^{2} |u(\boldsymbol{p}_\mathrm{f}, s)\gamma_0 u(\boldsymbol{p}_\mathrm{i}, r)|^2 .
\end{aligned} \tag{10.27}
$$

Because of

$$
(\bar{u}(\boldsymbol{p}_\mathrm{f}, s)\gamma_0 u(\boldsymbol{p}_\mathrm{i}, r))^* = \bar{u}(\boldsymbol{p}_\mathrm{i}, r)\gamma_0 u(\boldsymbol{p}_\mathrm{f}, s) ,
$$

we have:

$$\sum_{r,s=1}^{2} |\bar{u}(\boldsymbol{p}_{\rm f}, s)\gamma_0 u(\boldsymbol{p}_{\rm i}, r)|^2 = \sum_{r=1}^{2} \left(u_a(\boldsymbol{p}_{\rm f}, s)\bar{u}_b(\boldsymbol{p}_{\rm f}, s)\right) \gamma^0_{bc}$$
$$\times \sum_{r=1}^{2} \left(u_c(\boldsymbol{p}_{\rm i}, r)\bar{u}_d(\boldsymbol{p}_{\rm i}, r)\right) \gamma^0_{da}$$
$$= \frac{1}{4m^2}\mathrm{tr}[(\not{p}_{\rm f} + m)\gamma^0(\not{p}_{\rm i} + m)\gamma^0]$$
$$= \frac{1}{m^2}(E_{\rm i}E_{\rm f} + \boldsymbol{p}_{\rm i}\cdot\boldsymbol{p}_{\rm f} + m^2) \ . \qquad (10.28)$$

By plugging (10.28) into (10.27) one obtains

$$\langle |S_{\rm fi}|^2\rangle = \frac{e^2a^2\pi}{V^2E_{\rm i}E_{\rm f}}\left(\frac{\pi}{k^2}\right)^3 T\delta(E_{\rm i} - E_{\rm f})$$
$$\times\, \mathrm{e}^{-\frac{(\boldsymbol{p}_{\rm i} - \boldsymbol{p}_{\rm f})^2}{2k^2}}(E_{\rm i}E_{\rm f} + |\boldsymbol{p}_{\rm i}||\boldsymbol{p}_{\rm f}|\cos\theta + m^2) \ . \qquad (10.29)$$

By substituting (10.29) into the expression for the differential cross section,

$$\mathrm{d}\sigma = \frac{|S_{\rm fi}|^2}{T}\frac{VE_{\rm i}}{|\boldsymbol{p}_i|}\frac{V\mathrm{d}^3\boldsymbol{p}_{\rm f}}{(2\pi)^3} \ ,$$

one gets

$$\mathrm{d}\sigma = \frac{e^2a^2\pi}{8k^6}\left(E_{\rm i}E_{\rm f} + |\boldsymbol{p}_{\rm i}||\boldsymbol{p}_{\rm f}|\cos\theta + m^2\right)$$
$$\times \exp\left(-|\boldsymbol{p}_i|^2\frac{1-\cos\theta}{k^2}\right)\delta(E_{\rm f} - E_{\rm i})\frac{|\boldsymbol{p}_{\rm f}|}{|\boldsymbol{p}_{\rm i}|}\mathrm{d}E_{\rm f}\mathrm{d}\Omega \ .$$

The $E_{\rm f}$–integration gives

$$\frac{\mathrm{d}\sigma}{\mathrm{d}\Omega} = \frac{e^2a^2\pi}{8k^6}\left(E_{\rm i}^2 + |\boldsymbol{p}_{\rm i}|^2\cos\theta + m^2\right)\mathrm{e}^{-|\boldsymbol{p}|^2\frac{1-\cos\theta}{k^2}} \ .$$

(b) This problem is analogous to the previous one, so we shall discuss only the main steps. The transition amplitude is:

$$S_{\rm fi} = -\frac{2\mathrm{i}egm}{V\sqrt{E_{\rm i}E_{\rm f}}}(2\pi)\delta(E_{\rm f} - E_{\rm i})\frac{2\pi}{\boldsymbol{q}^2 + \frac{1}{a^2}}\bar{u}(\boldsymbol{p}_{\rm f}, s)\gamma^3 u(\boldsymbol{p}_{\rm i}, r) \ ,$$

where $\boldsymbol{q} = \boldsymbol{p}_{\rm f} - \boldsymbol{p}_{\rm i}$. The next step is to calculate the squared amplitude:

$$\sum_{r,s=1}^{2} |\bar{u}(\boldsymbol{p}_{\rm f}, s)\gamma^3 u(\boldsymbol{p}_{\rm i}, r)|^2 = \frac{1}{4m^2}\mathrm{tr}[(\not{p}_{\rm f} + m)\gamma^3(\not{p}_{\rm i} + m)\gamma^3]$$
$$= \frac{1}{m^2}(2p_{\rm i}^3p_{\rm f}^3 + p_{\rm i}\cdot p_{\rm f} - m^2)$$
$$= \frac{1}{m^2}(E_{\rm i}E_{\rm f} + |\boldsymbol{p}_{\rm i}||\boldsymbol{p}_{\rm f}|\cos\theta - m^2) \ .$$

The average value of the squared transition amplitude is:

$$\langle |S_{\rm fi}|^2 \rangle = \frac{16\pi^3 e^2 g^2 T}{V^2 E_{\rm i} E_{\rm f}} \frac{1}{\left(\boldsymbol{q}^2 + \frac{1}{a^2}\right)^2} \delta(E_{\rm f} - E_{\rm i})(E_{\rm i} E_{\rm f} + |\boldsymbol{p}_{\rm i}||\boldsymbol{p}_{\rm f}| \sin\theta - m^2) \ .$$

The differential cross section is:

$$\frac{{\rm d}\sigma}{{\rm d}\Omega} = 2e^2 g^2 \frac{(E^2 - m^2)(1 + \cos\theta)}{\left(\frac{1}{a^2} + 2(E^2 - m^2)(1 - \cos\theta)\right)^2} \ .$$

10.10 The initial state is vacuum $|0\rangle$, while the final state is

$$|{\rm f}\rangle = c^\dagger(\boldsymbol{p}_1, r) d^\dagger(\boldsymbol{p}_2, s) |0\rangle \ .$$

The transition amplitude is

$$S_{\rm fi} = \frac{{\rm i}e}{V} \int {\rm d}^4 x \sum_{r's'} \int {\rm d}^3 \boldsymbol{q}_1 {\rm d}^3 \boldsymbol{q}_2 \sqrt{\frac{m}{E_{q_1}}} \sqrt{\frac{m}{E_{q_2}}} \langle 0 | d(\boldsymbol{p}_2, s) c(\boldsymbol{p}_1, r)$$
$$\times \left(c^\dagger(\boldsymbol{q}_1, r') d^\dagger(\boldsymbol{q}_2, s') \bar{u}(\boldsymbol{q}_1, r') \gamma^\mu A_\mu(x) v(\boldsymbol{q}_2, s') {\rm e}^{{\rm i}q_1 \cdot x + {\rm i}q_2 \cdot x} + \ldots\right) |0\rangle \ ,$$

where we have dropped the vanishing terms. After reducing the last expression one obtains

$$\begin{aligned} S_{\rm fi} &= {\rm i}e \frac{ma}{V\sqrt{E_1 E_2}} \int {\rm d}^4 x \ \bar{u}(\boldsymbol{p}_1, r) \gamma_2 v(\boldsymbol{p}_2, s) {\rm e}^{{\rm i}(p_2 + p_1) \cdot x} {\rm e}^{-{\rm i}\omega t} \\ &= {\rm i}e (2\pi)^4 \frac{ma}{V\sqrt{E_1 E_2}} \\ &\times \bar{u}(\boldsymbol{p}_1, r) \gamma_2 v(\boldsymbol{p}_2, s) \delta^{(3)}(\boldsymbol{p}_1 + \boldsymbol{p}_2) \delta(E_1 + E_2 - \omega) \ . \end{aligned}$$

The average value of the squared transition amplitude is

$$\begin{aligned} \langle |S_{\rm fi}|^2 \rangle &= (2\pi)^4 T V \delta^{(3)}(\boldsymbol{p}_1 + \boldsymbol{p}_2) \delta(E_1 + E_2 - \omega) \\ &\times \frac{e^2 a^2}{4V^2 E_1 E_2} {\rm tr}[(\not{p}_1 + m) \gamma_2 (\not{p}_2 - m) \gamma_2] \\ &= (2\pi)^4 T \delta^{(3)}(\boldsymbol{p}_1 + \boldsymbol{p}_2) \delta(E_1 + E_2 - \omega) \frac{e^2 a^2}{V E_1 E_2} \\ &\times (E_1 E_2 + |\boldsymbol{p}_1||\boldsymbol{p}_2| - 2|\boldsymbol{p}_1||\boldsymbol{p}_2| \sin^2\theta \cos^2\phi + m^2) \ , \end{aligned}$$

since the four-momenta are:

$$p_1^\mu = (E_1, p_1 \sin\theta \cos\phi, p_1 \sin\theta \sin\phi, p_1 \cos\theta) \ ,$$

$$p_2^\mu = (E_2, -p_2 \sin\theta \cos\phi, -p_2 \sin\theta \sin\phi, -p_2 \cos\theta) \ .$$

The differential cross section is:

$${\rm d}\sigma = \frac{\langle |S_{\rm fi}|^2 \rangle}{T} \frac{V {\rm d}^3 p_1}{(2\pi)^3} \frac{V {\rm d}^3 p_2}{(2\pi)^3} \ .$$

By integrating over $\boldsymbol{p}_2$ and $\boldsymbol{p}_1$ one obtains the scattering cross section (per unit volume)

$$\sigma = \frac{e^2a^2}{3\pi\omega}(\omega^2+2m^2)\sqrt{\frac{\omega^2}{4}-m^2} \ .$$

10.11 The transition amplitude is

$$S_{\rm fi} = \frac{{\rm i}eam}{V}\frac{1}{\sqrt{E_{\rm i}E_{\rm f}}}\bar{u}(\boldsymbol{p}_{\rm f},s)\gamma_3(1-\gamma_5)u(\boldsymbol{p}_{\rm i},r)\int {\rm d}^4x{\rm e}^{-{\rm i}p_{\rm i}\cdot x+{\rm i}p_{\rm f}\cdot x}{\rm e}^{-k^2\boldsymbol{x}^2} \ .$$

By integrating over t and $\boldsymbol{x}$ we get

$$\begin{aligned} S_{\rm fi} &= {\rm i}ea\sqrt{\frac{m}{E_{\rm i}V}}\sqrt{\frac{m}{E_{\rm f}V}}\left(\frac{\pi}{k^2}\right)^{3/2}{\rm e}^{-\frac{(\boldsymbol{p}_{\rm i}-\boldsymbol{p}_{\rm f})^2}{4k^2}} \\ &\times 2\pi\delta(E_{\rm i}-E_{\rm f})\bar{u}(\boldsymbol{p}_{\rm f},s)\gamma_3(1-\gamma_5)u(\boldsymbol{p}_{\rm i},r) \ . \end{aligned}$$

The average value of the squared transition amplitude is:

$$\langle|S_{\rm fi}|^2\rangle = \frac{e^2a^2m^2}{V^2E_{\rm i}E_{\rm f}}2\pi T\delta(E_{\rm i}-E_{\rm f})\left(\frac{\pi}{k^2}\right)^3{\rm e}^{-\frac{(\boldsymbol{p}_{\rm i}-\boldsymbol{p}_{\rm f})^2}{2k^2}}\langle|\mathcal{M}|^2\rangle \ ,$$

where

$$\begin{aligned} \langle|\mathcal{M}|^2\rangle &= \frac{1}{2}\sum_{r,s=1}^{2}|\bar{u}(\boldsymbol{p}_{\rm f},s)\gamma_3(1-\gamma_5)u(\boldsymbol{p}_{\rm i},r)|^2 \\ &= \frac{1}{2}\frac{1}{4m^2}{\rm tr}\left[(\not{p}_{\rm f}+m)\gamma_3(1-\gamma_5)(\not{p}_{\rm i}+m)(1+\gamma_5)\gamma_3\right] \\ &= \frac{1}{m^2}(2p_{\rm f}^3p_{\rm i}^3+p_{\rm i}\cdot p_{\rm f}) \ . \end{aligned}$$

The differential cross section is:

$$\frac{{\rm d}\sigma}{{\rm d}\Omega} = \frac{e^2a^2\pi}{4k^6}\left(E_{\rm i}^2+|\boldsymbol{p}_{\rm i}|^2\cos\theta\right){\rm e}^{-\frac{1}{k^2}|\boldsymbol{p}_{\rm i}|^2(1-\cos\theta)} \ .$$

10.12 We shall present the expression for the transition amplitude and final result for the differential cross section only:

$$S_{\rm fi} = {\rm i}e\frac{m}{V\sqrt{E_{\rm i}E_{\rm f}}}\bar{v}(\boldsymbol{p}_{\rm i},s)v(\boldsymbol{p}_{\rm f},r)\int {\rm d}^4x({\rm i}E_{\rm f})\frac{g}{|\boldsymbol{x}|}{\rm e}^{-{\rm i}(p_{\rm i}-p_{\rm f})\cdot x} \ ,$$

$$\frac{{\rm d}\sigma}{{\rm d}\Omega} = \frac{e^2g^2E^2(E^2+m^2-\boldsymbol{p}^2\cos\theta)}{2|\boldsymbol{p}|^4(1-\cos\theta)^2} \ .$$

10.13 The transition amplitude $S_{\rm fi}$ is

$$\begin{aligned} S_{\rm fi} &= {\rm i}ea\sqrt{\frac{m}{VE_{\rm i}}}\sqrt{\frac{m}{VE_{\rm f}}}\bar{u}(\boldsymbol{p}_{\rm f},s_{\rm f})\gamma^0u(\boldsymbol{p}_{\rm i},s_{\rm i})\int {\rm d}^4x\delta^{(3)}(\boldsymbol{x}){\rm e}^{-{\rm i}(p_{\rm i}-p_{\rm f})\cdot x} \\ &= {\rm i}ea\frac{m}{V\sqrt{E_{\rm i}E_{\rm f}}}(2\pi)\delta(E_{\rm i}-E_{\rm f})\bar{u}(\boldsymbol{p}_{\rm f},s_{\rm f})\gamma^0u(\boldsymbol{p}_{\rm i},s_{\rm i}) \ , \end{aligned}$$

where $s_{\rm i}$ i $s_{\rm f}$ are initial and final electron polarizations. In order to calculate $|S_{\rm fi}|^2$ it is necessary to compute squared spin-part of the amplitude. Since

$$u(\boldsymbol{p},s)\bar{u}(\boldsymbol{p},s) = \frac{1+\gamma_5 \not{s}}{2}\frac{\not{p}+m}{2m} ,$$

we have

$$\begin{aligned}
|\bar{u}(\boldsymbol{p}_{\rm f},s_{\rm f})\gamma^0 u(\boldsymbol{p}_{\rm i},s_{\rm i})|^2 &= \frac{1}{16m^2}{\rm tr}\left[(1+\gamma_5\not{s}_{\rm f})(\not{p}_{\rm f}+m)\gamma_0(1+\gamma_5\not{s}_{\rm i})(\not{p}_{\rm i}+m)\gamma_0\right] \\
&= \frac{1}{16m^2}\left({\rm tr}[\not{p}_{\rm f}\gamma_0\not{p}_{\rm i}\gamma_0] + m^2{\rm tr}[1]\right. \\
&\quad \left. - \,{\rm tr}[\not{s}_{\rm f}\not{p}_{\rm f}\gamma_0\not{s}_{\rm i}\not{p}_{\rm i}\gamma_0] + m^2{\rm tr}[\not{s}_{\rm f}\gamma_0\not{s}_{\rm i}\gamma_0]\right) , \qquad (10.30)
\end{aligned}$$

where we have kept only the nonvanishing traces. The components of momenta and polarization vectors are:

$$\begin{aligned}
p_{\rm i}^\mu &= (E_{\rm i},0,0,|\boldsymbol{p}_{\rm i}|) , \\
p_{\rm f}^\mu &= (E_{\rm f},|\boldsymbol{p}_{\rm f}|\sin\theta\cos\phi,|\boldsymbol{p}_{\rm f}|\sin\theta\sin\phi,|\boldsymbol{p}_{\rm f}|\cos\theta) , \\
s_{\rm i}^\mu &= (|\boldsymbol{p}_{\rm i}|/m,0,0,E_{\rm i}/m), \\
s_{\rm f}^\mu &= (|\boldsymbol{p}_{\rm f}|/m,(E_{\rm f}/m)\sin\theta\cos\phi,(E_{\rm f}/m)\sin\theta\sin\phi,(E_{\rm f}/m)\cos\theta) .
\end{aligned}$$

The traces in the sum (10.30) are:

$$\begin{aligned}
{\rm tr}[\not{s}_{\rm f}\not{p}_{\rm f}\gamma_0\not{s}_{\rm i}\not{p}_{\rm i}\gamma_0] &= -4m^2\cos\theta , \\
{\rm tr}{\rm I} &= 4 , \\
{\rm tr}[\not{s}_{\rm f}\gamma_0\not{s}_{\rm i}\gamma_0] &= 4\left(\frac{k^2}{m^2}+\frac{E^2}{m^2}\cos\theta\right) , \\
{\rm tr}[\not{p}_{\rm f}\gamma_0\not{p}_{\rm i}\gamma_0] &= 4(E^2+k^2\cos\theta) ,
\end{aligned}$$

where $E_{\rm i}=E_{\rm f}=E$ while $k=|\boldsymbol{p}_{\rm i}|=|\boldsymbol{p}_{\rm f}|$. By summing all the terms we get

$$|\bar{u}(\boldsymbol{p}_{\rm f},s_{\rm f})\gamma^0 u(\boldsymbol{p}_{\rm i},s_{\rm i})|^2 = \frac{E^2}{m^2}\cos^2\left(\frac{\theta}{2}\right) . \qquad (10.31)$$

The differential cross section for the scattering is computed in the usual way. The result is:

$$\frac{{\rm d}\sigma}{{\rm d}\Omega} = \frac{e^2a^2}{4\pi^2}E^2\cos^2(\theta/2) .$$

10.14 The amplitude for this process is (see Fig. 10.2)

$$i\mathcal{M} = \frac{ie^2}{k^2}\bar{u}(\boldsymbol{p}_2,r)\gamma^\mu u_2(\boldsymbol{p}_1)\bar{v}_2(\boldsymbol{q}_1)\gamma_\mu v(\boldsymbol{q}_2,s) ,$$

where subscript 2 in u and v spinors indicates that these are negative helicity particles. The squared Feynman amplitude module is

$$\langle|\mathcal{M}|^2\rangle = \frac{e^4}{64m_e^2m_\mu^2k^4}\text{tr}[(\not{p}_2+m_e)\gamma^\mu(\not{p}_1+m_e)(1-\gamma_5\not{s}_1)\gamma_\nu]$$
$$\times\ \text{tr}[(\not{q}_1-m_\mu)(1-\gamma_5\not{s}_2)\gamma_\mu(\not{q}_2-m_\mu)\gamma^\nu]\ ,$$

where we have summed over polarization states of the final particles in the process. Here s_1 and s_2 are polarization vectors of the initial electron and muon which are going to be evaluated later. By applying corresponding identities from Problem 3.6 and corresponding expression for contractions of two ϵ symbols from Problem 1.5 we get

$$\begin{aligned}\langle|\mathcal{M}|^2\rangle = \frac{e^4}{2m_e^2m_\mu^2k^4}&[(p_2\cdot q_1)(p_1\cdot q_2)+(p_2\cdot q_2)(p_1\cdot q_1)-\\&-m_\mu^2(p_2\cdot p_1)-m_e^2(q_1\cdot q_2)+2m_e^2m_\mu^2+\\&+m_em_\mu\,((s_1\cdot s_2)(p_2\cdot q_2)-(s_1\cdot s_2)(p_2\cdot q_1)-\\&-(s_1\cdot s_2)(p_1\cdot q_2)+(s_1\cdot s_2)(p_1\cdot q_1)-\\&-(s_1\cdot q_2)(s_2\cdot p_2)+(s_1\cdot q_1)(s_2\cdot p_2)+\\&+\ (s_1\cdot q_2)(s_2\cdot p_1)-(s_1\cdot q_1)(s_2\cdot p_1))]\ .\end{aligned}\tag{10.32}$$

Since $m_\mu \approx 200m_e$ we will neglect the electron mass. In the center–of–mass frame four momenta are

$$\begin{aligned}p_1^\mu &= (E,0,0,p)\ ,\\ q_1^\mu &= (E',0,0,-p)\ ,\\ p_2^\mu &= (E,p\sin\theta\cos\phi,p\sin\theta\sin\phi,p\cos\theta)\ ,\\ q_2^\mu &= (E',-p\sin\theta\cos\phi,-p\sin\theta\sin\phi,-p\cos\theta)\ .\end{aligned}$$

Polarization vectors s_1 and s_2 are

$$\begin{aligned}s_1^\mu &= (\frac{p}{m_e},0,0,\frac{E}{m_e})\ ,\\ s_2^\mu &= (\frac{p}{m_\mu},0,0,-\frac{E'}{m_\mu})\ .\end{aligned}$$

After finding scalar products between four-vectors in (10.32) and reducing the obtained expression one gets

$$\begin{aligned}\langle|\mathcal{M}|^2\rangle = \frac{e^4}{32m_e^2m_\mu^2p^4\sin^4(\frac{\theta}{2})}&\left[(EE'+p^2)^2-2p^2(m_e^2+m_\mu^2)\sin^2\left(\frac{\theta}{2}\right)\right.\\&\left.+(EE'+p^2\cos\theta)^2+p^2\left(4p^2\sin^2\left(\frac{\theta}{2}\right)+EE'\sin^2\theta\right)\right]\ ,\end{aligned}\tag{10.33}$$

hence the differential cross section is

$$\begin{aligned}\frac{\mathrm{d}\sigma}{\mathrm{d}\Omega} = \frac{e^4}{128\pi^2(E+E')^2p^4\sin^4(\theta/2)}&\left[(EE'+p^2)^2-2p^2(m_e^2+m_\mu^2)\sin^2\left(\frac{\theta}{2}\right)\right.\\&\left.+(EE'+p^2\cos\theta)^2+p^2\left(4p^2\sin^2\left(\frac{\theta}{2}\right)+EE'\sin^2\theta\right)\right]\ .\end{aligned}\tag{10.34}$$

10.15 The interaction Hamiltonian is

$$H_{\rm int} = g \int \mathrm{d}^3\boldsymbol{x} \bar{\psi}\gamma_5\psi\phi \ ,$$

where the field operators are written in the interaction picture. In the lowest ("tree–level") order of the perturbation theory the transition amplitude is:

$$S_{\rm fi} = \frac{1}{2}(-\mathrm{i}g)^2 \langle \boldsymbol{p}'\boldsymbol{k}'| \int \mathrm{d}^4x\mathrm{d}^4y T\{ : (\bar{\psi}\gamma_5\psi\phi)_x :: (\bar{\psi}\gamma_5\psi\phi)_y : \} |\boldsymbol{pk}\rangle \ . \quad (10.35)$$

Because of

$$\psi(x)\,|\boldsymbol{p},r\rangle = \sqrt{\frac{m}{VE_p}}u(\boldsymbol{p},r)\mathrm{e}^{-\mathrm{i}p\cdot x} \ ,$$

$$\langle \boldsymbol{p},r|\,\bar{\psi}(x) = \sqrt{\frac{m}{VE_p}}\bar{u}(\boldsymbol{p},r)\mathrm{e}^{\mathrm{i}p\cdot x} \ ,$$

from the expression (10.35) we conclude that there are four ways to make contractions which correspond to the given process. In that way we obtain (note that there are two couples containing two identical terms)

$$\begin{aligned} S_{\rm fi} = &-g^2\frac{m^2}{V^2\sqrt{E_1E_2E_1'E_2'}}\int \mathrm{d}^4x\mathrm{d}^4y\mathrm{i}\Delta_{\rm F}(x-y) \\ &\times \Big[-\bar{u}(\boldsymbol{k}',s')\gamma_5 u(\boldsymbol{k},s)\bar{u}(\boldsymbol{p}',r')\gamma_5 u(\boldsymbol{p},r)\mathrm{e}^{\mathrm{i}(p'-p)\cdot y+\mathrm{i}(k'-k)\cdot x} \\ &+ \bar{u}(\boldsymbol{p}',r')\gamma_5 u(\boldsymbol{k},s)\bar{u}(\boldsymbol{k}',s')\gamma_5 u(\boldsymbol{p},r)\mathrm{e}^{\mathrm{i}(k'-p)\cdot y+\mathrm{i}(p'-k)\cdot x}\Big] \ . \end{aligned} \quad (10.36)$$

The minus sign in the first term is a consequence of the Wick theorem for fermions. After integrating the last expression and having in mind that

$$\mathrm{i}\Delta_{\rm F}(x-y) = \frac{\mathrm{i}}{(2\pi)^4}\int \mathrm{d}^4q\frac{\mathrm{e}^{-\mathrm{i}q\cdot(x-y)}}{q^2-M^2+\mathrm{i}\epsilon} \ ,$$

one obtains

$$\begin{aligned} S_{\rm fi} = &\,\mathrm{i}\frac{(2\pi)^4g^2m^2}{V^2\sqrt{E_1E_2E_1'E_2'}}\delta^{(4)}(p'+k'-p-k) \\ &\times \Big[\frac{1}{(p'-p)^2-M^2+\mathrm{i}\epsilon}\bar{u}(\boldsymbol{k}',s')\gamma_5 u(\boldsymbol{k},s)\bar{u}(\boldsymbol{p}',r')\gamma_5 u(\boldsymbol{p},r) - \\ &- \frac{1}{(p'-k)^2-M^2+\mathrm{i}\epsilon}\bar{u}(\boldsymbol{p}',r')\gamma_5 u(\boldsymbol{k},s)\bar{u}(\boldsymbol{k}',s')\gamma_5 u(\boldsymbol{p},r)\Big] \ . \end{aligned}$$

Feynman diagrams for the scattering are represented in the figure.

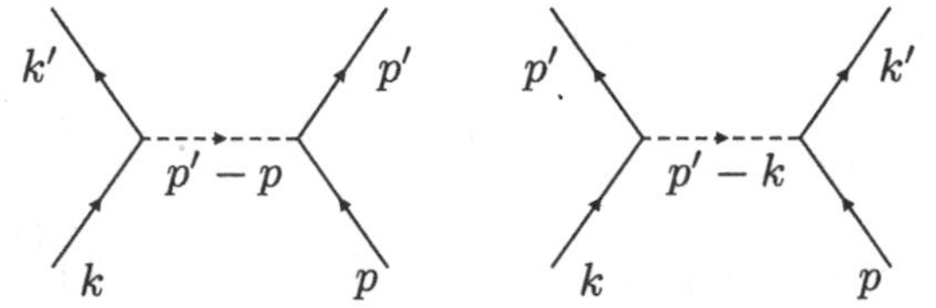

The squared amplitude is

$$
\begin{aligned}
\langle |S_{\mathrm{fi}}|^2\rangle = & \frac{g^4(2\pi)^4 T\delta^{(4)}(p'+k'-p-k)}{4V^3E_1E_2E_1'E_2'} \\
& \times \left[\frac{(k\cdot k')(p\cdot p')-(k\cdot k')m^2-(p\cdot p')m^2+m^4}{((p'-p)^2-M^2)^2}+\right. \\
& +\frac{(p\cdot k')(k\cdot p')-(p\cdot k')m^2-(k\cdot p')m^2+m^4}{((p'-k)^2-M^2)^2} \\
& -\frac{1}{2}\frac{1}{(p'-p)^2-M^2}\frac{1}{(p'-k)^2-M^2}\mathrm{Re}\left[(\mathrm{k}\cdot\mathrm{k}')(\mathrm{p}\cdot\mathrm{p}')\right. \\
& -(p'\cdot k')(k\cdot p)+(p\cdot k')(k\cdot p') \\
& -(k\cdot k')m^2-(p\cdot p')m^2-(k\cdot p')m^2 \\
& \left.\left. -(p\cdot k')m^2+(k\cdot p)m^2+(k'\cdot p')m^2+m^4\right]\right] .
\end{aligned}
$$

The squared amplitude per unit time as viewed from the center–of–mass frame is:

$$
\begin{aligned}
\frac{\langle |S_{\mathrm{fi}}|^2\rangle}{T} = & \frac{g^4(2\pi)^4\delta^{(4)}(p'+k'-p-k)}{4V^3E^4}|\boldsymbol{p}|^4 \\
& \times \left[\frac{(1-\cos\theta)^2}{(2|\boldsymbol{p}|^2(\cos\theta-1)-M^2)^2}+\frac{(1+\cos\theta)^2}{(2|\boldsymbol{p}|^2(\cos\theta+1)+M^2)^2}\right. \\
& \left. -\frac{\sin^2\theta}{(2|\boldsymbol{p}|^2(\cos\theta-1)-M^2)(2|\boldsymbol{p}|^2(\cos\theta+1)+M^2)}\right] , \qquad (10.37)
\end{aligned}
$$

where $E_1 = E_2 = E_1' = E_2' = E$ are the energies of the initial and final particles. All four fermions carry the momenta of the identical intensity $|\boldsymbol{p}|$. In the high energy limit from (10.37) one obtains

$$
\frac{\langle |S_{\mathrm{fi}}|^2\rangle}{T} = \frac{3g^4(2\pi)^4\delta^{(4)}(p'+k'-p-k)}{16V^3E^4} . \qquad (10.38)
$$

The total cross section for the scattering is

$$
\begin{aligned}
\sigma &= \int\int \frac{\langle |S_{\mathrm{fi}}|^2\rangle}{T}\frac{VE}{2|\boldsymbol{p}_1|}\frac{V\mathrm{d}^3\boldsymbol{p}_1'}{(2\pi)^3}\frac{V\mathrm{d}^3\boldsymbol{p}_2'}{(2\pi)^3} \\
&= \frac{3g^4}{4\pi^2}\frac{\delta(2E-2E')}{16E}\frac{\mathrm{d}E_1'\mathrm{d}\Omega_1'}{2E} \\
&= \frac{3g^4}{64\pi E^2} .
\end{aligned}
$$

10.16 By direct application of the Feynman rules we obtain the expression for the corresponding amplitudes. In the following expressions we drop external lines.

(a)

$\mathrm{i}\mathcal{M} =$ 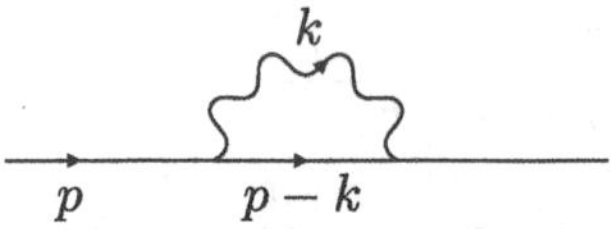

$$= (\mathrm{ie})^2 \int \frac{\mathrm{d}^4 k}{(2\pi)^4} \left(\gamma_\nu \frac{1}{\not{p} - \not{k} - m + \mathrm{i}\epsilon} \gamma_\mu \frac{g^{\mu\nu}}{k^2 + \mathrm{i}\epsilon} \right)$$

(b)

$\mathrm{i}\mathcal{M} =$ 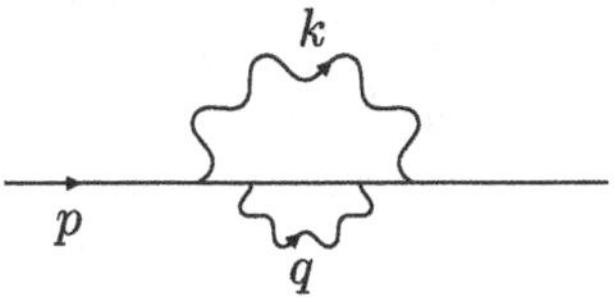

$$= \mathrm{i}(\mathrm{ie})^4 \int\int \frac{\mathrm{d}^4 k}{(2\pi)^4} \frac{\mathrm{d}^4 q}{(2\pi)^4} \left(\gamma^\mu \frac{1}{\not{p} - \not{k} - m + \mathrm{i}\epsilon} \gamma^\sigma \right.$$

$$\times \frac{1}{\not{p} - \not{k} - \not{q} - m + \mathrm{i}\epsilon} \gamma_\sigma \frac{1}{\not{p} - \not{k} - m + \mathrm{i}\epsilon}$$

$$\left. \times \gamma_\mu \frac{1}{k^2 + \mathrm{i}\epsilon} \frac{1}{q^2 + \mathrm{i}\epsilon} \right)$$

(c)

$\mathrm{i}\mathcal{M} =$ 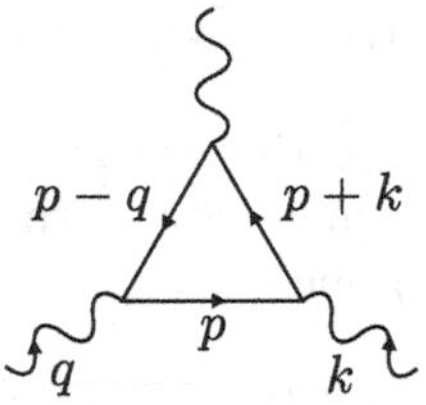

$$= -(\mathrm{ie})^3 \mathrm{i}^3 \int \frac{\mathrm{d}^4 p}{(2\pi)^4} \mathrm{tr} \left[\gamma^\nu \frac{1}{\not{p} - \not{q} - m + \mathrm{i}\epsilon} \gamma^\rho \right.$$

$$\left. \times \frac{1}{\not{p} + \not{k} - m + \mathrm{i}\epsilon} \gamma^\mu \frac{1}{\not{p} - m + \mathrm{i}\epsilon} \right]$$

(d)

$\mathrm{i}\mathcal{M} =$ 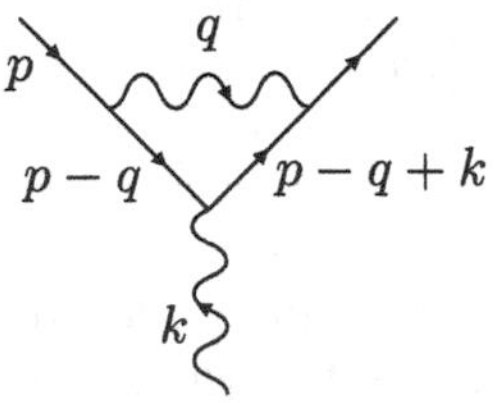

$$= \mathrm{i}(\mathrm{ie})^3 \int \frac{\mathrm{d}^4 p}{(2\pi)^4} \left(\gamma^\nu \frac{1}{\not{p} + \not{k} - \not{q} - m + \mathrm{i}\epsilon} \right.$$

$$\times\ \gamma^\rho \frac{1}{\not p - \not q - m + \mathrm{i}\epsilon}\gamma_\nu \frac{1}{q^2 + \mathrm{i}\epsilon}\Bigg)$$

(e)

$\mathrm{i}\mathcal{M} =$ 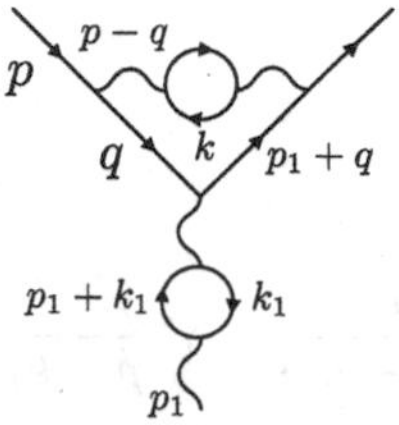

$$= (\mathrm{i}e)^7 \mathrm{i}^6 (-\mathrm{i})^3 \int\int\int \frac{\mathrm{d}^4k_1}{(2\pi)^4}\frac{\mathrm{d}^4q}{(2\pi)^4}\frac{\mathrm{d}^4k}{(2\pi)^4}$$
$$\times \left[\gamma^\nu \frac{1}{\not p_1 + \not q - m + \mathrm{i}\epsilon}\gamma^\alpha \frac{1}{\not q - m + \mathrm{i}\epsilon}\gamma^\mu\right.$$
$$\times \frac{g^{\mu\rho}}{(p-q)^2 + \mathrm{i}\epsilon}\frac{g^{\sigma\nu}}{(p-q)^2 + \mathrm{i}\epsilon}$$
$$\times \mathrm{tr}\left(\frac{1}{\not k - m + \mathrm{i}\epsilon}\gamma^\sigma \frac{1}{\not p - \not q + \not k - m + \mathrm{i}\epsilon}\gamma^\rho\right)$$
$$\left.\times \frac{g^{\alpha\beta}}{p_1^2 + \mathrm{i}\epsilon}\mathrm{tr}\left(\frac{1}{\not p_1 + \not k_1 - m + \mathrm{i}\epsilon}\gamma^\delta \frac{1}{\not k_1 - m + \mathrm{i}\epsilon}\gamma^\beta\right)\right]$$

(f)

$-\mathrm{i}\Pi^{\mu\nu}(k) =$ 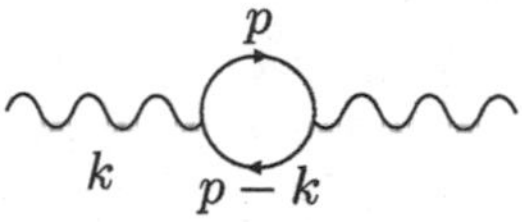

$$= (\mathrm{i}e)^2 \int \frac{\mathrm{d}^4p}{(2\pi)^4}\mathrm{tr}\left[\frac{1}{\not p - \not k - m + \mathrm{i}\epsilon}\gamma^\nu \frac{1}{\not p - m + \mathrm{i}\epsilon}\gamma^\mu\right]$$

(g)

$-\mathrm{i}\mathcal{M} =$ 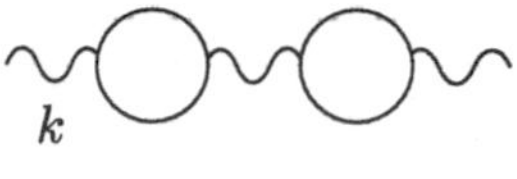

$= (-\mathrm{i})\Pi^{\mu\nu}(k)\dfrac{-\mathrm{i}g_{\nu\rho}}{k^2 + \mathrm{i}\epsilon}(-\mathrm{i})\Pi^{\rho\sigma}(k)$

(h)

$-\mathrm{i}\mathcal{M} =$ 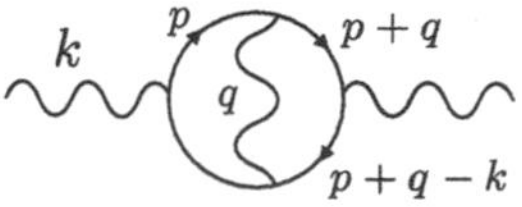

$$= -\mathrm{i}^4(-\mathrm{i})(\mathrm{i}e)^4 \int \frac{\mathrm{d}^4p}{(2\pi)^4}\frac{\mathrm{d}^4q}{(2\pi)^4}\mathrm{tr}\left[\frac{1}{\not p - \not k - m + \mathrm{i}\epsilon}\gamma^\sigma\right.$$
$$\times \frac{1}{\not p + \not q - \not k - m + \mathrm{i}\epsilon}\gamma^\nu \frac{1}{\not p + \not q - m + \mathrm{i}\epsilon}\gamma^\rho$$
$$\left.\times \frac{1}{\not p - m + \mathrm{i}\epsilon}\gamma^\mu\right]\frac{g_{\rho\sigma}}{q^2 + \mathrm{i}\epsilon}$$

(i)

$\mathrm{i}\mathcal{M} =$

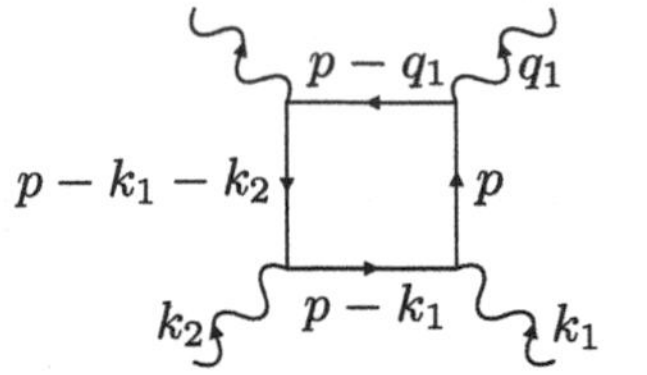

$$= -(\mathrm{i}e)^4 \int \frac{\mathrm{d}^4 p}{(2\pi)^4} \mathrm{tr}\left[\frac{1}{\not{p} - \not{k}_1 - m + \mathrm{i}\epsilon}\gamma^\mu \frac{1}{\not{p} - \not{k}_1 - \not{k}_2 - m + \mathrm{i}\epsilon}\right.$$

$$\left. \times\ \gamma^\sigma \frac{1}{\not{p} - \not{q}_1 - m + \mathrm{i}\epsilon}\gamma^\rho \frac{1}{\not{p} - m + \mathrm{i}\epsilon}\gamma^\nu\right]$$

11

Renormalization and regularization

11.1 In order to prove the Feynman formula we shall use mathematical induction. For $n = 2$ we have

$$
\begin{aligned}
I_2 &= \int_0^1 \mathrm{d}x_1 \int_0^1 \mathrm{d}x_2 \delta(x_1 + x_2 - 1) \frac{1}{[x_1 A_1 + x_2 A_2]^2} \\
&= \int_0^1 \mathrm{d}x_1 \frac{1}{[x_1 A_1 + (1 - x_1) A_2]^2} \\
&= \frac{1}{A_1 A_2} \ .
\end{aligned} \tag{11.1}
$$

By taking n-th derivative of (11.1) we get the useful identity

$$
\frac{1}{AB^n} = \int_0^1 \mathrm{d}x \int_0^1 \mathrm{d}y \delta(x + y - 1) \frac{n y^{n-1}}{[xA + yB]^{n+1}} \ . \tag{11.2}
$$

Now we shall assume that the Feynman formula is valid for $n = k$ and show that it holds for $n = k + 1$

$$
\begin{aligned}
\frac{1}{A_1 \ldots A_k A_{k+1}} &= \int_0^1 \mathrm{d}z_1 \ldots \mathrm{d}z_k \delta(z_1 + \ldots + z_k - 1) \frac{(k-1)!}{[z_1 A_1 + \ldots + z_k A_k]^k A_{k+1}} \\
&= \int_0^1 \mathrm{d}z_1 \ldots \mathrm{d}z_k \mathrm{d}y \, k! \, \delta(z_1 + \ldots + z_k - 1) \\
&\times \frac{y^{k-1}}{[y z_1 A_1 + \ldots + y z_k A_k + (1-y) A_{k+1}]^{k+1}} \ .
\end{aligned} \tag{11.3}
$$

By using substitution $x_1 = y z_1, \ldots, x_k = y z_k, x_{k+1} = 1 - y$ and a well known property of the δ–function

$$
\delta(ax) = \frac{1}{|a|} \delta(x) \ ,
$$

we obtain

$$\frac{1}{A_1...A_kA_{k+1}} = \int \mathrm{d}x_1...\mathrm{d}x_k\mathrm{d}x_{k+1}\ \delta(x_1+...+x_k+x_{k+1}-1)$$
$$\times \frac{k!}{[x_1A_1+...+x_{k+1}A_{k+1}]^{k+1}}\ , \tag{11.4}$$

which concludes the proof.

11.2 By introducing a new variable $q = k + p$, the integral I becomes

$$I = \int \mathrm{d}^D q \quad \frac{1}{(q^2 - m^2 - p^2 + \mathrm{i}\epsilon)^n}\ . \tag{11.5}$$

If we do a Wick rotation to the Euclidian space, $q^0 = \mathrm{i}q_{\mathrm{E}}^0$, $\boldsymbol{q} = \boldsymbol{q}_{\mathrm{E}}$, the integral I becomes

$$I = \mathrm{i}\int \mathrm{d}^D q_{\mathrm{E}} \quad \frac{1}{(-q_{\mathrm{E}}^2 - m^2 - p^2 + \mathrm{i}\epsilon)^n}\ . \tag{11.6}$$

The contour of the integration along the real axis can be rotated to the imaginary axis without passing through the poles. Transition from Minkowski space to Euclidian space is so–called Wick rotation.

The relation between the Cartesian and the spherical coordinates in the D dimensional space is

$$x_1 = r\sin\theta_{D-2}\sin\theta_{D-3}\dots\sin\theta_1\sin\phi\ ,$$
$$x_2 = r\sin\theta_{D-2}\sin\theta_{D-3}\dots\sin\theta_1\cos\phi\ ,$$
$$x_3 = r\sin\theta_{D-2}\sin\theta_{D-3}...\sin\theta_2\cos\theta_1\ ,$$
$$\vdots$$
$$x_D = r\cos\theta_{D-2}\ ,$$

where $0 < \phi < 2\pi,\ 0 < \theta_1, \dots, \theta_{D-2} < \pi$. The volume element, $\mathrm{d}V_D$ is

$$\mathrm{d}V_D = r^{D-1}\mathrm{d}r\,\mathrm{d}\phi\prod_1^{D-2}(\sin\theta_m)^m\mathrm{d}\theta_m\ .$$

Therefore

$$I = \frac{\mathrm{i}}{(-1)^n}2\pi\prod_{m=1}^{D-2}\int_0^\pi \mathrm{d}\theta_m\,(\sin\theta_m)^m\int_0^\infty \mathrm{d}r\frac{r^{D-1}}{(r^2+m^2+p^2)^n}\ . \tag{11.7}$$

If we use [9]

$$\int_0^\pi \mathrm{d}\theta\,(\sin\theta)^m = \sqrt{\pi}\frac{\Gamma\left(\frac{m+1}{2}\right)}{\Gamma\left(\frac{m+2}{2}\right)}\ ,$$

and

$$\int_0^\infty \mathrm{d}x\frac{x^b}{(x^2+M)^a} = \frac{\Gamma\left(\frac{1+b}{2}\right)\Gamma\left(a-\frac{1+b}{2}\right)}{2M^{a-\frac{1+b}{2}}\Gamma(a)}\ ,$$

we obtain

$$I = \mathrm{i}(-1)^n \pi^{\frac{D}{2}} \frac{\Gamma\left(n - \frac{D}{2}\right)}{\Gamma(n)} \frac{1}{(m^2 + p^2)^{n - \frac{D}{2}}} .$$

11.3 As we know, the Gamma–function is defined by

$$\Gamma(z) = \int_0^\infty \mathrm{d}t \mathrm{e}^{-t} t^{z-1} . \tag{11.8}$$

From the property $\Gamma(z) = \Gamma(z+1)/z$ follows that

$$\Gamma(z) = \Gamma(z+n+1) \prod_{k=0}^{n} \frac{1}{z+k} . \tag{11.9}$$

By using the definition of number e, the integral (11.8) becomes

$$\Gamma(z) = \lim_{n\to\infty} \int_0^n \mathrm{d}t \; t^{z-1}(1 - t/n)^n .$$

By introducing a new variable, $t/n = x$ the last integral is

$$\begin{aligned} \Gamma(z) &= \lim_{n\to\infty} n^z \int_0^1 \mathrm{d}x \; x^{z-1}(1-x)^n \\ &= \lim_{n\to\infty} n^z B(n+1, z) \\ &= \lim_{n\to\infty} n^z \frac{\Gamma(n+1)\Gamma(z)}{\Gamma(n+z+1)} \\ &= \lim_{n\to\infty} n^z \frac{\Gamma(n+1)}{z(z+1)\dots(z+n)} \\ &= \frac{1}{z} \lim_{n\to\infty} n^z \frac{1}{(1+z)(1+\frac{z}{2})\dots(1+\frac{z}{n})} , \end{aligned} \tag{11.10}$$

where we used (11.9).

Euler-Mascheroni constant, γ is defined by

$$\gamma = \lim_{n\to\infty} \left(1 + \frac{1}{2} + \frac{1}{3} + \dots + \frac{1}{n} - \ln n \right) .$$

Then

$$\mathrm{e}^{-\gamma z} = \lim_{n\to\infty} n^z \mathrm{e}^{-z(1+\frac{1}{2}+\dots+\frac{1}{n})} . \tag{11.11}$$

From (11.10) and (11.11) follows

$$\Gamma(z) = \mathrm{e}^{-z\gamma} \frac{1}{z} \prod_{n=1}^{\infty} \frac{\mathrm{e}^{z/n}}{1 + \frac{z}{n}} .$$

By taking the logarithm of the previous formula we get

$$\ln \Gamma(z) = -\gamma z - \ln z + \sum_{n=1}^{\infty} \left(\frac{z}{n} - \ln(1 + \frac{z}{n}) \right) .$$

Hence

$$\psi(z) = \frac{\mathrm{d} \ln \Gamma(z)}{\mathrm{d}z} = \frac{\Gamma'(z)}{\Gamma(z)} = -\gamma - \frac{1}{z} + \sum_{k=1}^{\infty} \left(\frac{1}{k} - \frac{1}{k+z} \right) . \tag{11.12}$$

For $z = n$ from the previous expression we get

$$\psi(n) = -\gamma + 1 + \frac{1}{2} + \frac{1}{3} + \ldots + \frac{1}{n-1} . \tag{11.13}$$

Expanding $\Gamma(1+\epsilon)$ according the Taylor formula we obtain

$$\begin{aligned} \Gamma(1+\epsilon) &= \Gamma(1) + \epsilon \Gamma'(1) + \ldots \\ &= 1 - \gamma\epsilon + o(\epsilon) . \end{aligned} \tag{11.14}$$

By using (11.9) and the previous expression we have

$$\begin{aligned} \Gamma(-n+\epsilon) &= \frac{\Gamma(1+\epsilon)}{\epsilon(\epsilon-1)\ldots(\epsilon-n)} \\ &= \frac{(-1)^n(1-\epsilon\gamma+o(\epsilon))}{n!\epsilon(1-\epsilon)(1-\epsilon/2)\ldots(1-\epsilon/n)} \\ &= \frac{(-1)^n}{n!}\left(\frac{1}{\epsilon}-\gamma\right)\left(1+\epsilon\left(1+\frac{1}{2}+\ldots+\frac{1}{n}\right)\right)+o(\epsilon) \\ &= \frac{(-1)^n}{n!}\left(\frac{1}{\epsilon}-\gamma+1+\frac{1}{2}+\ldots+\frac{1}{n}+o(\epsilon)\right) \\ &= \frac{(-1)^n}{n!}\left(\frac{1}{\epsilon}+\psi(n+1)+o(\epsilon)\right) . \end{aligned} \tag{11.15}$$

11.4 By applying the Feynman parametrization (11.G), the integral becomes

$$I = \int_0^1 \mathrm{d}x \int \mathrm{d}^4 k \frac{1}{[(k+px)^2 - \Delta]^2} ,$$

where $\Delta = p^2(x^2 - x) + m^2 x$. By making change of variable $l = k + px$ and going to Euclidian space ($l^0 = \mathrm{i} l_{\mathrm{E}}^0$, $\boldsymbol{l} = \boldsymbol{l}_{\mathrm{E}}$) we get

$$I = \mathrm{i} \int_0^1 \mathrm{d}x \int \mathrm{d}^4 l_{\mathrm{E}} \frac{1}{[l_{\mathrm{E}}^2 + \Delta]^2} .$$

In order to compute the integral we introduce spherical coordinates. The angular integration can be done immediately

$$I = \mathrm{i}\int_0^1 \mathrm{d}x \int_0^{2\pi} \mathrm{d}\phi \int_0^{\pi} \mathrm{d}\theta_1 \sin\theta_1 \int_0^{\pi} \mathrm{d}\theta_2 \sin^2\theta_2 \int_0^{\infty} \mathrm{d}l_\mathrm{E} \frac{l_\mathrm{E}^3}{(l_\mathrm{E}^2+\Delta)^2}$$
$$= \mathrm{i}\pi^2 \int_0^1 \mathrm{d}x \int_0^{\infty} \mathrm{d}l_\mathrm{E}^2 l_\mathrm{E}^2 \frac{1}{(l_\mathrm{E}^2+\Delta)^2} = \mathrm{i}\pi^2 \int_0^1 \mathrm{d}x \left[\ln(l_\mathrm{E}^2+\Delta)|_0^\infty - 1\right] .$$

The previous integral diverges logarithmically. Performing the Pauli–Villars regularization the propagator $1/k^2$ in the integral I becomes

$$\frac{1}{k^2} \to \frac{1}{k^2} - \frac{1}{k^2-\Lambda^2} ,$$

where Λ is a large parameter. A contribution of the second term in the previous expression to the integral is

$$I_\Lambda = \mathrm{i}\pi^2 \int_0^1 \mathrm{d}x \left[\ln(l_\mathrm{E}^2+\Delta_\Lambda)|_0^\infty - 1\right] ,$$

where we introduced

$$\Delta_\Lambda = \Lambda^2 + p^2(x^2-x) + x(m^2-\Lambda^2).$$

By subtracting these two results we get

$$I - I_\Lambda = \mathrm{i}\pi^2 \int_0^1 \mathrm{d}x \ln\left(\frac{\Lambda^2 + p^2(x^2-x) + x(m^2-\Lambda^2)}{p^2(x^2-x)+m^2x}\right)$$
$$= \mathrm{i}\pi^2 \int_0^1 \mathrm{d}x \ln\left(\frac{\Lambda^2(1-x)}{p^2(x^2-x)+m^2x}\right) .$$

11.5 The integrand is symmetric with respect to any two indices and therefore $I_{\alpha\beta\mu\nu\rho\sigma}$ is of the form

$$\begin{aligned} I_{\alpha\beta\mu\nu\rho\sigma} &= C\,[g_{\alpha\beta}(g_{\mu\nu}g_{\rho\sigma} + g_{\mu\rho}g_{\nu\sigma} + g_{\mu\sigma}g_{\nu\rho}) \\ &+ g_{\alpha\mu}(g_{\beta\nu}g_{\rho\sigma} + g_{\beta\rho}g_{\nu\sigma} + g_{\beta\sigma}g_{\nu\rho}) \\ &+ g_{\alpha\nu}(g_{\beta\mu}g_{\rho\sigma} + g_{\beta\rho}g_{\mu\sigma} + g_{\beta\sigma}g_{\mu\rho}) \\ &+ g_{\alpha\rho}(g_{\beta\mu}g_{\nu\sigma} + g_{\beta\nu}g_{\mu\sigma} + g_{\beta\sigma}g_{\nu\mu}) \\ &+ g_{\alpha\sigma}(g_{\beta\mu}g_{\nu\rho} + g_{\beta\nu}g_{\mu\rho} + g_{\beta\rho}g_{\mu\nu})] , \end{aligned}$$

where C is a constant. In order to determine C we will compute the contraction $g^{\alpha\beta}g^{\mu\nu}g^{\rho\sigma}I_{\alpha\beta\mu\nu\rho\sigma}$. It is easy to get

$$g^{\alpha\beta}g^{\mu\nu}g^{\rho\sigma}I_{\alpha\beta\mu\nu\rho\sigma} = C(D^3+6D^2+8D) .$$

On the other hand

$$g^{\alpha\beta}g^{\mu\nu}g^{\rho\sigma}I_{\alpha\beta\mu\nu\rho\sigma} = \int \frac{\mathrm{d}^Dk}{(k^2)^{n-3}} = \lim_{\mu\to 0}\int \frac{\mathrm{d}^Dk}{(k^2-\mu^2)^{n-3}}$$
$$= \lim_{\mu\to 0} \mathrm{i}(-1)^{n-3}\pi^2 \frac{\Gamma(n-3-\frac{D}{2})}{\Gamma(n-3)}(\mu^2)^{3-n+\frac{D}{2}} ,$$

where μ is a infrared parameter. Comparing these results we get

$$C = \frac{1}{D^3 + 6D^2 + 8D} \lim_{\mu \to 0} \mathrm{i}(-1)^{n-3} \pi^2 \frac{\Gamma(n-3-\frac{D}{2})}{\Gamma(n-3)} (\mu^2)^{3-n+\frac{D}{2}} .$$

Specially, for $n = 5$ the divergent part of the integral $I_{\alpha\beta\mu\nu\rho\sigma}$ is

$$\begin{aligned} I_{\alpha\beta\mu\nu\rho\sigma}\big|_{\text{div}} = \frac{\mathrm{i}\pi^2}{96\epsilon} [& g_{\alpha\beta}(g_{\mu\nu}g_{\rho\sigma} + g_{\mu\rho}g_{\nu\sigma} + g_{\mu\sigma}g_{\nu\rho}) \\ & + g_{\alpha\mu}(g_{\beta\nu}g_{\rho\sigma} + g_{\beta\rho}g_{\nu\sigma} + g_{\beta\sigma}g_{\nu\rho}) \\ & + g_{\alpha\nu}(g_{\beta\mu}g_{\rho\sigma} + g_{\beta\rho}g_{\mu\sigma} + g_{\beta\sigma}g_{\mu\rho}) \\ & + g_{\alpha\rho}(g_{\beta\mu}g_{\nu\sigma} + g_{\beta\nu}g_{\mu\sigma} + g_{\beta\sigma}g_{\nu\mu}) \\ & + g_{\alpha\sigma}(g_{\beta\mu}g_{\nu\rho} + g_{\beta\nu}g_{\mu\rho} + g_{\beta\rho}g_{\mu\nu}) . \end{aligned}$$

11.6 In D–dimensional space the interaction term takes the form $-g\mu^{\epsilon/2}\chi\phi^2$.

(a) The self–energy of the χ particle is determined by the diagram

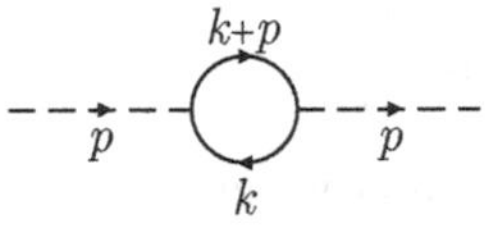

from which we read

$$-\mathrm{i}\Pi(p^2) = 2g^2\mu^\epsilon \int \frac{\mathrm{d}^D k}{(2\pi)^D} \frac{1}{k^2 - m^2 + \mathrm{i}0} \frac{1}{(k+p)^2 - m^2 + \mathrm{i}0} . \qquad (11.16)$$

By introducing the Feynman parametrization (11.G) and integrating over the momentum k we get:

$$\begin{aligned} -\mathrm{i}\Pi(p^2) &= \frac{\mathrm{i}g^2}{8\pi^2} \left(\frac{2}{\epsilon} - \gamma - \int_0^1 \mathrm{d}x \ln \frac{m^2 + p^2 x(x-1) - \mathrm{i}0}{4\pi\mu^2} \right) \\ &= \frac{\mathrm{i}g^2}{8\pi^2} \left[\frac{2}{\epsilon} - \gamma - \ln \frac{m^2}{4\pi\mu^2} \right. \\ & \left. - \int_0^1 \mathrm{d}x \ln \left(1 + \frac{p^2}{m^2} x(x-1) - \mathrm{i}0 \right) \right] . \end{aligned} \qquad (11.17)$$

As we know from the complex analysis the logarithm function, $w = \ln z$ has a branch cut along the positive x–axis which starts at the branch point $z = 0$. This branch cut is necessary if we want that branches of logarithm function to be single valued and holomorphic functions. Let us find the branch point for function

$$\ln[1 + \frac{p^2}{m^2} x(x-1)] .$$

It is the smallest value of p^2 for which the argument of logarithm function vanishes:

$$1+\frac{p^2}{m^2}(x^2-x)=0\ ,$$

i.e.

$$\frac{\partial p^2}{\partial x}=m^2\frac{2x-1}{(x^2-x)^2}=0\ ,$$

from which we get $x=\frac{1}{2}$. The point $p^2=4m^2$, which is step energy for the decay $\chi\to 2\phi$, is the branch point. A branch cut starts at this point and goes along x–axis in the positive direction to the infinity. Let us introduce the following notation

$$I=\frac{g^2}{8\pi^2}\int_0^1 \mathrm{d}x \ln\left(1+\frac{p^2}{m^2}x(x-1)-\mathrm{i}\delta\right)\ .$$

We shall calculate first this integral in the case $p^2>4m^2$. For $X>0$ we have

$$\log[-X-\mathrm{i}0]=\log|X|-\mathrm{i}\pi\ .$$

The zeroes of $1+\frac{p^2}{m^2}x(x-1)$ are

$$x_{1,2}=\frac{1\pm\sqrt{1-\frac{4m^2}{p^2}}}{2}\ .$$

For $x_1<x<x_2$ the expression $1+\frac{p^2}{m^2}(x^2-x)$ is negative, otherwise it is positive. Then

$$\begin{aligned}I=\frac{g^2}{8\pi^2}&\left[\int_0^{x_1}\mathrm{d}x\ln\left(1+\frac{p^2}{m^2}x(x-1)\right)\right.\\&+\int_{x_2}^1\mathrm{d}x\ln\left(1+\frac{p^2}{m^2}x(x-1)\right)\\&\left.+\int_{x_1}^{x_2}\mathrm{d}x\ln\left(-1-\frac{p^2}{m^2}x(x-1)\right)-\mathrm{i}\pi(x_2-x_1)\right]\ .\end{aligned}\qquad(11.18)$$

By doing partial integration we have

$$\begin{aligned}I=\frac{g^2}{8\pi^2}&\left[x\ln\left(1+\frac{p^2}{m^2}x(x-1)\right)\Big|_0^{x_1}-\frac{p^2}{m^2}\int_0^{x_1}\mathrm{d}x\frac{x(2x-1)}{1+p^2(x^2-x)/m^2}\right.\\&+x\ln\left(1+\frac{p^2}{m^2}x(x-1)\right)\Big|_{x_2}^{1}-\frac{p^2}{m^2}\int_{x_2}^{1}\mathrm{d}x\frac{x(2x-1)}{1+p^2(x^2-x)/m^2}\\&+x\ln\left(-1-\frac{p^2}{m^2}x(x-1)\right)\Big|_{x_1}^{x_2}-\frac{p^2}{m^2}\int_{x_1}^{x_2}\mathrm{d}x\frac{x(2x-1)}{1+p^2(x^2-x)/m^2}\\&\left.-\ \mathrm{i}\pi(x_2-x_1)\right]\ .\end{aligned}\qquad(11.19)$$

Combining the terms in the previous formula we get

$$I=\frac{g^2}{8\pi^2}\left[-\mathrm{i}\pi(x_2-x_1)-\frac{p^2}{m^2}\int_0^1\mathrm{d}x\frac{x(2x-1)}{1+p^2(x^2-x)/m^2}\right]\,. \tag{11.20}$$

The integral in the previous formula can be simplified by introducing the new variable $t=2x-1$. The result is (see [9])

$$I=-\mathrm{i}\frac{g^2}{8\pi}\sqrt{1-\frac{4m^2}{p^2}}-\frac{g^2}{4\pi^2}\left[1+\frac{1}{2}\sqrt{1-\frac{4m^2}{p^2}}\ln\frac{1-\sqrt{1-\frac{4m^2}{p^2}}}{1+\sqrt{1-\frac{4m^2}{p^2}}}\right]\,.$$

For $0<p^2<4m^2$ we get [9]

$$I=\frac{g^2}{4\pi^2}\left[-1+\sqrt{\frac{4m^2}{p^2}-1}\arcsin\sqrt{\frac{p^2}{4m^2}}\right]\,.$$

The final result for the vacuum polarization, $-\mathrm{i}\Pi(p^2)$ is

$$-\mathrm{i}\Pi(p^2)=\frac{\mathrm{i}g^2}{8\pi^2}\left(\frac{2}{\epsilon}-\gamma-\ln\frac{m^2}{4\pi\mu^2}+2\right)+\pi(p^2)\,, \tag{11.21}$$

where

$$\pi(p^2)=-\frac{\mathrm{i}g^2}{4\pi^2}\sqrt{\frac{4m^2}{p^2}-1}\arcsin\sqrt{\frac{p^2}{4m^2}}$$

for $0<p^2<4m^2$ and

$$\pi(p^2)=\frac{\mathrm{i}g^2}{8\pi^2}\left(\mathrm{i}\sqrt{1-\frac{4m^2}{p^2}}+\sqrt{1-\frac{4m^2}{p^2}}\ln\frac{1-\sqrt{1-\frac{4m^2}{p^2}}}{1+\sqrt{1-\frac{4m^2}{p^2}}}\right)$$

for $p^2>4m^2$.

(b) In the lowest order of the perturbation theory the transition amplitude is given by

$$\begin{aligned}S_{fi}&=-\mathrm{i}g\int\mathrm{d}^4x\,\langle\boldsymbol{p}_1,\boldsymbol{p}_2|\,\chi(x)\phi(x)\phi(x)\,|M,\boldsymbol{p}=0\rangle\\&=(2\pi)^4\delta^{(4)}(p-p_1-p_2)\sqrt{\frac{1}{2VM}}\sqrt{\frac{1}{2VE_1}}\sqrt{\frac{1}{2VE_2}}(-2\mathrm{i}g)\,,\end{aligned}$$

where $\boldsymbol{p}_{1,2}$ are the momenta of the decay products. Also we take that χ particle is in the rest. The decay rate is

$$\mathrm{d}\Gamma=\frac{|S_{fi}|^2}{T}\frac{V^2\mathrm{d}^3p_1\mathrm{d}^3p_2}{(2\pi)^6}\,.$$

By integrating over the momentum $\boldsymbol{p}_2$ we get:

$$\Gamma = \frac{4g^2}{(2\pi)^2} \int \mathrm{d}E pE \frac{1}{8ME^2} \delta(M - 2E) \int_0^\pi \mathrm{d}\theta \int_0^\pi \mathrm{d}\phi \, ,$$

and the space angle integration gives 2π (not 4π, because the final particles are identical). The final result is given by:

$$\Gamma = \frac{g^2}{4\pi M^2} \sqrt{\frac{M^2}{4} - m^2} \, .$$

(c) The imaginary part of $\Pi(p^2)$ can be read off the part (a):

$$\mathrm{Im}\Pi(p^2) = -\frac{g^2}{8\pi} \sqrt{1 - \frac{4m^2}{p^2}} \theta(p^2 - 4m^2) \, . \tag{11.22}$$

This result also can be obtained using *Cutkosky rule.* The expression (11.16) can be rewritten in the following form

$$-\mathrm{i}\Pi(p^2) = 2g^2 \int \frac{\mathrm{d}^4 k}{(2\pi)^4} \frac{1}{(-k)^2 - m^2 + \mathrm{i}0} \frac{1}{(k+p)^2 - m^2 + \mathrm{i}0} \, . \tag{11.23}$$

The discontinuity of the amplitude

$$\mathrm{Disc}\, \Pi(p^2) = \Pi(p^2 + \mathrm{i}\epsilon) - \Pi(p^2 - \mathrm{i}\epsilon) \, ,$$

is obtained by making the substitution

$$\frac{1}{p^2 - m^2} \to (-2\mathrm{i}\pi)\delta^{(4)}(p^2 - m^2)\theta(p^0) \, ,$$

in the expression (11.23). Since $\Pi(p^2)$ is a Lorentz scalar we shall take that $p^\mu = (p_0, \boldsymbol{p} = 0)$ i.e. we shall calculate it in the rest frame of the particle χ. In this way we obtain

$$\begin{aligned} \mathrm{Disc}\Pi(p^2) &= 2\mathrm{i}g^2(-2\mathrm{i}\pi)^2 \int \frac{\mathrm{d}^4 k}{(2\pi)^4} \delta^{(4)}(k^2 - m^2) \\ &\quad \times \delta^{(4)}((k+p)^2 - m^2)\theta(-k_0)\theta(k_0 + p_0) \\ &= -\frac{g^2 \mathrm{i}}{8\pi^2} \int \mathrm{d}^4 k \frac{1}{\omega_k^2} \delta(k_0 + \omega_k)\delta(k_0 + p_0 - \omega_k) \\ &= -\frac{\mathrm{i}g^2}{8\pi^2} \int \mathrm{d}^3 \boldsymbol{k} \frac{\delta(p_0 - 2\omega_k)}{\omega_k^2} \, . \end{aligned} \tag{11.24}$$

By performing the integration over the momentum $\boldsymbol{k}$ we get

$$\mathrm{Disc}\, \Pi(p^2) = -\frac{\mathrm{i}g^2}{4\pi} \sqrt{1 - \frac{4m^2}{p^2}} \, .$$

Since

$$\mathrm{Im}\,\Pi(\mathrm{p}^2) = \frac{1}{2\mathrm{i}}\mathrm{Disc}\,\Pi(\mathrm{p}^2) \ ,$$

we again obtain the result (11.22). From the expressions for Γ and $\Pi(M^2)$ we immediately see that the relation which was given in problem is valid. This relation is a consequence of *the optic theorem.*

11.7 In $D = 4 - \epsilon$ dimensional spacetime the dimension of a scalar field is $D/2 - 1$, while the dimensions of the coupling constants are the same as in four dimensions: $[\lambda] = 0$, $[g] = 1$. The dimension of the Lagrangian density must be $[\mathcal{L}] = D$, so it is given by

$$\mathcal{L} = \frac{1}{2}(\partial_\mu \phi)^2 - \frac{m^2}{2}\phi^2 - \frac{g\mu^{\epsilon/2}}{3!}\phi^3 - \frac{\lambda\mu^\epsilon}{4!}\phi^4 \ ,$$

where we introduced the parameter μ which has the dimension of mass. The self–energy is determined by diagrams shown in Fig. 11.1.

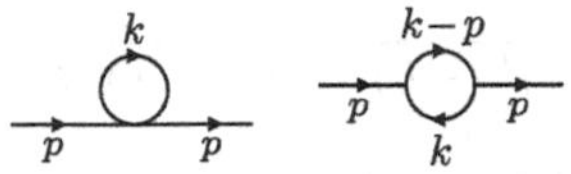

Fig. 11.1. The one-loop contribution to the self–energy of ϕ field

The contribution of the first one is

$$-\mathrm{i}\Sigma_1 = -\frac{\mathrm{i}\lambda}{2}\mu^\epsilon \int \frac{\mathrm{d}^D k}{(2\pi)^D} \frac{\mathrm{i}}{k^2 - m^2} \ .$$

By applying the formula (11.A) we get

$$-\mathrm{i}\Sigma_1 = -\frac{\mathrm{i}\lambda m^2}{32\pi^2}\left(\frac{4\pi\mu^2}{m^2}\right)^{\epsilon/2} \Gamma\left(-1 + \frac{\epsilon}{2}\right) \ ,$$

which, using (11.F), gives

$$\begin{aligned} -\mathrm{i}\Sigma_1 &= \frac{\mathrm{i}\lambda m^2}{32\pi^2}\left(1 + \frac{\epsilon}{2}\ln\left(\frac{4\pi\mu^2}{m^2}\right) + o(\epsilon)\right)\left(\frac{2}{\epsilon} + 1 - \gamma + o(\epsilon)\right) \\ &= \frac{\mathrm{i}\lambda m^2}{32\pi^2}\left(\frac{2}{\epsilon} + 1 - \gamma + \ln\left(\frac{4\pi\mu^2}{m^2}\right) + o(\epsilon)\right) \ . \end{aligned}$$

The second integral is

$$-\mathrm{i}\Sigma_2(p) = \frac{(-\mathrm{i}g)^2}{2}\mu^\epsilon \int \frac{\mathrm{d}^D k}{(2\pi)^D} \frac{\mathrm{i}}{k^2 - m^2} \frac{\mathrm{i}}{(k-p)^2 - m^2} \ .$$

By using the Feynman parametrization formula (11.G) the last expression becomes

$$-\mathrm{i}\Sigma_2(p) = -\frac{(-\mathrm{i}g)^2}{2}\mu^\epsilon \int_0^1 \mathrm{d}x \int \frac{\mathrm{d}^D k}{(2\pi)^D} \frac{1}{[k^2 - 2k\cdot px + p^2x - m^2]^2} .$$

The integration over the momentum k gives

$$\begin{aligned}
-\mathrm{i}\Sigma_2(p) &= \frac{\mathrm{i}}{2}\mu^\epsilon g^2 \frac{1}{(4\pi)^{2-\epsilon/2}}\Gamma\left(\frac{\epsilon}{2}\right)\int_0^1 \mathrm{d}x\,(m^2 - p^2x + p^2x^2)^{-\epsilon/2} \\
&= \frac{\mathrm{i}g^2(4\pi\mu^2)^{\epsilon/2}}{2(4\pi)^2}\left(\frac{2}{\epsilon} - \gamma + o(\epsilon)\right) \\
&\times \left[1 - \frac{\epsilon}{2}\int_0^1 \mathrm{d}x\,\left(\ln m^2 + \ln(1 + \frac{p^2}{m^2}x(x-1))\right)\right] .
\end{aligned}$$

Finally, the integration over the Feynman parameter x gives (for $p^2 < 4m^2$)

$$-\mathrm{i}\Sigma_2(p) = \frac{\mathrm{i}g^2}{32\pi^2}\left[\frac{2}{\epsilon} - \gamma + 2 + \ln\frac{4\pi\mu^2}{m^2} - 2\sqrt{\frac{4m^2}{p^2} - 1}\arcsin\sqrt{\frac{p^2}{4m^2}}\right] .$$

The self–energy of the particle is

$$-\mathrm{i}\Sigma(p) = -\mathrm{i}\Sigma_1(p) - \mathrm{i}\Sigma_2(p) .$$

The mass shift is $\delta m^2 = \Sigma(m^2) = \Sigma_1(m^2) + \Sigma_2(m^2)$.

11.8 The vertices in this theory are shown in Fig. 11.2.

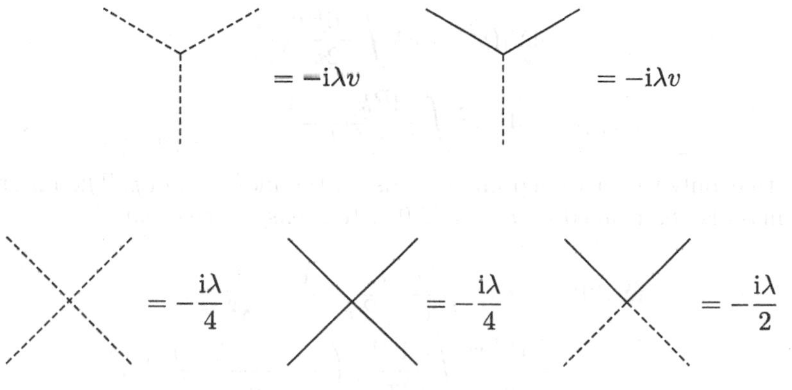

Fig. 11.2. Vertices in σ–model

The self–energy of the π particle is determined by the diagrams given in Fig. 11.3. The full line depict the π field, while the dashed line depict σ.

The first diagram is one of the terms in the second order of the perturbation theory

$$\frac{1}{2}(-\mathrm{i}\lambda v)^2 2\int \mathrm{d}x_1 \mathrm{d}x_2\,\langle 0|\,T(\pi(y_1)\pi(y_2)\sigma^3(x_1)\sigma(x_2)\pi^2(x_2))\,|0\rangle \ , \qquad (11.25)$$

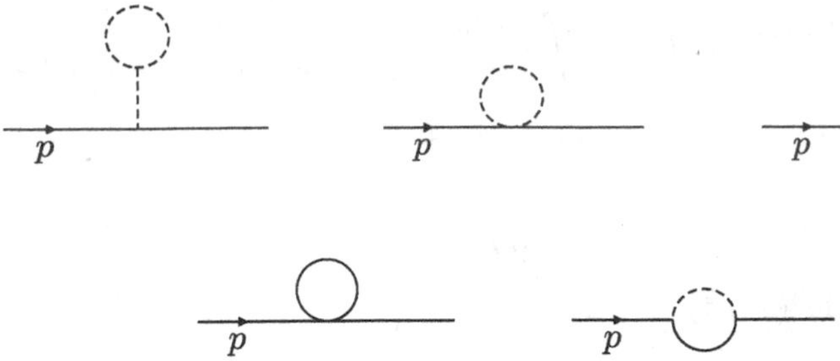

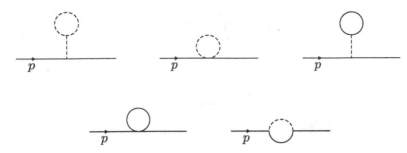

Fig. 11.3. The one-loop correction to the π propagator

so that

$$-\mathrm{i}\Sigma_1(p^2) = 6(-\mathrm{i}v\lambda)^2 \frac{\mathrm{i}}{-m^2} \int \frac{d^D k}{(2\pi)^D} \frac{\mathrm{i}}{k^2 - m^2} \; .$$

The symmetry factor of this diagram is 6, since one π field can be contracted to π field from $\pi\pi\sigma$-vertex in two ways, while $\sigma\sigma$ contraction in the vertex $\sigma\sigma\sigma$ can be done in 3 ways. Other diagrams are:

$$-\mathrm{i}\Sigma_2(p^2) = \lambda \int \frac{\mathrm{d}^D k}{(2\pi)^D} \frac{1}{k^2 - m^2} \; ,$$

$$-\mathrm{i}\Sigma_3(p^2) = -\frac{2v^2\lambda^2}{m^2} \int \frac{\mathrm{d}^D k}{(2\pi)^D} \frac{1}{k^2} \; ,$$

$$-\mathrm{i}\Sigma_4(p^2) = 3\lambda \int \frac{\mathrm{d}^D k}{(2\pi)^D} \frac{1}{k^2} \; ,$$

$$-\mathrm{i}\Sigma_5(p^2) = 4\lambda^2 v^2 \int \frac{\mathrm{d}^D k}{(2\pi)^D} \frac{1}{k^2 - m^2} \frac{1}{(k+p)^2} \; .$$

Note that only the last diagram depends on the momentum p. The renormalized mass is determined by $m_R^2 = \Sigma(0)$. It is easy to see that

$$\begin{aligned}
-\mathrm{i}\Sigma_5(0) &= 4\lambda^2 v^2 \int \frac{\mathrm{d}^D k}{(2\pi)^D} \frac{1}{k^2 - m^2} \frac{1}{k^2} \\
&= \frac{4\lambda^2 v^2}{m^2} \int \frac{\mathrm{d}^D k}{(2\pi)^D} \left(\frac{1}{k^2 - m^2} - \frac{1}{k^2} \right) \; .
\end{aligned}$$

By summing all diagrams we obtain

$$\Sigma(0) = \Sigma_1(0) + \Sigma_2(0) + \Sigma_3(0) + \Sigma_4(0) + \Sigma_5(0) = 0 \; ,$$

so $m_R = 0$.

11.9 The amplitude for the diagram

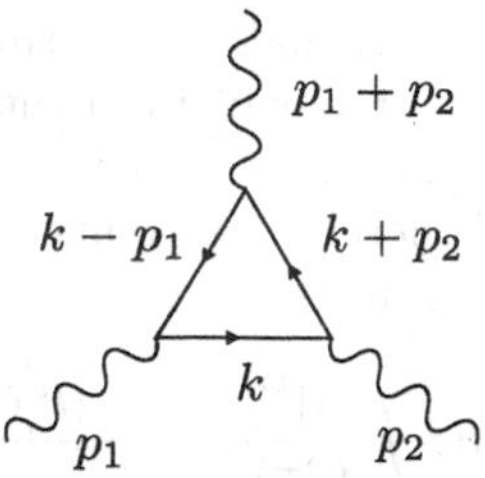

is

$$i\mathcal{M} = e^3 \int \frac{d^D k}{(2\pi)^D} \frac{\mathrm{tr}[\gamma^\mu(\not k - \not p_1 + m)\gamma^\nu(\not k + \not p_2 + m)\gamma^\rho(\not k + m)]}{((k-p_1)^2 - m^2)((k+p_2)^2 - m^2)(k^2 - m^2)} . \tag{11.26}$$

By applying the Feynman parametrization (11.H) we get

$$\begin{aligned}
&\frac{1}{((k-p_1)^2 - m^2)((k+p_2)^2 - m^2)(k^2 - m^2)} \\
&= 2\int_0^1 dx \int_0^{1-x} dz \frac{1}{[k^2 - m^2 + (p_2^2 + 2k\cdot p_2)x + (p_1^2 - 2k\cdot p_1)z]^3} \\
&= 2\int_0^1 dx \int_0^{1-x} dz \frac{1}{[(k + p_2 x - p_1 z)^2 - \Delta]^3} ,
\end{aligned}$$

where we introduce the notation

$$\Delta = (p_2 x - p_1 z)^2 - p_2^2 x - p_1^2 z + m^2 .$$

The numerator of the integrand in (11.26) is

$$\begin{aligned}
&\mathrm{tr}[\gamma^\mu(\not k - \not p_1 + m)\gamma^\nu(\not k + \not p_2 + m)\gamma^\rho(\not k + m)] \\
&= \mathrm{tr}[\gamma^\mu(\not l + \not A + m)\gamma^\nu(\not l + \not B + m)\gamma^\rho(\not l + \not C + m)] ,
\end{aligned} \tag{11.27}$$

where

$$\begin{aligned}
l &= k + p_2 x - p_1 z , \\
A &= p_1 z - p_2 x - p_1 , \\
B &= p_1 z - p_2 x + p_2 , \\
C &= p_1 z - p_2 x .
\end{aligned}$$

Since the trace of the odd number of γ–matrices is zero, (11.27) becomes

$$\begin{aligned}
&\mathrm{tr}[\gamma^\mu(\not l + \not A + m)\gamma^\nu(\not l + \not B + m)\gamma^\rho(\not l + \not C + m)] \\
&= \mathrm{tr}[\gamma^\mu \not l \gamma^\nu \not l \gamma^\rho \not l] + \mathrm{tr}[\gamma^\mu \not l \gamma^\nu \not l \gamma^\rho \not C] + \mathrm{tr}[\gamma^\mu \not l \gamma^\nu \not B \gamma^\rho \not l] + \\
&+ \mathrm{tr}[\gamma^\mu \not l \gamma^\nu \not B \gamma^\rho \not C] + \mathrm{tr}[\gamma^\mu \not A \gamma^\nu \not l \gamma^\rho \not l] + \mathrm{tr}[\gamma^\mu \not A \gamma^\nu \not l \gamma^\rho \not C] + \\
&+ \mathrm{tr}[\gamma^\mu \not A \gamma^\nu \not B \gamma^\rho \not l] + \mathrm{tr}[\gamma^\mu \not A \gamma^\nu \not B \gamma^\rho \not C] + m^2 \mathrm{tr}[\gamma^\mu \not l \gamma^\nu \gamma^\rho] + \\
&+ m^2 \mathrm{tr}[\gamma^\mu \not A \gamma^\nu \gamma^\rho] + m^2 \mathrm{tr}[\gamma^\mu \gamma^\nu \not l \gamma^\rho] + \\
&+ m^2 \mathrm{tr}[\gamma^\mu \gamma^\nu \not B \gamma^\rho] + m^2 \mathrm{tr}[\gamma^\mu \gamma^\nu \gamma^\rho \not l] + m^2 \mathrm{tr}[\not C \gamma^\mu \gamma^\nu \gamma^\rho] .
\end{aligned} \tag{11.28}$$

To calculate the integral (11.26) we make substitution of variable $k \to l$. Terms in (11.28) which contain odd number of momenta l after integration vanish. The terms which are proportional to m^2 as well as the term proportional to $\mathrm{tr}[\gamma^\mu \not{A}\gamma^\nu \not{B}\gamma^\rho \not{C}]$ are finite, and therefore we consider only the remaining terms. The first of the divergent integrals is

$$\mathrm{i}\mathcal{M}_1 = 8e^3 \int_0^1 \mathrm{d}x \int_0^{1-x} \mathrm{d}z \int \frac{\mathrm{d}^D l}{(2\pi)^D} \left[\frac{2l^\nu (l^\mu C^\rho - g^{\mu\rho} C \cdot l + l^\rho C^\mu)}{(l^2 - \Delta)^3} - \right.$$
$$\left. - \frac{l^2 (g^{\mu\nu} C^\rho - g^{\mu\rho} C^\nu + g^{\nu\rho} C^\mu)}{(l^2 - \Delta)^3} \right] ,$$

since

$$\mathrm{tr}[\gamma^\mu \not{l}\gamma^\nu \not{l}\gamma^\rho \not{C}] = 2l^\nu \mathrm{tr}[\gamma^\mu \not{l}\gamma^\rho \not{C}] - l^2 \mathrm{tr}[\gamma^\mu \gamma^\nu \gamma^\rho \not{C}] .$$

By integrating over l (using (11.C)) we get

$$\mathrm{i}\mathcal{M}_1 = \frac{4\mathrm{i}e^3}{(4\pi)^{D/2}} \Gamma\left(\frac{\epsilon}{2}\right) \int_0^1 \mathrm{d}x \int_0^{1-x} dz \left[1 - \frac{\epsilon}{2} \ln \Delta + o(\epsilon^2) \right]$$
$$\times (1 - \frac{D}{2})(g^{\mu\nu} C^\rho - g^{\mu\rho} C^\nu + g^{\nu\rho} C^\mu) .$$

The divergent part of this integral is

$$\mathrm{i}\mathcal{M}_1|_{\mathrm{div}} = -\frac{\mathrm{i}e^3}{2\pi^2 \epsilon} \int_0^1 \mathrm{d}x \int_0^{1-x} \mathrm{d}z (g^{\mu\nu} C^\rho - g^{\mu\rho} C^\nu + g^{\nu\rho} C^\mu) .$$

The other two integrals can be evaluated in the same way. The final result is

$$\mathrm{i}\mathcal{M}|_{\mathrm{div}} = -\frac{\mathrm{i}e^3}{2\pi^2 \epsilon} \left[\frac{1}{6} (g^{\mu\nu}(p_1 - p_2)^\rho + g^{\mu\rho}(p_1 - p_2)^\nu + g^{\rho\nu}(p_1 - p_2)^\mu) + \right.$$
$$\left. + \frac{1}{2} (g^{\mu\nu}(p_1 + p_2)^\rho + g^{\mu\rho}(p_2 - p_1)^\nu - g^{\rho\nu}(p_1 + p_2)^\mu) \right] .$$

The diagram where the orientation in the loop is opposite is shown in the following figure.

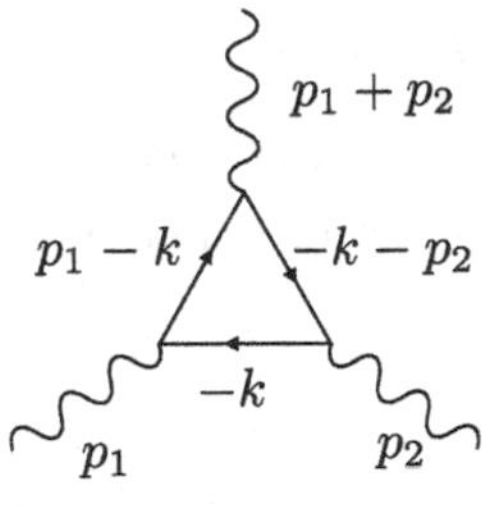

The amplitude is the same as in (11.26) except that the trace in (11.26) should be replaced by

$$\mathrm{tr}[\gamma^\rho(-\not{k} - \not{p}_2 + m)\gamma^\nu(\not{p}_1 - \not{k} + m)\gamma^\mu(-\not{k} + m)] .$$

By putting $C^{-1}C$ in the previous expression, where matrix C is the charge conjugation matrix (4.K), we get

$$\begin{aligned}&\mathrm{tr}[C\gamma^{\rho}C^{-1}C(-\not{k}-\not{p}_2+m)C^{-1}C\gamma^{\nu}C^{-1}C\\&\times(\not{p}_1-\not{k}+m)C^{-1}C\gamma^{\mu}C^{-1}C(-\not{k}+m)C^{-1}].\end{aligned}$$

By using (4.K) we have

$$\begin{aligned}&\mathrm{tr}[\gamma^{\rho}(-\not{k}-\not{p}_2+m)\gamma^{\nu}(\not{p}_1-\not{k}+m)\gamma^{\mu}(-\not{k}+m)]\\&=(-)^3\mathrm{tr}[\gamma^{\rho}(\not{k}+m)\gamma^{\mu}(\not{k}-\not{p}_1+m)\gamma^{\nu}(\not{k}+\not{p}_2+m)]\,,\end{aligned}$$

from which the we get the requested result. The statement is valid for all diagrams of this type with the odd number of vertices and this is called *the Furry theorem.*

11.10 The vacuum polarization in QED is

$$-\mathrm{i}\Pi_{\mu\nu}(q)=-e^2\int\frac{\mathrm{d}^4k}{(2\pi)^4}\frac{\mathrm{tr}[(\not{k}+m)\gamma_{\mu}(\not{k}+\not{q}+m)\gamma_{\nu}]}{(k^2-m^2)((k+q)^2-m^2)}\,. \tag{11.29}$$

From the Ward identity we know that this expression has the following form

$$-\mathrm{i}\Pi_{\mu\nu}(q)=-(q_{\mu}q_{\nu}-q^2g_{\mu\nu})\mathrm{i}\Pi(q^2)\,.$$

By multiplying the previous expression by $g^{\mu\nu}$ and using (11.29) we get

$$\begin{aligned}\mathrm{i}\Pi(q^2)&=-\frac{1}{3q^2}\mathrm{i}g^{\mu\nu}\Pi_{\mu\nu}\\&=-\frac{4e^2}{3q^2}\int\frac{\mathrm{d}^4k}{(2\pi)^4}\frac{-2k\cdot(k+q)+4m^2}{(k^2-m^2)((k+q)^2-m^2)}\,.\end{aligned} \tag{11.30}$$

Discontinuity in the expression $\Pi(q^2)$ can be calculated by applying *the Cutkosky rule.* Then

$$\begin{aligned}\mathrm{Disc}\,\Pi(q^2)=&\frac{4\mathrm{i}e^2}{3q^2}\frac{1}{(2\pi)^4}(-2\pi\mathrm{i})^2\int\mathrm{d}^4k(4m^2-2k\cdot(k+q))\delta^{(4)}(k^2-m^2)\\&\times\delta^{(4)}((k+q)^2-m^2)\theta(-k_0)\theta(k_0+q_0).\end{aligned} \tag{11.31}$$

By using

$$\delta(x^2-a^2)=\frac{1}{2|a|}(\delta(x-a)+\delta(x+a))$$

and taking $q^{\mu}=(q_0,\mathbf{0})$ we get

$$\begin{aligned}\mathrm{Disc}\,\Pi(q^2)=&-\frac{16\mathrm{i}\pi^2e^2}{3q^2}\frac{1}{(2\pi)^4}\int\mathrm{d}^4k(4m^2-2k\cdot(k+q))\\&\times\frac{1}{4\omega_k^2}\delta(k_0+\omega_k)\delta(k_0+q_0-\omega_k)\,.\end{aligned} \tag{11.32}$$

Integration over k_0 gives

$$\mathrm{Disc}\,\Pi(q^2) = -\frac{4\mathrm{i}\pi^2 e^2}{3q^2}\frac{1}{(2\pi)^4}\int \mathrm{d}^3\boldsymbol{k}(2m^2 + 2q_0\omega_k)\frac{1}{\omega_k^2}\delta(q_0 - 2\omega_k)\ . \quad (11.33)$$

Since $\mathrm{d}^3\boldsymbol{k} = |\boldsymbol{k}|\omega_k \mathrm{d}\omega_k \sin\theta \mathrm{d}\phi \mathrm{d}\theta$ we have

$$\mathrm{Disc}\,\Pi(q^2) = -\frac{\mathrm{i}e^2}{3\pi q^2}\int_m^\infty \mathrm{d}\omega_k \frac{2m^2 + 2q_0\omega_k}{\omega_k}\sqrt{\omega_k^2 - m^2}\delta(q_0 - 2\omega_k)\ . \quad (11.34)$$

Integration over ω_k gives

$$\mathrm{Disc}\,\Pi(q^2) = \frac{e^2}{6\pi\mathrm{i}}\left(1 + \frac{2m^2}{q^2}\right)\sqrt{1 - \frac{4m^2}{q^2}}\theta(q^2 - 4m^2)\ . \quad (11.35)$$

Finally

$$\begin{aligned}\mathrm{Im}\Pi(\mathrm{q}^2 + \mathrm{i}\epsilon) &= \frac{1}{2\mathrm{i}}\mathrm{Disc}\,\Pi(p^2) \\ &= -\frac{e^2}{12\pi}\left(1 + \frac{2m^2}{q^2}\right)\sqrt{1 - \frac{4m^2}{q^2}}\theta(q^2 - 4m^2)\ . \quad (11.36)\end{aligned}$$

11.11 Scalar electrodynamics has two vertices:

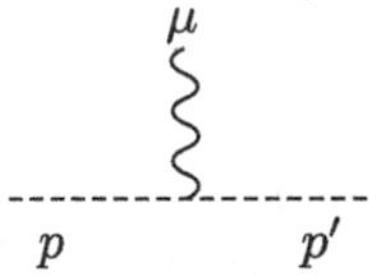

$= -\mathrm{i}e(p + p')_\mu$

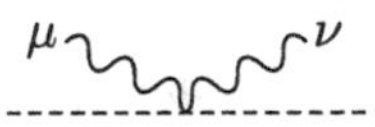

$= 2\mathrm{i}e^2 g_{\mu\nu}$

The Feynman rules are standard except that for every closed photon loop we have an extra factor 1/2. The photon self-energy is determined by the diagrams:

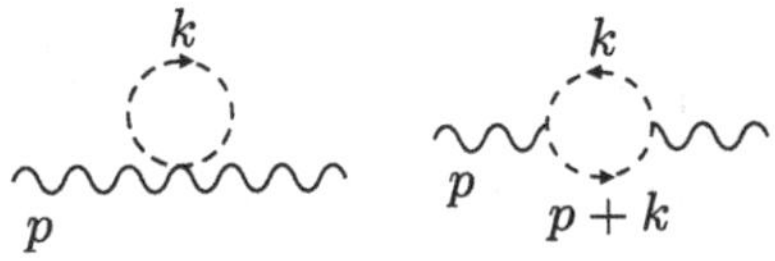

The first one is

$$-\mathrm{i}\Pi_{\mu\nu}^{(1)} = 2\mathrm{i}e^2 g_{\mu\nu}\int \frac{\mathrm{d}^D k}{(2\pi)^D}\frac{\mathrm{i}}{k^2 - m^2}\ .$$

By applying (11.A) and (11.F) we obtain:

$$-\mathrm{i}\Pi_{\mu\nu}^{(1)} = -\frac{\mathrm{i}e^2}{4\pi^2\epsilon}m^2 g_{\mu\nu} + \text{fin. part}\ . \quad (11.37)$$

The second diagram is

$$-\mathrm{i}\Pi^{(2)}_{\mu\nu} = e^2 \int \frac{\mathrm{d}^D k}{(2\pi)^D} \frac{(2k+p)_\mu (2k+p)_\nu}{(k^2-m^2)((k+p)^2-m^2)} \; .$$

By using the Feynman parametrization in the previous integral we get

$$-\mathrm{i}\Pi^{(2)}_{\mu\nu} = e^2 \int_0^1 \mathrm{d}x \int \frac{\mathrm{d}^D k}{(2\pi)^D} \frac{4k_\mu k_\nu + 2k_\mu p_\nu + 2k_\nu p_\mu + p_\mu p_\nu}{[k^2 + 2xk\cdot p + p^2 x - m^2]^2} \; .$$

Applying the formulae (11.A–C) it follows that :

$$\begin{aligned} -\mathrm{i}\Pi^{(2)}_{\mu\nu} = & \frac{\mathrm{i}e^2\pi^{D/2}}{(2\pi)^D} \int_0^1 \mathrm{d}x \left[\Gamma\left(\frac{\epsilon}{2}\right) \frac{1}{(m^2 + p^2x^2 - p^2x)^{\epsilon/2}} (4x^2 - 4x + 1) p_\mu p_\nu \right. \\ & \left. - \; 2g_{\mu\nu} \frac{\Gamma\left(\frac{\epsilon}{2} - 1\right)}{(m^2 + p^2x^2 - p^2x)^{\epsilon/2-1}} \right] \; , \end{aligned}$$

which is equal to

$$-\mathrm{i}\Pi^{(2)}_{\mu\nu} = \frac{\mathrm{i}e^2}{16\pi^2} \left(\frac{2}{3\epsilon}(p_\mu p_\nu - p^2 g_{\mu\nu}) + \frac{4m^2}{\epsilon} g_{\mu\nu} \right) + \text{fin. part} \; . \qquad (11.38)$$

Adding the divergent parts of the expressions (11.37) and (11.38) we get the requested result. Note that the terms proportional to m^2 cancel. So, the final result is gauge invariant, as expected.

11.12

(a) Let us introduce the following notation:
N_f– the number of external fermionic lines
N_s– the number of external scalar lines
P_f– the number of internal fermionic lines
P_s– the number of internal scalar lines
V_3– the number of $\bar{\psi}\gamma_5\psi\phi$ vertices
V_4– the number of ϕ^4 vertices
L– the number of loops.
Then the superficial degree of divergence for a diagram is

$$\mathcal{D} = 4L - 2P_s - P_f \; .$$

On the other hand, L can be expressed as

$$L = P_s + P_f - (V-1) \; ,$$

since it is a number of independent internal momenta. By combining the previous formulae with

$$2V_3 = N_f + 2P_f \; ,$$

$$V_3 + 4V_4 = N_s + 2P_s \; ,$$

we get

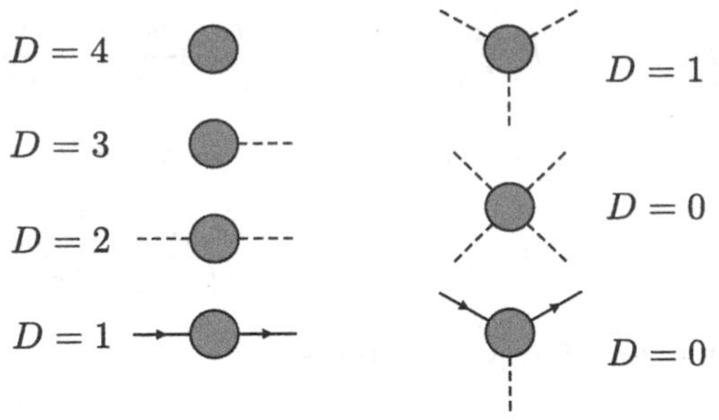

Fig. 11.4. Superficially divergent diagrams in the Yukawa theory

$$\mathcal{D} = 4 - N_s - \frac{3}{2}N_f \ .$$

Superficially divergent amplitudes are shown in Fig. 11.4.
The first diagram is the vacuum one and it can be ignored; the second and fifth are equal to zero. The bare Lagrangian density is

$$\mathcal{L}_0 = \frac{1}{2}(\partial\phi_0)^2 - \frac{m_0^2}{2}\phi_0^2 + \bar{\psi}_0(\mathrm{i}\gamma_\mu\partial^\mu - M_0)\psi_0 - \mathrm{i}g_0\bar{\psi}_0\gamma_5\psi_0\phi_0 - \frac{\lambda_0}{4!}\phi_0^4 \ . \quad (11.39)$$

If we rescale the fields as

$$\phi_0 = \sqrt{Z_\phi}\phi = \sqrt{1+\delta Z_\phi}\phi \ ,$$
$$\psi_0 = \sqrt{Z_\psi}\psi = \sqrt{1+\delta Z_\psi}\psi \ ,$$

and introduce a new set of variables:

$$\begin{aligned} Z_\phi m_0^2 &= m^2 + \delta m^2 \\ Z_\psi M_0 &= M + \delta M \\ Z_\psi\sqrt{Z_\phi}g_0 &= \mu^{\epsilon/2}(g+\delta g) \\ Z_\phi^2\lambda_0 &= \mu^\epsilon(\lambda+\delta\lambda) \ , \end{aligned}$$

the bare Lagrangian density becomes

$$\begin{aligned} \mathcal{L}_0 = \frac{1}{2}(1+\delta Z_\phi)(\partial\phi)^2 - \frac{m^2+\delta m^2}{2}\phi^2 + \mathrm{i}(1+\delta Z_\psi)\bar{\psi}\partial\!\!\!/\psi \\ - (M+\delta M)\bar{\psi}\psi - \mathrm{i}(g+\delta g)\mu^{\epsilon/2}\bar{\psi}\gamma_5\psi\phi - \frac{(\lambda+\delta\lambda)\mu^\epsilon}{4!}\phi^4 \ . \end{aligned}$$

The Feynman rules are given in the Fig. 11.5

(b) The one–loop fermionic propagator correction is represented in Fig. 11.6. The first diagram is

$$-\mathrm{i}\Sigma_2(p) = -g^2\mu^\epsilon\int\frac{\mathrm{d}^Dk}{(2\pi)^D}\frac{1}{k^2-m^2+\mathrm{i}0}\gamma_5\frac{p\!\!\!/ - k\!\!\!/ + M}{(p-k)^2-M^2+\mathrm{i}0}\gamma_5 \ .$$

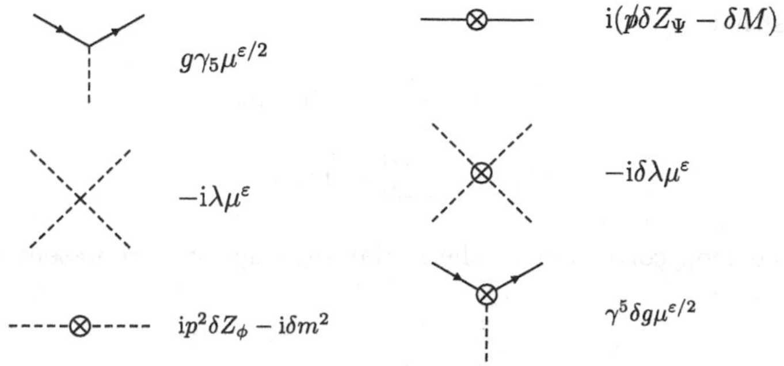

Fig. 11.5. Feynman rules in renormalized Yukawa theory

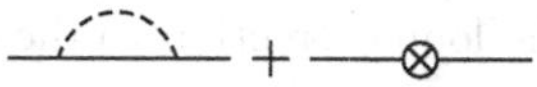

Fig. 11.6. The one–loop correction to fermionic propagator

Since $\gamma_5\not{a}\gamma_5 = -\not{a}$ and $(\gamma_5)^2 = 1$ we have

$$\begin{aligned}
-\mathrm{i}\Sigma_2(p) &= -\frac{g^2\mu^\epsilon}{(2\pi)^D}\int \mathrm{d}^D k \frac{-\not{p}+\not{k}+M}{(k^2-m^2+\mathrm{i}0)((p-k)^2-M^2+\mathrm{i}0)} \\
&= -\frac{g^2\mu^\epsilon}{(2\pi)^D}\int \mathrm{d}^D k \int_0^1 \mathrm{d}x \frac{-\not{p}+\not{k}+M}{(k-px)^2-\Delta+\mathrm{i}0)^2} \\
&= -\frac{g^2\mu^\epsilon}{(2\pi)^D}\mathrm{i}\pi^{D/2}\Gamma\left(\frac{\epsilon}{2}\right)\int_0^1 \mathrm{d}x \frac{\not{p}(x-1)+M}{\Delta^{\epsilon/2}} , \qquad (11.40)
\end{aligned}$$

where $\Delta = M^2 x + m^2(1-x) - p^2 x + p^2 x^2$. Since

$$\frac{\mu^\epsilon}{2^D\pi^{D/2}} = \frac{1}{16\pi^2}(4\pi\mu^2)^{\epsilon/2} = \frac{1}{16\pi^2}\left(1 + \frac{\epsilon}{2}\ln(4\pi\mu^2) + \dots\right) ,$$

we have

$$\begin{aligned}
-\mathrm{i}\Sigma_2(p) &= -\frac{\mathrm{i}g^2}{16\pi^2}\left[\frac{2}{\epsilon} - \gamma + o(\epsilon)\right]\int_0^1 \mathrm{d}x\,[M + (x-1)\not{p}]\left[1 - \frac{\epsilon}{2}\ln\frac{\Delta}{4\pi\mu^2}\right] \\
&= -\frac{\mathrm{i}g^2}{8\pi^2\epsilon}(M - \frac{1}{2}\not{p}) + \text{fin. part} . \qquad (11.41)
\end{aligned}$$

The full one–loop correction to the fermionic propagator is

$$-\mathrm{i}\Sigma(p) = -\frac{\mathrm{i}g^2}{8\pi^2\epsilon}(M - \frac{1}{2}\not{p}) - \mathrm{i}\delta M + \mathrm{i}\delta Z_\psi\not{p} + \text{fin. part} .$$

From the renormalization conditions:

$$\begin{aligned}
\Sigma(\not{p} = M) &= 0 , \\
\left.\frac{\mathrm{d}\Sigma}{\mathrm{d}\not{p}}\right|_{\not{p}=M} &= 0 , \qquad (11.42)
\end{aligned}$$

follows that

$$\begin{aligned}\delta Z_\psi &= -\frac{g^2}{16\pi^2\epsilon} + \text{fin. part} \;, \\ \delta M &= -\frac{g^2 M}{8\pi^2\epsilon} + \text{fin. part} \;. \end{aligned} \tag{11.43}$$

(c) The one–loop correction to the scalar propagator is represented in Fig. 11.7.

Fig. 11.7. The one-loop correction to the scalar propagator

The first diagram is

$$\begin{aligned}-\mathrm{i}\Pi_1(p^2) &= -\frac{\mathrm{i}^2 g^2\mu^\epsilon}{(2\pi)^D}\int \mathrm{d}^D k \frac{\mathrm{tr}[\gamma_5(\not k + M)\gamma_5(\not p + \not k + M)]}{(k^2 - M^2 + \mathrm{i}0)((p+k)^2 - M^2 + \mathrm{i}0)} \\ &= \frac{g^2\mu^\epsilon}{(2\pi)^D}\int \mathrm{d}^D k \int_0^1 \mathrm{d}x \frac{\mathrm{tr}[(-\not k + M)(\not p + \not k + M)]}{(k^2 + 2k\cdot px - M^2 + p^2 x)^2} \\ &= \frac{g^2\mu^\epsilon}{(2\pi)^D}\int_0^1 \mathrm{d}x \int \mathrm{d}^D k \frac{4(-k\cdot p - k^2 + M^2)}{(k^2 + 2k\cdot px - M^2 + p^2 x)^2} \;,\end{aligned}$$

where we use the Feynman parametrization formula (11.G). Introducing a new variable $l = k + px$ we further have

$$\begin{aligned}-\mathrm{i}\Pi_1(p^2) &= 4g^2\mu^\epsilon \int_0^1 \mathrm{d}x \int \frac{\mathrm{d}^D l}{(2\pi)^D}\frac{2M^2 - \Delta - l^2}{(l^2 - \Delta + \mathrm{i}0)^2} \\ &= \frac{\mathrm{i}g^2}{4\pi^2}\int_0^1 \mathrm{d}x \left(1 - \frac{\epsilon}{2}\ln\frac{\Delta}{4\pi\mu^2}\right) \\ &\times \Big((M^2 - p^2(x^2 - x))(\frac{2}{\epsilon} - \gamma + o(\epsilon)) + \\ &+ \frac{D}{2}(-\frac{2}{\epsilon} - 1 + \gamma + o(\epsilon))(M^2 + p^2(x^2 - x))\Big) \\ &= \frac{\mathrm{i}g^2}{2\pi^2\epsilon}\left(\frac{p^2}{2} - M^2\right) + \text{fin. part} \;,\end{aligned}$$

where $\Delta = M^2 + p^2(x^2 - x)$. The second diagram is

$$-\mathrm{i}\Pi_2 = \frac{\mathrm{i}\lambda m^2}{16\pi^2\epsilon} + \text{fin. part} \;. \tag{11.44}$$

Summing, we obtain

$$-\mathrm{i}\Pi(p^2) = \frac{\mathrm{i}g^2}{2\pi^2\epsilon}\left(\frac{p^2}{2} - M^2\right) + \frac{\mathrm{i}\lambda m^2}{16\pi^2\epsilon} + \mathrm{i}\delta Z_\phi p^2 - \mathrm{i}\delta m^2 + \text{fin.part}\,. \quad (11.45)$$

Using the renormalization conditions:

$$\Pi(p^2 = m^2) = 0$$
$$\left.\frac{\mathrm{d}\Pi}{\mathrm{d}p^2}\right|_{p^2=m^2} = 0\,, \quad (11.46)$$

we get

$$\delta Z_\phi = -\frac{g^2}{4\pi^2\epsilon} + \text{fin. part}$$
$$\delta m^2 = \frac{\lambda m^2}{16\pi^2\epsilon} - \frac{g^2 M^2}{2\pi^2\epsilon} + \text{fin. part}\,. \quad (11.47)$$

(d) The amplitude of the diagram

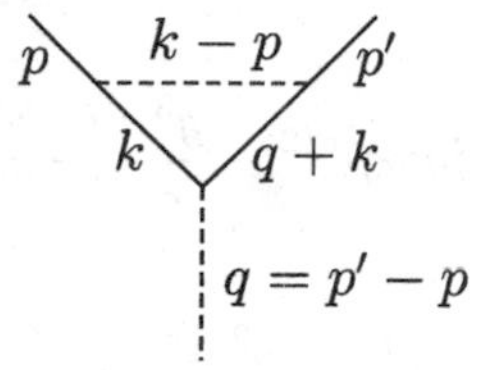

is

$$\begin{aligned}
\mathrm{i}\mathcal{M}_3 &= (\mathrm{i}g)^3\mu^{3\epsilon/2}\int\frac{\mathrm{d}^D k}{(2\pi)^D}\frac{\gamma_5(\not{k}+\not{q}+M)\gamma_5(\not{k}+M)\gamma_5}{((k+q)^2-M^2)(k^2-M^2)((k-p)^2-m^2)} \\
&= -\frac{2\mathrm{i}g^3\mu^{3\epsilon/2}}{(2\pi)^D}\gamma_5\int_0^1\mathrm{d}x\int_0^{1-x}\mathrm{d}z\int\mathrm{d}^D k\frac{M^2-\not{q}\not{k}+M\not{q}-k^2}{((k+qx-pz)^2-\Delta)^3} \\
&= -\frac{2\mathrm{i}g^3\mu^{3\epsilon/2}}{(2\pi)^D}\gamma_5\int_0^1\mathrm{d}x\int_0^{1-x}\mathrm{d}z\int\mathrm{d}^D l\frac{N}{(l^2-\Delta)^3}\,,
\end{aligned}$$

where

$$\Delta = x^2q^2 + z^2p^2 + (1-z)M^2 - xq^2 + zm^2 - p^2z - 2xzq\cdot p$$

and

$$N = M^2 - (l - xq + zp)^2 + M\not{q} - \not{q}(\not{l} - x\not{q} + z\not{p})\,.$$

In the previous formulae we introduced a variable $l = k + xq - zp$. As we are interested to find only the divergent part of $\mathrm{i}\mathcal{M}_3$, it is useful to note that only l^2–term in the numerator of the integrand is divergent. So, by using (11.C) we get:

$$
\begin{aligned}
i\mathcal{M}_3 &= 2ig^3\mu^{3\epsilon/2}\gamma_5 \int \frac{d^D l}{(2\pi)^D} \int_0^1 dx \int_0^{1-x} dz \frac{l^2}{(l^2-\Delta)^3} + \dots \\
&= -\frac{g^3\mu^{\epsilon/2}(4-\epsilon)}{32\pi^2}\gamma_5 \left(\frac{2}{\epsilon} - \gamma + \dots\right) \int_0^1 dx \\
&\times \int_0^{1-x} dz \left(1 - \frac{\epsilon}{2}\ln\frac{\Delta}{4\pi\mu^2}\right) .
\end{aligned}
$$

Finally

$$
i\mathcal{M}_3 = -\frac{g^3\mu^{\epsilon/2}}{8\pi^2\epsilon}\gamma_5 + \text{fin. part} . \tag{11.48}
$$

The vertex correction is

so, from

$$
iV_3 = \left(g\gamma_5\mu^{\epsilon/2} + \delta g\gamma_5\mu^{\epsilon/2} - \frac{g^3\mu^{\epsilon/2}}{8\pi^2\epsilon}\gamma_5 + \text{fin.part} \right)\Big|_{q^2=0} = g\gamma_5
$$

follows

$$
\delta g = \frac{g^3}{8\pi^2\epsilon} + \text{fin. part} .
$$

(e) Let us first calculate the following diagram

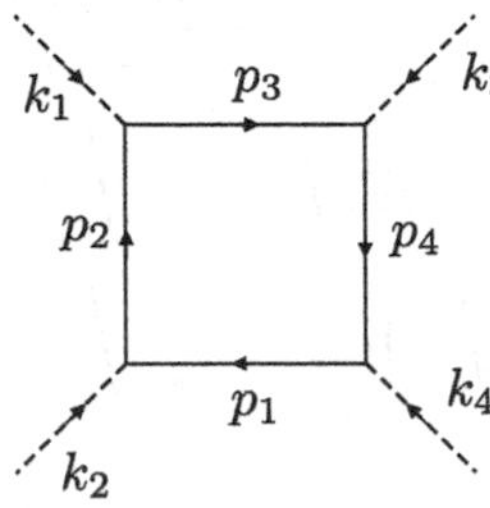

Since we have to find the divergent part of this diagram we can put that the external momenta are equal to zero. Then,

$$
i\mathcal{M}_4(k_1 = k_2 = k_3 = k_4 = 0) = -g^4\mu^{2\epsilon} \int \frac{d^D p}{(2\pi)^D} \frac{\mathrm{tr}[\gamma_5(\not{p} + M)]^4}{(p^2 - M^2)^4} . \tag{11.49}
$$

Since

$$
\gamma_5(\not{p} + M)\gamma_5(\not{p} + M) = (-\not{p} + M)(\not{p} + M) = M^2 - p^2
$$

we have

$$\begin{aligned}
\mathrm{i}\mathcal{M}_4(k_1 = k_2 = k_3 = k_4 = 0) &= -4g^4\mu^{2\epsilon}\int \frac{\mathrm{d}^D p}{(2\pi)^D}\frac{1}{(p^2-M^2)^2} \\
&= -\frac{\mathrm{i}g^4\mu^\epsilon}{4\pi^2}\left(\frac{2}{\epsilon}-\gamma\right)\left(1-\frac{\epsilon}{2}\ln\frac{M^2}{4\pi\mu^2}\right) \\
&= -\frac{\mathrm{i}g^4\mu^\epsilon}{2\pi^2\epsilon} + \text{fin. part} . \qquad (11.50)
\end{aligned}$$

The previous result should be multiplied by a factor 6 as there are six diagrams of this type.
The complete four vertex is

$$\begin{aligned}
\mathrm{i}V_4 &= \left(-\mathrm{i}\lambda\mu^\epsilon - \mathrm{i}\delta\lambda\mu^\epsilon - \frac{6\mathrm{i}g^4\mu^\epsilon}{2\pi^2\epsilon} + \frac{3\mathrm{i}\lambda^2\mu^\epsilon}{16\pi^2\epsilon} + \text{fin. part}\right)\Big|_{s=4m^2,t=u=0} \\
&= -\mathrm{i}\lambda , \qquad (11.51)
\end{aligned}$$

and finally

$$\delta\lambda = -\frac{3g^4}{\pi^2\epsilon} + \frac{3\lambda^2}{16\pi^2\epsilon} + \text{fin. part} . \qquad (11.52)$$

11.13 In this problem dimension of spacetime is $D = 2-\epsilon$.

(a) The polarization of vacuum is given by:

$$-\mathrm{i}\Pi_{\mu\nu}(p) = (\mathrm{i}e)^2(-\mathrm{i}^2)\int \frac{\mathrm{d}^D q}{(2\pi)^D}\frac{\mathrm{tr}[(\not{q}-\not{p})\gamma_\nu\not{q}\gamma_\mu]}{q^2(q-p)^2} . \qquad (11.53)$$

In D-dimensional space trace identities necessary to calculate the previous expression read:

$$\mathrm{tr}(\gamma_\mu\gamma_\nu) = f(D)g_{\mu\nu},$$

$$\mathrm{tr}(\gamma_\mu\gamma_\nu\gamma_\rho\gamma_\sigma) = f(D)(g_{\mu\nu}g_{\rho\sigma} - g_{\mu\rho}g_{\nu\sigma} + g_{\mu\sigma}g_{\rho\nu}) ,$$

where $f(D)$ is any analytical function which satisfies the condition $f(2) = 2$. Instead of $f(D)$ we will write 2 as we did in the previous problems (of course, there $f(D) = 4$). The Feynman parametrization gives

$$\begin{aligned}
-\mathrm{i}\Pi_{\mu\nu}(p) &= -\frac{2e^2}{(2\pi)^D}\int_0^1 \mathrm{d}x\int \mathrm{d}^D q \\
&\times \frac{2q_\mu q_\nu - q^2 g_{\mu\nu} - p_\mu q_\nu - p_\nu q_\mu + (p\cdot q)g_{\mu\nu}}{(q^2-2p\cdot qx+p^2x)^2} . \qquad (11.54)
\end{aligned}$$

By using (11.A–C) in (11.54) we obtain

$$\begin{aligned}
-\mathrm{i}\Pi_{\mu\nu} &= -\frac{2\mathrm{i}e^2\pi^{D/2}}{(2\pi)^D}\int_0^1 \mathrm{d}x\left[2\left(\frac{x^2p_\mu p_\nu}{(-p^2x+p^2x^2)^{1+\epsilon/2}}\Gamma(1+\frac{\epsilon}{2})\right.\right. \\
&\left.-\frac{1}{2}\frac{g_{\mu\nu}}{(-p^2x+p^2x^2)^{\epsilon/2}}\Gamma(\frac{\epsilon}{2})\right)
\end{aligned}$$

$$
\begin{aligned}
&- g_{\mu\nu}\left(\frac{x^2p^2}{(-p^2x+p^2x^2)^{1+\epsilon/2}}\Gamma(1+\frac{\epsilon}{2})\right.\\
&\left.- \frac{2-\epsilon}{2}\frac{1}{(-p^2x+p^2x^2)^{\epsilon/2}}\Gamma(\frac{\epsilon}{2})\right)\\
&- 2\frac{xp_\mu p_\nu}{(-p^2x+p^2x^2)^{1+\epsilon/2}}\Gamma(1+\frac{\epsilon}{2})\\
&\left.+ g_{\mu\nu}\frac{p^2x}{(-p^2x+p^2x^2)^{1+\epsilon/2}}\Gamma(1+\frac{\epsilon}{2})\right] .
\end{aligned}
$$

From the previous expression (for $D \to 2$ i.e. $\epsilon \to 0$) we obtain

$$
\begin{aligned}
-\mathrm{i}\Pi_{\mu\nu}(p) &= -\mathrm{i}(p_\mu p_\nu - p^2 g_{\mu\nu})\Pi(p^2)\\
&= -\frac{\mathrm{i}e^2}{\pi p^2}(p_\mu p_\nu - p^2 g_{\mu\nu}) , \qquad (11.55)
\end{aligned}
$$

from which we see that the polarization of vacuum is a finite quantity.

(b) The full photon propagator is obtained by summing the diagrams in the Figure

$$
\begin{aligned}
\mathrm{i}D_{\mu\nu}(p) &= \frac{-\mathrm{i}g_{\mu\nu}}{p^2+\mathrm{i}0} + \frac{-\mathrm{i}g_{\mu\rho}}{p^2+\mathrm{i}0}[p^2g^{\rho\sigma} - p^\rho p^\sigma]\mathrm{i}\Pi(p^2)\frac{-\mathrm{i}g_{\sigma\nu}}{p^2+\mathrm{i}0} + \dots\\
&= -\frac{\mathrm{i}}{p^2+\mathrm{i}0}(g_{\mu\nu} - \frac{p_\mu p_\nu}{p^2})(1+\Pi(p^2)+\Pi^2(p^2)+\dots) - \frac{\mathrm{i}p_\mu p_\nu}{p^4}\\
&= -\frac{\mathrm{i}(g_{\mu\nu} - \frac{p_\mu p_\nu}{p^2})}{p^2(1-\Pi(p^2)+\mathrm{i}0)} , \qquad (11.56)
\end{aligned}
$$

were we discarded the $\mathrm{i}p_\mu p_\nu/p^4$-term in the last line since the propagator is coupled to a conserved current. Then the photon propagator is

$$
\mathrm{i}D_{\mu\nu}(p) = -\frac{\mathrm{i}(g_{\mu\nu} - \frac{p_\mu p_\nu}{p^2})}{p^2 - \frac{e^2}{\pi}} . \qquad (11.57)
$$

Photon mass is $e/\sqrt{\pi}$.

11.14 The dimension of spacetime is $D = 6 - \epsilon$.

(a) The renormalized Lagrangian density is

$$
\mathcal{L}_{\mathrm{ren}} = \mathcal{L} + \mathcal{L}_{\mathrm{ct}} , \qquad (11.58)
$$

where

$$
\mathcal{L} = \frac{1}{2}(\partial\phi)^2 - \frac{m^2}{2}\phi^2 - \frac{g\mu^{\epsilon/2}}{3!}\phi^3 - h\mu^{-\epsilon/2}\phi , \qquad (11.59)
$$

$$\mathcal{L}_{\rm ct} = \frac{1}{2}\delta Z(\partial\phi)^2 - \frac{\delta m^2}{2}\phi^2 - \frac{\mu^{\epsilon/2}\delta g}{3!}\phi^3 - \mu^{-\epsilon/2}\delta h\phi \,. \tag{11.60}$$

By introducing new quantities

$$Z = 1 + \delta Z \,, \tag{11.61}$$
$$m_0^2 Z = m^2 + \delta m^2 \,, \tag{11.62}$$
$$g_0 Z^{3/2} = (g + \delta g)\mu^{\epsilon/2} \,, \tag{11.63}$$
$$h_0 Z^{1/2} = (h + \delta h)\mu^{-\epsilon/2} \,, \tag{11.64}$$

and rescaling the field, $\phi_0 = \sqrt{Z}\phi$, the renormalized Lagrangian density becomes

$$\mathcal{L}_{\rm ren} = \frac{1}{2}(\partial\phi_0)^2 - \frac{m_0^2}{2}\phi_0^2 - \frac{g_0}{3!}\phi_0^3 - h_0\phi_0 \,.$$

The quantities with index 0 are called bare. The Feynman rules are given in Figure 11.8.

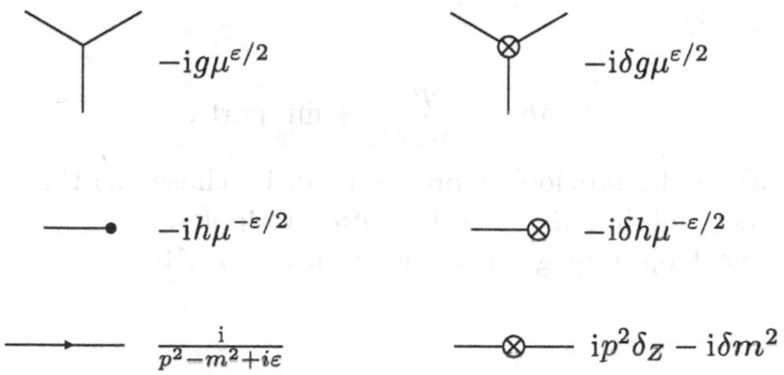

Fig. 11.8. Feynman rules in ϕ^3 theory

Superficially divergent amplitudes are:

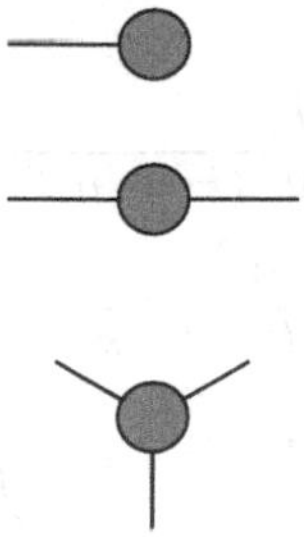

Fig. 11.9. Divergent amplitudes in ϕ^3 theory

(b) The tadpole diagram in one–loop order is shown in the following figure.

The second term is

$$
\begin{aligned}
& - \mathrm{i}g\mu^{\epsilon/2} \int \frac{\mathrm{d}^D k}{(2\pi)^D} \frac{\mathrm{i}}{k^2 - m^2 + \mathrm{i}0} \\
&= -\mathrm{i} \frac{g\mu^{\epsilon/2}}{(2\pi)^D} \frac{\pi^{D/2}}{(m^2)^{-2+\epsilon/2}} \Gamma\left(-2 + \frac{\epsilon}{2}\right) \\
&= -\frac{\mathrm{i}gm^4\mu^{-\epsilon/2}}{128\pi^3} \left(\frac{2}{\epsilon} + \ln\left(\frac{4\pi\mu^2}{m^2}\right) + \frac{3}{2} - \gamma\right) \\
&= -\frac{\mathrm{i}gm^4\mu^{-\epsilon/2}}{64\pi^3\epsilon} + \text{fin. part} ,
\end{aligned}
$$

and it does not depend on momentum. Summing all diagrams we get

$$
\mathrm{i}H = -\mathrm{i}h\mu^{-\epsilon/2} - \frac{\mathrm{i}gm^4\mu^{-\epsilon/2}}{64\pi^3\epsilon} - \mathrm{i}\delta h\mu^{-\epsilon/2} + \text{fin. part} . \tag{11.65}
$$

Hence,

$$
\delta h = -\frac{gm^4}{64\pi^3\epsilon} + \text{fin. part} . \tag{11.66}
$$

Finite part in the previous expression can be chosen so that $H = 0$ and we can ignore all diagrams which contain tadpoles.

(c) The full one–loop propagator is shown in Fig. 11.10.

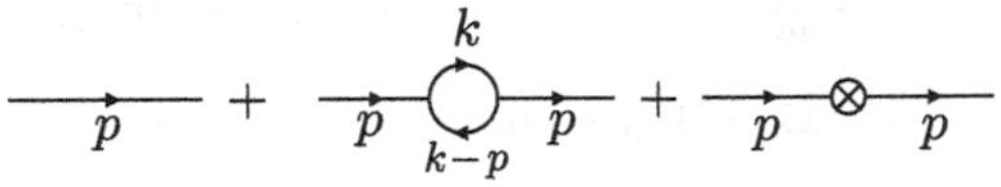

Fig. 11.10. The one–loop propagator in ϕ^3 theory

The second diagram is

$$
\begin{aligned}
-\mathrm{i}\Pi_2 &= \frac{(\mathrm{i}g)^2\mu^\epsilon}{2} \int \frac{\mathrm{d}^D k}{(2\pi)^D} \frac{\mathrm{i}^2}{(k^2 - m^2 + \mathrm{i}0)((k-p)^2 - m^2 + \mathrm{i}0)} \\
&= \frac{g^2\mu^\epsilon}{2} \int_0^1 \mathrm{d}x \int \frac{\mathrm{d}^D k}{(2\pi)^D} \frac{1}{(k^2 - 2k \cdot px + p^2 x - m^2 + \mathrm{i}0)^2} \\
&= -\frac{\mathrm{i}g^2}{128\pi^3} \left(\frac{2}{\epsilon} + 1 - \gamma + o(\epsilon)\right) \\
&\times \int_0^1 \mathrm{d}x (m^2 + p^2 x(x-1)) \left(1 - \frac{\epsilon}{2} \ln \frac{m^2 + p^2 x(x-1)}{4\pi\mu^2}\right) \\
&= -\frac{\mathrm{i}g^2}{64\pi^3\epsilon} \left(m^2 - \frac{p^2}{6}\right) + \text{fin. part} .
\end{aligned} \tag{11.67}
$$

Propagator correction is

$$-\mathrm{i}\Pi(p^2) = -\frac{\mathrm{i}g^2}{64\pi^3\epsilon}\left(m^2 - \frac{p^2}{6}\right) + \mathrm{i}p^2\delta Z - \mathrm{i}\delta m^2 + \text{fin. part} . \quad (11.68)$$

From the condition $-\mathrm{i}\Pi(p^2) = \text{finite}$ we get

$$\delta Z = -\frac{g^2}{384\pi^3\epsilon} + \text{fin. part} , \quad (11.69)$$

$$\delta m^2 = -\frac{m^2 g^2}{64\pi^3\epsilon} + \text{fin. part} . \quad (11.70)$$

In MS scheme the finite parts in (11.69) and (11.70) are zero.

(d) The vertex correction is given in Fig 11.11.

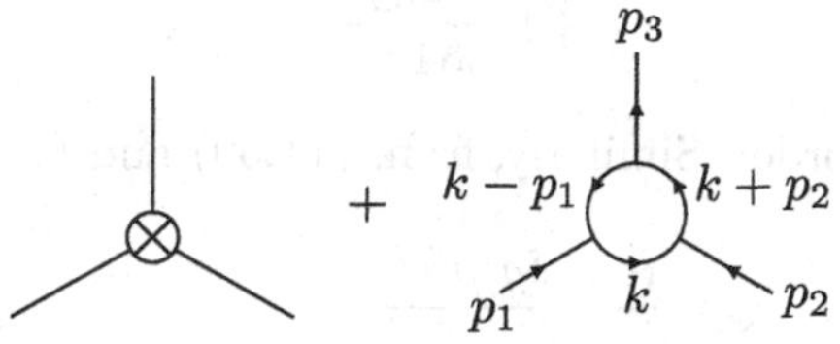

Fig. 11.11. Vertex correction in ϕ^3 theory

The second diagram is

$$\mathrm{i}\Gamma = (-\mathrm{i}g)^3\mu^{3\epsilon/2}\int\frac{\mathrm{d}^D k}{(2\pi)^D}\frac{\mathrm{i}^3}{(k^2-m^2)((k+p_2)^2-m^2)((k-p_1)^2-m^2)} . \quad (11.71)$$

By applying (11.H) and integrating over the momentum k we get

$$\begin{aligned}\mathrm{i}\Gamma = &-(-\mathrm{i}g)^3\mu^{3\epsilon/2}\frac{\pi^{D/2}}{(2\pi)^D}\Gamma\left(\frac{\epsilon}{2}\right)\int_0^1\mathrm{d}x\int_0^{1-x}\mathrm{d}z\\ &\times\frac{1}{(m^2-p_2^2x-p_1^2z+p_2^2x^2+p_1^2z^2)^{\epsilon/2}-2xzp_1\cdot p_2}\\ = &-\frac{\mathrm{i}g^3\mu^{\epsilon/2}}{2^{6-\epsilon}\pi^{3-\epsilon/2}}\left(\frac{2}{\epsilon}+\dots\right)\int_0^1\mathrm{d}x\int_0^{1-x}\mathrm{d}z\\ &\times\left(1-\frac{\epsilon}{2}\ln\frac{m^2-p_2^2x-p_1^2z+p_2^2x^2+p_1^2z^2-2xzp_1\cdot p_2}{\mu^2}\right) \quad (11.72)\end{aligned}$$

From the last formula we find that the divergent part of $\mathrm{i}\Gamma$ is given by

$$-\frac{\mathrm{i}g^3\mu^{\epsilon/2}}{64\pi^3\epsilon} . \quad (11.73)$$

The full one–loop vertex in the renormalized theory is

$$iV_3 = -ig\mu^{\epsilon/2} - i\delta g\mu^{\epsilon/2} + i\Gamma \ .$$

In minimal subtraction scheme δg is

$$\delta g = -\frac{g^3}{64\pi^3\epsilon} \ . \tag{11.74}$$

(e) From (11.61), (11.69) and (11.70) follows

$$Z = 1 - \frac{g^2}{384\pi^3\epsilon} \ , \tag{11.75}$$

$$\begin{aligned} m^2 &= m_0^2\left(1 - \frac{g^2}{384\pi^3\epsilon}\right) + \frac{m^2 g^2}{64\pi^3\epsilon} \\ &= m_0^2 + \frac{5m_0^2 g^2}{384\pi^3\epsilon} \ , \end{aligned} \tag{11.76}$$

in the one–loop order. Similarly, from (11.69) and (11.74) we have

$$g_0 = \frac{(g+\delta g)\mu^{\epsilon/2}}{Z^{3/2}} \tag{11.77}$$

$$= g\mu^{\epsilon/2}\left(1 - \frac{g^2}{64\pi^3\epsilon} + \frac{g^2}{256\pi^3\epsilon}\right) \tag{11.78}$$

$$= g\mu^{\epsilon/2}\left(1 - \frac{3g^2}{256\pi^3\epsilon}\right) \ . \tag{11.79}$$

The last expression is important for calculation of the β function.

References

1. D. Bailin and A. Love, *Introduction to Gauge Field Theory*, Adam Hilger, Bristol, 1986
2. J. Bjorken and S. Drell, *Relativistic Quantum Mechanics*, McGraw-Hill, New York, 1964
3. J. Bjorken and S. Drell, *Relativistic Quantum Fields*, McGraw-Hill, New York, 1965
4. N. N. Bogoljubov and D.V. Shirkov, *Introduction to the Theory of Quantized Fields*, Wiley-Interscience, New York, 1980
5. M. Blagojević, *Gravitation and Gauge Symmetries*, IOP Publishing, Bristol, 2002
6. T.P. Cheng and L.F. Li, *Gauge Theory of Elementary Particle Physics*, Oxford University Press, New York, 1984
7. T.P. Cheng and L.F. Li, *Gauge Theory of Elementary Particle Physics, Problems and Solutions*, Oxford University Press, New York, 2000
8. M. Damnjanović, *Hilbert spaces and group theory*, Faculty of Physics, Beograd, 2000 (in Serbian)
9. I.S. Gradshteyn and I.M. Ryzhnik, *Table of Integrals, Series and Products*, (trans. and ed. by Alan Jeffrey), Academic Press, Orlando, Florida, 1980
10. W. Greiner and J, Reinhardt, *Quantum Electrodinamics*, Springer, Berlin, Heidelberg, New York, 1996
11. W. Greiner and J, Reinhardt, *Field Quantization*, Springer, Berlin, Heidelberg, New York, 1996
12. F. Gross, *Relativistic Quantum Mechanics and Field Theory*, Wiley, New York, 1993
13. C. Itzykson and J.B. Zuber, *Quantum Field Theory*, McGraw-Hill, New York, 1980
14. M. Kaku, *Quantum Field Theory: A Modern Introduction*, Oxford University Press, New York, 1993
15. F. Mandl and G. Show, *Quantum Field Theory*, New York, 1999
16. M. E. Peskin and D. V. Schroeder, *An Introduction to Quantum Field Theory*, Addison Wesley, 1995
17. P. Ramond, *Field Theory: A Modern Primer* (second edition), Addison-Wesley, RedwoodCity, California, 1989

18. L. Rayder, *Quantum Field Theory*, Cambridge University Press, Cambridge, 1985
19. J. J. Sakurai, *Advanced Quantum Mechanics*, Addison-Wesley, Reading, 1967
20. S. S. Schweber, *An Introduction to Relativistic Quantum Field Theory*, Harpen and Row, New York, 1962
21. A. G. Sveshnikov and A. N. Tikhonov, *The Theory of Functions of a Complex Variable*, Mir Publisher, Moscow, 1978
22. G. Sterman, *Introduction to Quantum Field Theory*, Cambridge University Press, Cambridge, 1993
23. S. Weinberg, *The Quantum Theory of Fields I and II*, Cambridge University Press, New York, 1996

Index

Action, 25
  Einstein–Hilbert, 27
Advanced Green function
  Dirac equation, 138
  Klein–Gordon equation, 132
Angular momentum tensor
  Dirac field, 44, 45, 164
  electromagnetic field, 52, 183–185
  Klein–Gordon field, 36, 37, 144
Anticommutation relations
  Dirac field, 43

Baker–Hausdorff formula, 91, 144
Bianchi identity, 49

Casimir effect, 53, 187–190
Casimir operator, 7
Charge
  Dirac field, 45, 162
  Klein–Gordon field, 37, 142
Charge conjugation
  Dirac equation, 18
    bilinears, 23–24, 115–118
  Dirac field, 45
    bilinears, 47, 175–177
  scalar field, 41, 159
Chiral transformations, 28
Coherent states, 40, 156–158
Commutation relations
  electromagnetic field, 50
  scalar field, 35
Conformal group, 75
Conformal transformations, 7
Continuity equation, 10
Cross section, 55
Cutkosky rule, 62, 225

Decay rate, 218
Differential cross section, 192
Dilatations
  Dirac field, 30, 46, 129, 168
  scalar field, 29, 38, 129, 148–150
Dimensional regularization, 63
Dirac equation, 17
  helicity, 99, 118
  helicity basic, 20, 95
  plane wave solutions, 17, 18, 93–95
  spinor basic, 20
Dirac field
  quantization, 43
Dirac particle
  in a hole, 22, 110–111
  in a magnetic field, 23, 113
Dyson Green function
  Klein–Gordon equation, 133

Electromagnetic field
  quantization, 49
Energy–momentum tensor, 26, 126
  symmetric or Belinfante tensor, 29, 127
Euler–Lagrange equations, 25, 121

Feynman parametrization, 62, 211
Feynman propagator
  Dirac equation, 138, 139
  Dirac field, 44

Klein–Gordon equation, 31, 33, 132, 136
Klein–Gordon field, 36, 153
Foldy–Wouthuysen transformation, 24, 118–119
Functional derivative, 25, 121
Furry theorem, 225

Galilean algebra, 39, 156
Gamma matrices, 13
contraction identities, 14, 86–87
Dirac representation, 13
Majorana representation, 13
trace identities, 15, 87–89
Weyl representation, 13
Gamma–function, 62, 213
γ_5–matrix, 13, 86, 102
$\gamma_5 \not{s}$–operator, 98
gauge transformations, 49
Gordon identity, 21, 104
Grassmann variable, 173
Green function
Dirac equation, 31, 33
Klein-Gordon equation, 31
massive vector field, 33, 140
massless vector field, 34, 140
Schrödinger equation, 154
Gupta–Bleuler quantization, 50

Hamiltonian
Dirac field, 44, 45, 162
Klein–Gordon field, 36, 37, 142
Helicity, 94, 165, 181

Klein paradox
Dirac particle, 109
scalar particle, 82
Klein–Gordon equation, 9
plane wave solutions, 77
Klein–Gordon particle
in a hole, 10, 79
in a magnetic field, 10, 81
in the Coulomb potential, 10, 83

Lagrangian density
Dirac field, 43
massive vector field, 27
massless vector field, 49
Schrödinger field , 39
sigma model, 28
Left/right spinors, 102–103
Levi-Civita tensor, 4, 5, 68
Little group, 74
Lorentz group, 5, 67
generators in defining repr., 69
Lorentz transformations
Dirac equation, 17
bilinears, 23–24, 115–118
Dirac field, 44, 170
bilinears, 47, 174–177
scalar field, 158–159

Majorana spinor, 47, 173
Maxwell equations, 49
Metric tensor, 3
Minkowski space, 3
Momentum
Dirac field, 44, 45
Klein–Gordon field, 36, 37, 142
MS scheme, 237

Noether theorem, 26
Normal ordering
Dirac field, 44, 47, 172
Klein–Gordon field, 36

Optic theorem, 220

Parity
Dirac equation, 18
bilinears, 23–24, 115–118
Dirac field, 44
bilinears, 47, 174–177
scalar field, 41, 159
Pauli matrices, 5
Pauli–Lubanski vector, 7, 19, 72–74, 98
Pauli–Villars regularization, 62, 215
Phase transformations, 28, 125
ϕ^3 theory in 4D, 58
ϕ^3 theory in 6D, 64, 234–238
Poincaré algebra, 6, 71, 72
Poincaré group, 4, 6
Poincaré transformations, 4
scalar field, 40
Projection operators
energy, 19, 95–96
spin, 100

QED processes

scattering in an external electromagnetic field, 202
QED processes
$\mu^- \mu^+ \to e^- e^+$, 58, 196–198
$e^- \mu^+ \to e^- \mu^+$, 58
$e^- \mu^+ \to e^- \mu^+$, 198
Compton scattering, 58, 199
scattering in an external electromagnetic field, 58, 200

Reflection and transmission coefficients
Dirac equation, 22
Klein–Gordon equation, 10
Reiman ζ–function, 53
Retarded Green function
Klein–Gordon equation, 132, 137

S–matrix, 55
Scalar electrodynamics, 64, 226
Scalar field
quantization, 35
Scalar product, 4
Scattering of polarized particles, 59, 203–205
Schrödinger equation, 153
Schwinger model, 64, 233
$\boldsymbol{\Sigma}$–vector, 96
$\sigma_{\mu\nu}$–matrices, 14, 85, 87
SL(2, C) group, 5
Superficial degree of divergence, 64, 227
Symmetry factor in ϕ^4 theory, 57, 194–195

Tensor of rank (m, n), 4
Time reversal
Dirac equation, 18
bilinears, 23–24, 115–118
Dirac field, 44
bilinears, 47, 175–178
scalar field, 41, 159

Vacuum polarization, 63, 225
Vector, 3
contravariant components, 3
covariant components, 4
dual vector or one–form, 4
Vertex correction, 231–232, 237
Virasora algebra, 38

Weyl fields, 20
Wick rotation, 212
Wick theorem, 55, 57, 152, 172, 193–196

Yukawa theory, 64, 206, 227–233